Ableton Live 9

Ableton Live 9

Create, Produce, Perform

Keith Robinson

Focal Press
Taylor & Francis Group

NEW YORK AND LONDON

First published 2014
by Focal Press
70 Blanchard Road, Suite 402, Burlington, MA 01803

and by Focal Press
2 Park Square, Milton Park, Abingdon, Oxon OX14 4RN

Focal Press is an imprint of the Taylor & Francis Group, an informa business

Library of Congress Cataloging in Publication Data
Robinson, Keith, 1977–
Ableton Live 9: create, produce, perform/Keith Robinson.
  pages cm
  1. Ableton Live. 2. Digital audio editors. 3. Software synthesizers.
  I. Title.
  ML74.4.A23R63 2014
  781.3'4536—dc23   2013030324

ISBN: 978–0–240–81789–7 (pbk)
ISBN: 978–0–240–81790–3 (ebk)

Typeset in Myriad and DIN
By Florence Production Ltd, Stoodleigh, Devon, UK

Bound to Create

You are a creator.

Whatever your form of expression — photography, filmmaking, animation, games, audio, media communication, web design, or theatre — you simply want to create without limitation. Bound by nothing except your own creativity and determination.

Focal Press can help.

For over 75 years Focal has published books that support your creative goals. Our founder, Andor Kraszna-Krausz, established Focal in 1938 so you could have access to leading-edge expert knowledge, techniques, and tools that allow you to create without constraint. We strive to create exceptional, engaging, and practical content that helps you master your passion.

Focal Press and you.

Bound to create.

We'd love to hear how we've helped you create. Share your experience:
www.focalpress.com/boundtocreate

Focal Press
Taylor & Francis Group

Contents

About the Author

Keith Robinson is a composer, sound designer, and producer living in New York City. He is also part of the adjunct faculty at The Clive Davis Institute of Recorded Music, New York University, where he teaches Producing Music with Software and MIDI (Pro Tools, Live, Logic, Reason). Keith's music has been used in numerous commercials and television shows, and he has composed additional music and sound design for multiple feature films. Currently, Keith serves as vice president and co-developer of the sample library company Sample Logic LLC (www.samplelogic.com). His sample company's award winning products A.I.R. Expanded, The Elements EXP, Synergy X, Morphestra, Cinematic Guitars 1 & 2, Rumble, Fanfare, Acoustix, Metalix, Tronix, Assault, and Cyclone are heavily used throughout music for film, TV, and the gaming world.

Acknowledgments

Keith Robinson would like to thank his wife Amy Baer for all of her help, support, and encouragement; and Joe Trupiano at Sample Logic LLC, Brandon Rowan, Huston Singletary, Justin Hegburg, and Larry Wilson for showing Keith the possibilities of music technology.

Special thanks to Carlin Bowers for making this book a possibility and seeing it through to the end.

Live 9:
Create, Produce, Perform

1.1 Introduction

Never has there been music production software that so closely emulates the human mind and the demand for delivering music through a computer than Live. With an imaginative design and a forward-thinking mission, Ableton continues their legacy with Live 9, a software package that drives music production to the cutting edge while squarely meeting the needs of the composer, producer, performer, songwriter, DJ, and beyond. With such a progressive approach to its development, some of you may feel a bit disoriented or even intimidated at first sight of Live's unconventional design, especially those coming from a traditional Digital Audio Workstation (DAW) background. If you are new to DAWs, DJ style programs, or software music production in general, then you'll soon be right at home with the "parallel concept" of Live's Session and Arrangement Views. For the rest of you, you'll have to rethink your approach to composing, arranging, and producing music just a bit; but it will be a worthwhile adjustment. That is why this book has been written: to help reinvent the experienced software-based music producer and to unleash the new user. The goal here is to build and cultivate a strong understanding of Live 9's concepts and to provide material that will engage all DAW users alike. With this goal in mind, at the end of each reading you should feel that your current skills and knowledge base have been elevated to the next level. For the current Ableton Live user—yes, you—there is plenty here to unlock! After all, there is still a little "new user" inside us all.

Now it is time to learn how to *Create, Produce,* and *Perform* with Live 9—all you have to do is decide what your needs are, because it's all here.

1.2 Approaching Ableton Live in Three Intuitive Ways

Live naturally presents itself as a multipurpose music tool for the composer, producer, and performer. Which one are you? For this reason, this book provides an accessible way for both new and intermediate Live users of all musical styles to obtain the critical information that will act as a constant guide for all things relating to Live. The goal is to present a balance between creative techniques, tips, tricks, and the vital nuts and bolts of Live by concentrating on its three main facets: *Create, Produce, Perform*. Breaking Live down into these categories will help to educate, inspire, and propel you to new heights, synergizing creativity, production, and performance. So, use this book as a means to uncover specific features and gain conceptual and practical knowledge that you can apply in real-world situations. Rediscover yourself in the way that fits you best, whether through creation, production, or performance.

Now, let's take a look to see where you fit within the three content areas. Each area of focus will help you to identify your unique needs in order to maximize your time spent with Live. Use them as a guide to tailor-make your own experience while reading this book. Which one describes you?

- *Create*: composers, song writers, sound designers.
- *Produce*: producers, mixers, re-mixers, audio engineers.
- *Perform*: performing artists, studio, live stage, real-time performances and improvisations, interactive music installations, choreographies.

1.3 Immersing Yourself in Live

The heart and soul of this book is about communicating the idea of making music on-the-fly. This means an uninterrupted workflow that allows for multi-tasking the most complex musical ideas in real-time. Therefore, this text is all about immersing you into the world of Live. It's time to think like Live, reading and learning with the freedom to customize your learning experience. This is the ideology supporting every Scene (chapter). The content of the book should be used as a launching pad for navigating to and from related concepts, skills, and the unique features of Live 9. This being said, you'll find that the experience has been engineered in the same way that Live's own non-linear musical environment has been conceived. Not only can the book be read front-to-back in a traditional fashion, it can also, and should be, read and used as a non-linear resource for unleashing new and efficient skills for music creation, production, and performance. Simply put, read it in any order you desire!

In order to deliver the content in this way while staying true to Live, the key elements of the Live software have been integrated into the book itself. This includes the same creative features found in Live, transformed into a textbook format. The terminology, labels, icons, and features of the program have been explicitly adopted and literally implemented, treating the book as if it were part of Live's user interface. The concept is to have the book look and feel like the software. Think of it as a new way of using a computer program and guidebook in tandem to foster new and creative ways to use Live.

1.3.1 How the Book Has Been Constructed

This book does not use the traditional structural labeling system; rather, it incorporates Live's nomenclature and terminologies in order to draw a parallel to Live itself. The goal is that while reading, you will be learning the program without realizing it—thinking in Live's language. The following legend identifies the terms, labels, and images that are used to parallel Live—with some liberties, of course:

- *Categories* = contents.
- *Places* = index.
- *Scenes* = chapters.
- *Clips* = topic titles.
- *Info Boxes* = sidebar definitions, concepts, topic highlights. Note that these resemble Live's Info View located at the bottom left corner of the main Live screen.
- **Hot Tips** = a cool and quick tip or trick. Hot Tips appear in two different formats. Note that neither of these physically exists in Live, but they resemble its Info View:
  - Hot Tip icon;
  - Hot Tip box.

- **Launch Boxes** = launch points for linking Scene or Clip topics and concepts to other related Scenes or Clips in the book. They can be found throughout the book embedded in the text and in Info Boxes:
  - Scene Launch Boxes are teal (light blue) with a grey launch button.
    ▶ Scene
  - Clip Launch Boxes are green, blue, or red.
    ▶ Create ▶ Produce ▶ Perform

Also included are screenshots from Live 9. Be on the lookout for graphics and pictures that will help to solidify concepts. A picture is worth a thousand words!

1.3.2 General Scene Layout

Each Scene is constructed around the following general outline:

▶ *Concepts and Technology:*
- Features;
- Functions;
- Techniques;
- Tutorials/examples.

▶ *Create Concepts:*
- Composing/arranging;
- Song writing;
- Sound design.

▶ *Produce Concepts:*
- Producing;
- Mixing;
- Remixing.

▶ *Perform Concepts:*
- Real-time on-the-fly;
- On-stage performance;
- Improvisation.

▶ *Information/Hot Tips:*
- Hot Tips;
- Definitions.

▶ *Launch Points:*

- Contextual Clip Launch Boxes;
- Interconnected topics, techniques, features, tips, and tricks.

1.4 How to Use This Book

This book has been designed to be read in any order you choose, not necessarily from cover to cover. Customize your own path on-the-fly as you read. A good approach is to read straight through the first three Scenes. This will take you through the basic essentials. At that point you will be ready to take off on your own journey through the rest of the content. Feel free to approach your reading as an interlinked adventure, tying together the specific concepts you need to enhance your workflow and unlock Live's power. If you're new to Live, then you might choose to read straight through, or attack the areas that address your immediate needs. Here is a strategy to get you started:

- Locate information and concepts that interest you, then let the book take you on a journey through its related topics.
- For those of you looking for basic principles, the nuts and bolts, you can read straight through cover to cover. It's all here!
- If you haven't the time for a long read, dive right into the content you need, then launch from Scene to Scene, clip to clip as needed. Use the Categories or Places Browsers (Contents and Index) or Launch Boxes within a Scene.
- If you have a little more time on your hands, dig deep into the concepts, definitions, and Hot Tips that are tailored to creating, producing, and performing with Live.
- Many of the exercises, tutorials, walkthroughs, and examples reference specific media. This content has been made available for you on the companion website: www.focalpress.com/cw/Robinson. Download Sets, Audio, MIDI, Clips, and more from the site when recommended. You will also find supplemental chapters on Live 9.
- Visit www.ableton.com/en/packs for additional Packs (presets and sample content) that will enhance your experience with Live 9.

1.5 Summing It All Up

Provided for you is a customizable learning environment that will allow you to move freely throughout the text. The strategic pointers (launch points) open up pathways to new features and concepts, tying together other related features and concepts. Think of this as a "lateral learning" approach. In this way, the book becomes a tangible testament of Live, freeing you from constraining methods to better service your unique needs. Make reading it part of your daily routine and apply a new tool or concept to a composition, production, or a performance. The point is: you can focus in on one or all of Live's three facets—create, produce, perform—while navigating freely throughout. Choose where to start, stop, and go with the freedom to launch to and from any Scene or Clip in a non-linear fashion. This allows you to choose your own adventure and find answers and justifications that define how music and technology converge with Live in a customized way. Remember, launch points can be found inside Info Boxes, Hot Tips, and embedded within the text. They will guide you to related material that will help to build your practical knowledge of Live.

1.6 Taking Advantage of "Hot Tips"

Hot Tips are invaluable cool insider tips that save you time while enhancing your workflow. They have been crafted and designed to provide important information that will streamline your experience with Live. Some of them are hidden tricks of the trade, while others are newfound workarounds. Hot Tips are quick timesaving devices that can be put into practice right away! Some also act as launching points to interconnect skills, tips, tricks, and topics throughout the book where indicated. Keep an eye out for Hot Tip boxes **Hot Tip** and icons. They have been strategically placed throughout the book.

Live 9:
Overview

2.1 Introduction

First released in 2001, Ableton Live has quickly become a mainstay in the music industry due to its fresh approach to creating and producing music in the studio and onstage. When you think of a Digital Audio Workstation (DAW), you don't normally think of a musical instrument for performance and song creation, but Live can be used in precisely that way. It's the one thing that makes the program so unique. At the same time, Live's differences are why most of you are reading this book—that is, you have found yourself intrigued, yet out of your comfort zone and hesitant to take the leap. Let's change that, but first it should be pointed out that there are three editions of Live 9: *Intro*, *Standard*, and *Suite*. Live 9 Intro is a slimmed down version of Live with only the essential features: 3 Ableton software instruments, 26 effects, and 4GB of sound content. Live 9 Standard is the full-featured version: Live 9 and comes with 3 Ableton software instruments, 37 effects, and 11GB of sounds. Live 9 Suite is the complete studio package that bundles the full-featured version of Live 9 with 9 Ableton software instruments, 40 effects, Max for Live, and 54GB of sounds. Live 9 Suite will be discussed throughout, but for a dedicated overview, launch to ▶ **Web** .

Ok, now we can begin.

2.2 The Concept

Live 9 is a one-of-a-kind real-time musical instrument, sequencer, and production tool. It's all about non-stop creative flow. At first glance you will notice that Live is designed around two parallel views, or windows if you will: *Session View* and *Arrangement View*. Each view has its own unique musical purpose and functionality. They cannot be viewed simultaneously, but they do, in fact, work together in representing your music productions. There will be much more on this parallel concept throughout the book, but for now we'll focus on the general concept.

It is from one of these two views that you will create, produce, or perform your music. Generally speaking, you will use the Session View as a musical sketchpad and launching base for your musical ideas, and the Arrangement View as a traditional linear music timeline for recording audio and sequencing your music, the latter of which is like any other DAW or sequencer software. This is in no way set in stone, but more often than not it's the most common method of using Live. Whichever approach you choose, it can accommodate. More importantly, the real beauty is in Live's ability to realize and humanize the approach to creating music on-the-fly and to enable real-time improvisations. Of course, it also provides all of the traditional features of a DAW—powerful audio effects, instruments, intuitive editing tools and features; but it's Live's non-traditional real-time flexibility and creative workflow that sets it apart.

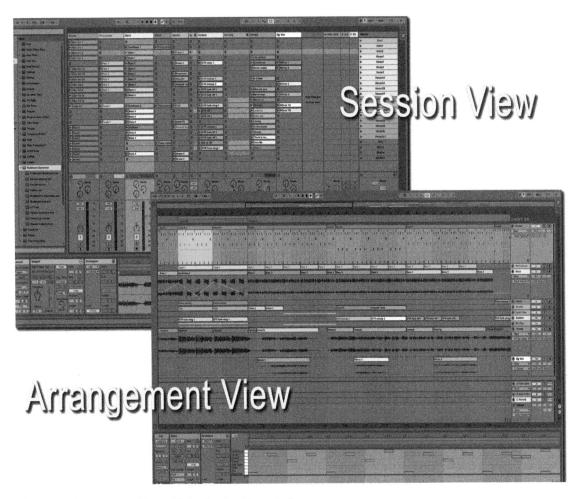

Figure 2.1 Arrangement View – Right; Session View – Left.

With the world of Live now fresh on your mind, you will be guided through a few quick steps to get optimized and running. Let's start out by addressing some software settings. Once your system is streamlined, we'll create a Live Set and get you on your way.

2.3 Important Preference Tweaks

A little knowledge about how Live and your computer interact can go a long way. The last thing anyone wants is to be performing on stage when suddenly playback pops, glitches, or cuts out. Even worse is experiencing a black screen, blue screen, or the dreaded "Pinwheel of Death". Although no one can foresee those tragic system crashes, optimizing your central processing unit (CPU), Audio, MIDI, and other performance-related settings are good pre-emptive measures. In a perfect world, we'd all own the latest and greatest computer with the fastest CPU, tons of RAM, and all of the disk space in the world. Unfortunately, this is never the case. That being said, you should be sure that your system at least exceeds the minimum recommended system requirements of any production software you use! Here's something that software developers don't tell you: they rate their particular system specifications as low as they can before the software will not run. That means that as soon as you load a third-party plug-in or multi-task, your system will stutter, pop, glitch, overload, error: you name it. In any case, keep an eye out for system requirements and if relying on Live for your bread and butter, you're better off exceeding Live's minimum requirements. Here are a few ways to smooth out your experience.

2.3.1 CPU Load

When it comes to CPU, you will want to keep an eye on the *CPU Load Meter* at the top right of Live's main screen.

Figure 2.2
CPU Load Meter.

This shows the amount of processing your computer is taking on at any given time. When this meter reaches 100 percent, it is maxed out. Fortunately, Live audio processes are the last to be sacrificed for processing power. Therefore, non-critical processes and functions are the first to be dropped so your music can remain uninterrupted during a power struggle—theoretically, at least.

What will cause strain on your CPU? The higher the track count and number of active devices—software instruments and effects processing—the more strain on your system resources; but the biggest offender will be third-party virtual instruments and plug-ins. They can eat up a lot of your CPU, hard disk, and RAM. Instruments, in general, can be CPU hogs and sample libraries are the worst. Fortunately, Live is pretty efficient in dealing with tracks and devices. Only active instruments and effects contribute to the CPU load. When they are deactivated there is little to no pull on resources.

2.3.2 Disk Load

The number of simultaneous audio channels used in a Set is directly related to disk load. Therefore, Live's performance is dependent upon your computer's hard drive speed. Most optimized desktop computers have sufficient hard drives (7,200 RPMs) for operation with Live. Laptops, on the other hand, generally use slower drives (5,400 RPMs). Rest assured that Live is laptop friendly and most systems work without a hiccup. In any case, track count and virtual instruments will dictate the disk performance of your Live set. The *Hard Disk Overload Indicator* will light up if hard disk overloading occurs. In this situation, you may need to reduce the track count in your Set, deactivate unnecessary effects, or use *RAM Mode* for certain clips. Launch to Clip ▶ **7.5.4** for more details on clip RAM Mode. If a third-party software sample player is the culprit, you could switch its internal sample loading to a RAM-based mode as opposed to direct-from-disk mode (DFD), when applicable. In short, in DFD mode, samples are read directly from the hard drive. Using a RAM-based mode will free up the hard disk by loading the samples into your computer's memory instead. The goal is to take the strain of reading and writing off the hard disk. Each plug-in handles this differently, so familiarize yourself with their unique options.

2.3.3 Track Freeze

Freezing tracks (*Edit Menu>Freeze Track*) is a great way to free up system resources, both CPU, RAM, and disk usage when you are simply pulling too much power with your Set. Once a track is frozen, it plays back from its Freeze File stored in the project sample folder. This file is a render of the entire track's contents, along with its effects. This allows you to use more instruments, effects, and tracks, even if your computer is low on processing power. It's a simple, quick, and practical way to improve your system performance, and any track that can hold clips may be frozen—Audio or MIDI. Track Freeze is located under the Edit Menu or the clip's contextual menu.

Although freezing is a very intelligent and helpful feature, it has a few minor limitations. It limits some editing features such as note and audio changes or manipulations. On the other hand, Mixer functions are still available, as well as launching clips, copy, paste, rearranging, volume, panning, sends, and so on. Ideally, you won't have to use the Track Freeze command for CPU/Disk relief, but if you do, you can unfreeze tracks at any time. Track Freeze is a non-destructive process, but with the convenient option to *Flatten* a frozen track permanently, making any edits to frozen audio clips destructive and converting a frozen MIDI track into an audio track with audio clips. The Flatten command is located just under the Freeze command in the Edit Menu. That being said, conserving CPU is not the only reason to use Track Freeze. There are a few very useful—actually mind boggling— ways to use Track Freeze with MIDI in which you can turn your MIDI clips into audio clips in a blink of an eye. What? Yes, it's true: MIDI clips converted into audio and in real-time. To be honest, it doesn't stop with MIDI. Any frozen audio clip can be instantly rendered when *dragged* from a frozen track to another audio track. To learn more about Freezing and Converting MIDI clips into audio clips, launch to ▶ 8.3.1 .

2.3.4 Audio Preferences

It's always a good idea to verify your audio interface setup. To do this, go to the *Preferences>Audio* tab. Make sure that your audio input and output device is configured correctly and efficiently. Any unused input/output channels should be turned off. This will reduce the load on your CPU. As for sample rate (SR), you will determine that yourself, but use High Quality SR and Pitch Conversion unless your system cannot handle the load. Change this to "Normal" if you must, but why settle for less? It only makes sense to choose the best quality. As for any audio latency issues, they should be addressed with the Buffer Size settings. For Plug-in Buffer Size, go to the CPU tab. The smaller the size, the higher the CPU load.

2.3.5 MIDI Preferences

To setup your external MIDI devices, go to the *Preferences>MIDI Sync* tab. From there you can configure Control Surfaces, MIDI interfaces, or other MIDI devices that are connected to your computer. In the top area of this tab you can choose up to six control surfaces and set its MIDI Input/Output ports via the available chooser menus. To the far right is the *Dump* button, short for preset dump. This feature will send preset information to a controller so that you can map its physical controllers (slider, faders, knobs, pads, etc.) correctly if and when needed. Note that Dump is only needed for certain hardware surface controllers/devices. You won't need to use the Dump feature for devices natively supported by Live.

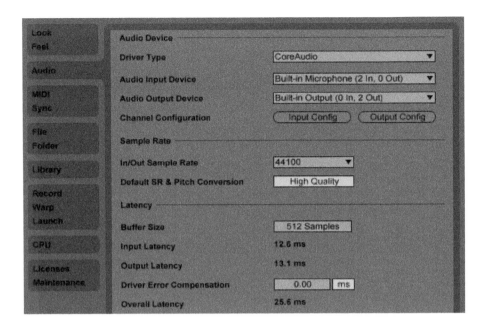

Figure 2.3
Audio Preferences
Tab.

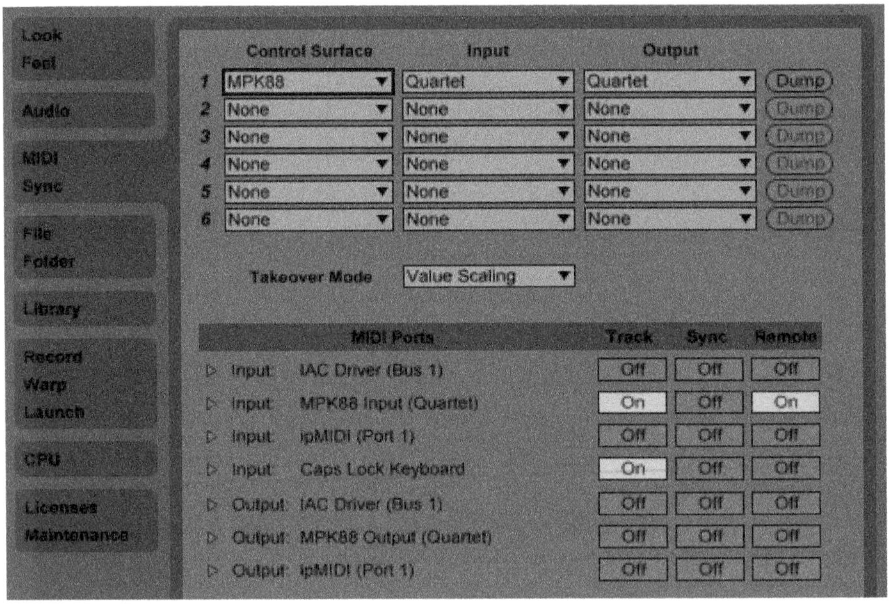

Figure 2.4
MIDI Preferences
Tab.

Below this you can set the *Takeover Mode* for your device. This is a very important preference! It determines how Live will respond to continuous controller data (MIDI values) sent from a physical knob or slider on your MIDI controller. There are three modes: *None*, *Pickup*, and *Values Scaling*. Each option determines how Live behaves and reacts to sudden jumps in values sent from a controller. Refer to "Setting up Control" in ▶ 17.1.1 .

The bottom section of the MIDI/Sync tab is the MIDI Ports List where you will find all of your available MIDI Input and Output ports. The left column lists the actual MIDI In/Out Ports. To the right of these are three columns and their associated "On/Off" switches: *Track*, *Sync*, and *Remote*. In order to send MIDI to Live's tracks, you will need to activate Track ("On") for the input port(s) you wish to use for sending MIDI. You can also send MIDI from Live tracks to external MIDI devices by activating the Track switch for the output port you wish to use for sending MIDI out. For synchronization of MIDI Clock (MBC) or MIDI Timecode (MTC) from external devices with Live, activate the Sync switch for MIDI input ports that are to be used for syncing Live. To send sync messages from Live, activate Sync for the desired output ports that are going to be used. Sync allows you to synchronize playback, tempo-based effects, Arpeggiators, low-frequency oscillators (LFOs), etc. If you wish to Remote Control Live with a controller using customized mapping assignments, then Remote should be activated for the input ports that are to be used for remote controlling Live. For control surfaces with motorized faders or real-time displays, Remote should be activated for the associated output ports. Live can then send control update messages to the control surface so that its controllers, faders, or display will mirror Live's related controls. Learn more about Remote Control. Launch to ▶ 17.1 .

2.3.6 Record, Warp, and Launch Preferences

Here you can set Record File Type (.wav or. aiff), Bit Depth (16, 24, or 32), and Record Count-in (1, 2, and 4 bar). File Type and Bit Depth should always be set prior to recording audio. The *Exclusive Arm* and *Solo* options establish arming (record-enable) and solo restrictions so that only one track at a time can be armed or soloed. *Click + hold* the (⌘ Mac/[ctrl] PC) key while *clicking* track Arm and Solo buttons to temporarily bypass these Exclusives when you're working in Live.

Another important preference is Warp/Fades and Launch, which help to streamline your creative workflow. They manage how Live handles imported audio samples and the behaviors associated with launching clips. If you are just starting off with Live, the default settings are fine. For more information on warp settings and concepts, refer to Clip ▶ 7.5.9 and ▶Scene13 . For launching clips, refer to ▶ 5.3.2 .

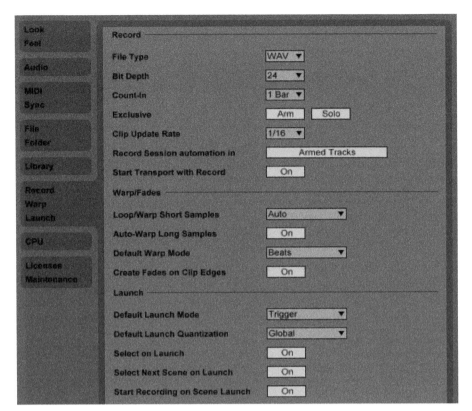

Figure 2.5
Record, Warp,
Launch
Preferences Tab.

2.4 The Live Browser

Before you dive into making music, you should first understand how Live's Browser system works, because it will literally jumpstart your creative flow. So, what is it? The Live Browser is an interactive window for accessing all of Live's built-in sounds, instruments, effects, installed third-party plug-ins, Packs, samples, presets, and any file folder located on your hard drive(s). Simply put, everything you need to make music can be accessed from the Browser. By design, you should never have to leave Live while working on your music. You can always *drag* and *drop* content from your Mac Finder or Explorer Window; but utilizing Live's built-in Browser makes for an intuitive way to quickly access any sounds and files. It is the one-stop shop for all of your production tools. You can also customize it by adding your own file folders directly to the Browser without having to copy your sounds to one central library location. Imagine how seamless a real-time impromptu performance will be.

2.4.1 Browser Layout

The Live Browser window is made up of two columns located on the far left side of the main Live screen. It can be shown or hidden by *clicking* the unfold button (triangle) at the upper left directly below the *TAP* button ([⌘+opt/alt+B] Mac/ [ctrl+opt/alt+B] PC).

Each icon or label in the *Browser Sidebar* (leftmost column) identifies and links to individual sub-browser items (content) that appear to the right in the *Content Pane*. The sidebar is divided into two main sections: *Categories* and *Places*. Categories is where you will generally access instruments, presets, effects, samples, etc. Places is where you'll find Live's pre-installed Core Library, Packs, User Library, Current Project content and any folders you have added to Live are located. Navigate subfolders and presets by *clicking* on the arrow left of the names.

Figure 2.6
Show/Hide Browser.

With the Live Browser open in view you can access:

- ♪ *Sounds:* Instrument presets.
- ⊞ *Drums:* Drum presets.
- ⋀ *Instruments:* Live Instrument devices and presets.
- ⫴ *Audio Effects:* Live Audio Effect devices and presets.
- ⌷ *MIDI Effects:* Live MIDI Effect devices and presets.
- ⬒ *Max for Live:* Max for Live devices and presets.
- ◁ *Plug-ins:* AU/VST third-party instruments and effects.
- ▶ *Clips:* all clips installed in Live.

Samples: Audio files.

Packs: Live Library content and installed content.

User Library: user content.

Add Folder … : Browser customization.

Instruments

The Instruments label provides access to all of your default Live Instrument devices (Ableton Live Software Instruments), Instrument Racks, and related factory and installed Pack presets organized by device type. Live Instrument devices—synths and sample players—come pre-installed with Live depending upon which version you own. As mentioned earlier, Live 9 Intro and Standard come with three Live Instruments and a humble set of presets. Live 9 Suite, on the other hand, includes all nine Live Instruments and thousands of presets, as shown in the example here.

It is strongly recommended that you consider upgrading to or buying Suite. Having all of the Live devices is essential in creating, producing, and performing with Live. For a detailed description of the instruments included in Live 9 Suite, launch to ▶ **Web** and also check out the Ableton website. For now, let's zero in the two core instruments included in all versions of Live 9 located from Instruments in the Live Browser: *Impulse* and *Simpler*.

Impulse is a sample player primarily used for triggering and playing back drums and percussive one-shot audio samples via its eight sample slots only. Simpler is also a basic sample player, but triggers samples via the standard 88-key chromatic keyboard map. A sample player is designed to play back any sample content, whether melodic, percussive or otherwise. These two devices will get you up and running in no time, making it very easy to begin creating music without investing in third-party virtual instruments. Use them as single standalone instruments or as part of an Instrument Rack or Drum Rack. Both devices come with presets, so feel free to start exploring their capabilities as soon as you can. For specific details on Impulse and Simpler, launch to ▶ **15.3** and for more on Device Racks launch to ▶**Scene16** .

Sounds

The Sounds label displays all of your installed Live instrument presets intelligently organized by sound type.

Drums

The Drums label displays all of your drum-based presets, including Drum Racks, Instruments Racks, and Drum Kits.

Figure 2.7
Live Instrument
devices.

Audio Effects

The Audio Effects label displays all of Live's native audio effects, presets, and Effect Racks. Depending on the version of Live, there are over 40 unique audio effects accessible that include everything from classic effects used for mixing and tracking to those that can be used to wildly and creatively enhance and manipulate your audio. A sound designer's dream!

The vast and creative quality of Live's audio effects will help you to quickly realize that there is not much reason to run out and buy a huge and costly bundle of third-party effects plug-ins, especially if you have Suite. Here is a short list of some of the Audio Effects included in Live 9: *Amp**, *Auto Filter*, *Cabinet**, *Corpus**, *Dynamic Tube*, *Glue*, *Multiband Dynamics*, *Vocoder*, and the standout classics such as *Beat Repeat* and *Filter Delay*. For a detailed description and explanation of how audio effects work and apply to creating music with Live, launch to ▶ 15.5 . If you already have a handle on audio effects and are feeling ambitious, launch to ▶Scene16 . There you will find out how these monster effects can be linked together, utilizing *Audio Effect Racks*. For **Audio Effects included in Live 9 Suite*, launch to ▶ 15.5 and ▶ Web .

MIDI Effects

The MIDI Effects label provides access to all of Live's native MIDI-based effects, including presets and MIDI Effect Racks. These effects are inspiring devices that can literally manipulate and grow your music into dynamically complex and rich arrangements. MIDI effects have a unique ability to alter your MIDI tracks and patterns. Start with the included presets to get a handle on their instant creative power. Just like audio effects, MIDI effects can also be turned into custom Effects Racks. Launch to ▶ 15.4 to learn more about MIDI effects.

Plug-ins

The Plug-ins Browse label displays all of your third-party instrument and effects plug-ins installed on your computer (Audio Units and VSTs). This conveniently gives your third-party plug-ins the same seamless integration as the native Live effects. Although Live includes many fantastic instruments and effects, once you have exhausted all of them, you may consider branching out and investing in some third-party instruments and effects plug-ins.

Figure 2.8 Live Audio Effects.

Figure 2.9 Plug-In Devices: Third-party AU and VST instruments and effects.

Why do you need more? These are your tools, and you can never have enough! There are new plug-ins made available every day from numerous developers, created by some amazing programmers and musicians who specialize and dedicate every waking hour to their development. So when you are ready and in need, search the web for third-party plug-ins that are compatible with Live, and don't forget to continually check the Ableton website for new *Packs* (downloadable content for Live).

Before you can use third-party Audio Units (Mac only) or virtual studio technology (VST) plug-ins with Live, you will need to turn them on in the Live Preferences menu (*Preferences>File Folder>Plug-in Sources*). Once activated, Live will scan and locate all of your audio unit (AU) and VST Plug-ins, thus making them accessible in the Browser.

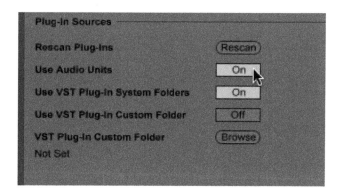

Figure 2.10 Activate AU and/or VST plug-ins for use in Live.

Keep in mind that this includes both virtual instruments and audio effects. They have all been placed together into one single content pane listed alphabetically. To learn how to use third-party plug-ins in Live, launch to Clip: ▶ 15.8 .

Clips

Clips is where you will be able to access all of your installed Live Clips—audio and MIDI clips. This includes Live Clips installed with a Pack or any Live Clips you save to the User Library or Current Project Folder. Live Clips consist of audio or MIDI events and store various clip, track, and device settings. Learn more about the Live Clip format. More about Live Clips: ▶ 15.2.4 .

Samples

Just as the title suggests, the Samples label is where you will find all of your installed audio samples as contained in Packs.

Packs

The Packs label is where you will find all of your installed content organized by Pack. This includes the pre-installed Core Library and any other expanded content. Packs are essentially Live-formatted containers or bundles of content designed by Ableton or other third-party developers that contain devices, samples, clips, grooves, etc. The Core Library itself includes an array of key elements to be used with all of your Live projects, but the content varies depending upon which edition of Live 9 you've installed. *Click* on a Pack and you will find folders in the Content Pane dedicated to Audio Effect Racks, Clips, Devices, Drums, Samples, Sounds, and Swing and Groove. You can download additional Packs from the Ableton website (www.ableton.com/en/packs).

User Library

The User Library is the default location where all of your own personally saved items can be accessed, such as presets, device settings, Racks, samples, Live Clips, and much more. Here you can store all of your own customized default settings, device presets, clips, and Sets, etc.

Current Project

This label provides a convenient way to view and access all files related to an open Live Project such as tracks, clips, and devices.

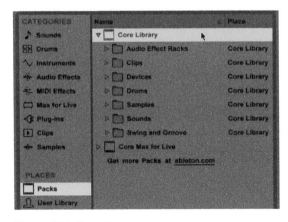

Figure 2.11 Core Library.

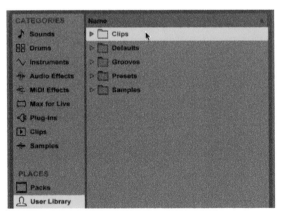

Figure 2.12 User Library.

Add Folder …

In addition to Packs and the User Library, you can customize the Live Browser for quick and efficient access to your own system file directories or external hard drives. Simply *click* on the *Add Folder* label and then add your own custom files, file folders, or entire hard drives right into Places. Not only are these like bookmarks, Live analyzes these folders and then adds their contents to the appropriate Categories in the Browser. With the Add Folder feature and the other labels under Places, the Live Browser becomes an intuitive and adjustable data hierarchy that gives you quick access to all of the information located on your internal and external hard drives for instant access and use in Live. This is an incredibly crafty way to view your common and most used items while simultaneously creating, producing, or performing. The concept is built around having your tools and content right in front of you ready to be accessed at the *click* of a button in both the Session and Arrangement Views.

Figure 2.13 Add a custom folder to Places (Mac).

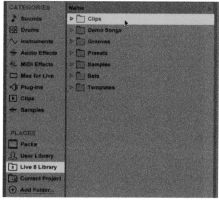

Figure 2.14 Live 8 library added to Places.

2.4.2 The Browser Concept

The Live Browser is simple and exciting! With a quick *drag, drop,* or *click* you can load instruments, effects, sounds, create tracks, save presets, or simply browse custom content all within Live and even during playback of a song. The design is dynamic, making a simple browser into an interactive tool not only for locating and accessing content, but also for auditioning sample content, devices, effects and presets. As far as browsing is concerned, that's self-explanatory; but loading and auditioning clips and devices is revolutionary. You'll notice in both the Session View and Arrangement View an area labeled *"Drop Files and Devices Here"*. You can insert a Live device into your Set by *dragging* and *dropping* it directly from the Browser to the Drop Area, to a track, or by *double-clicking* on a device name inside the Browser. When *dropping* a device outside of a track in one of the given Drop Areas or *double-clicking*, Live will automatically load up and create the necessary tracks. To learn all about Live Devices (instruments and effects), launch to ▶**Scene15**.

In addition to easy access, Live has the ability to change out instruments and effects on-the-fly with the *Hot-Swap* buttons. With *Hot-Swap Mode* activated, you can instantly change out an instrument, preset, or effect by *double-clicking* on a new one in the Browser. Grey = Off: 🔄. Orange = On: ❌. When "On", an orange title bar will also appear above the Content Pane indicating Hot-Swap Mode. Follow any one of the following launch links below to learn more about Hot-Swap Mode.

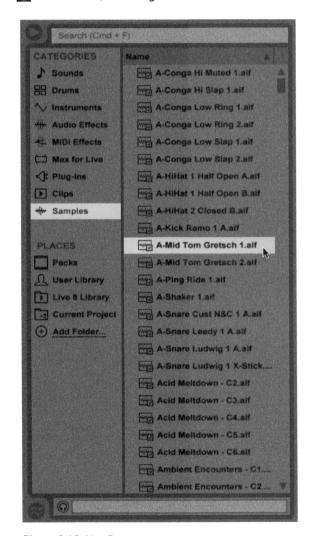

Hot-Swap buttons are incorporated for use with:

- Devices/Presets ▶ **15.2** .
- Grooves ▶**Scene12**.
- Sample-based Live Instruments/Device Chains ▶**Scene15**.
- Device Racks ▶**Scene16**.
- Replacing/Managing files referenced in a Set ▶ **3.7.3** .

Live's Browser is used to import a variety of media, such as audio files (samples), MIDI files, devices/presets (instruments and effects), and Live Clips (Sets, tracks, devices, clips) into a Set. For example, all of your audio/MIDI files and Live software instruments may be previewed directly from the Browser prior to importing or inserting. This is very useful for auditioning loops, samples, and instrument presets without having to commit them to your Set or increase your track count. This is also ideal for auditioning entire songs for live DJ sets since preview material can be routed for

Figure 2.15 Live Browser.

monitoring through a secondary audio output channel instead of the main audio outputs of Live—audio interface permitting. This is most commonly done through headphones, but could also occur through any secondary set of audio device/ interface outputs. Note that previewing is possible in real-time and is in sync with Live's current tempo. Test song against song on-the-fly without interrupting playback. To activate preview, *click* on the *Preview button* grey/blue headphones icon located at the bottom of the Live Browser. Grey = Off : 🎧. Blue = On: 🎧. This is the Preview Tab.

Figure 2.16
Preview Tab: Listen to and audition Audio/MIDI files and clips.

To expedite the browsing process, use the *Search* field at the top of the Browser to search within any specific category or place. Type in a keyword into the field and any results will instantly appear. If you want to search all Categories and Places, type ([⌘+F] Mac/[ctrl+F] PC). The results will appear under a new icon called All Results and list samples, presets, devices, etc.

Figure 2.17
Searching the Browser.

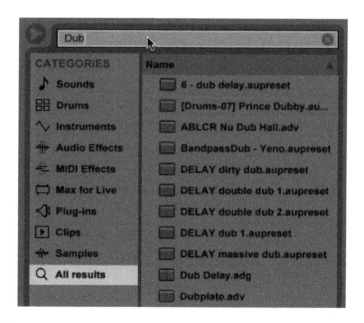

Once you become familiar with browsing around your content via Live's Browser, you'll stop using those extra Finder/Explorer windows that can clutter your desktop workspace. This is really efficient as you begin to rely on the Browser features to help you create and arrange your music without the distraction of locating files outside of Live that are on your desktop or remote drives.

Simplifying your workflow is a key part of working in any DAW. There will always be cumbersome tasks no matter which DAW you are using, but one way to take advantage of Live is to use the Browser search functionality. Taking this a step further, be sure to customize your browsing experience by adding user folders!

Now that the Live Browser system has been explained, let's create your own custom folder:

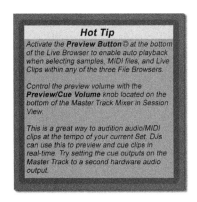

Hot Tip

Activate the **Preview Button** © at the bottom of the Live Browser to enable auto playback when selecting samples, MIDI files, and Live Clips within any of the three File Browsers.

Control the preview volume with the **Preview/Cue Volume** knob located on the bottom of the Master Track Mixer in Session View.

This is a great way to audition audio/MIDI clips at the tempo of your current Set. DJs can use this to preview and cue clips in real-time. Try setting the cue outputs on the Master Track to a second hardware audio output.

1. *Click* on the Add Folder ... button.
2. *Choose* a folder to add from the dialog box that appears. Or ...
3. *Drag* a folder directly from a location or directory on your computer and *drop* it into Places.
4. To *remove, right-click/ctrl + click* on the folder in the sidebar you want to delete and select "Remove from Sidebar".

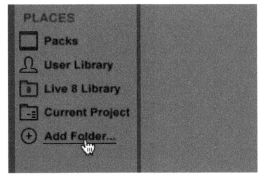

Figure 2.18
Click Add Folder…and select.

If you have a number of external drives, adding them to Places is the best way in which to view them. Two, three, or even a dozen external drives are not uncommon in a studio arsenal, which can get quite confusing. Using the Add Folders ... feature in the Browser will really sort things out for you. When you first add a whole volume, it will take a while for all its contents to populate the Content Browser.

Now you should be looking at all of the folders and files of that particular drive. Continue to build a custom Browser as you see fit. This makes it very easy to access, audition, and load your files in real-time.

Figure 2.19
External Volumes.

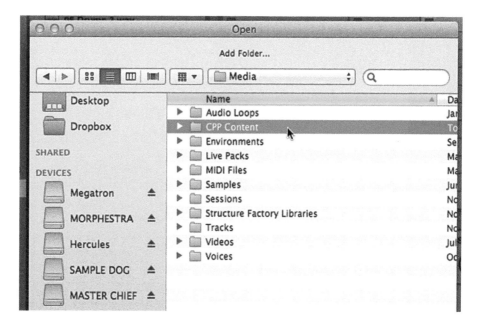

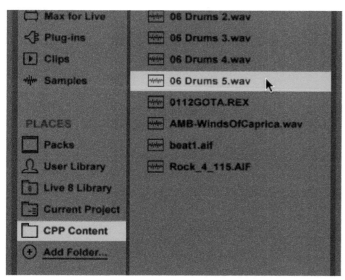

Hot Tip

Use your computer keyboard arrow keys to navigate through the list of sounds and files to preview.

Use the right-arrow key to open a folder.

Figure 2.20
Access your files in real-time.

The Quick Way to Start Making Music!

3.1 Introduction

So, you've installed and authorized Live 9 and now you're wondering how to get started. Whether or not you have experience with a sequencer or a Digital Audio Workstation (DAW), sometimes you just don't have the time to dig in and read the entire manual cover to cover. For that reason, this Scene is a quick start guide to get you into Live in the least amount of time possible. It underlines the general concepts without too much theoretical and technical jargon. When you are done with this Scene you will be able to execute the basic functions of Live with a general understanding of how it works. You won't always know how or why something does what it does, but that's ok. The goal here is to build a foundation that you can rely on later while saving you time now. If you feel like you still need more help as you're reading, then be sure to launch to the various locations highlighted within this Scene. There you will find dedicated and in-depth information pertaining to each topic. You can also take a look at the interactive *Live Lessons* that come pre-installed with Live. Lessons can be found in *Help View* located on the right-hand side of your Live Set. If it's hidden from view, select *Help View* from the View Menu or *drag* it open with your mouse from the right edge of the main Live screen.

Now, here's a little tip that can really help you to get out of a rut while learning Live. You will find helpful information displayed inside the *Info View* in the lower left corner of Live's main screen. *Click* on the arrow button in the lower left corner of the main Live screen to open it or use the shortcut *shift + /*. Nearly every feature, button, and object in Live is methodically tagged with a brief description of its function.

As you mouse over and around any area of Live's interface, descriptions will automatically pop up in the Info View. This is a true time saver in discovering or rediscovering feature functions and parameters.

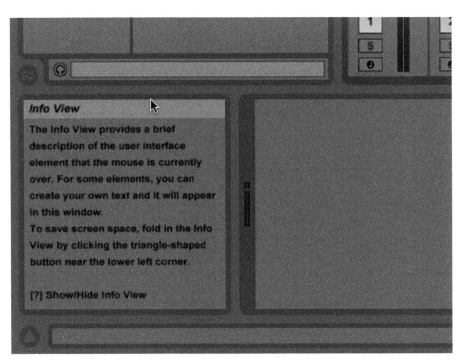

Figure 3.1
Info View.

Hot Tip If you're the really organized producer or engineer, add your own info text to items in Live, such as tracks, Scenes, clips, and devices. Place your mouse over any one of these items and right click or ctrl + click to bring up a contextual menu, and then select "Edit Info Text". Whatever you type will then be viewable in the Info View box when you hover your mouse over the item you named. Think about using these as comments or notes. You could also use them to leave a friend a little surprise.

3.2 Starting a Project

All right, it's time to get you on your way to creating and producing music, so open up Live 9 and let's begin. When you open Live 9 the default *Set* will appear in Session View consisting of two audio tracks, two MIDI tracks, two Return tracks (reverb and delay), and the Master track—the same default setup occurs when you create a new Set. This is just for starters; you can always add or delete tracks as needed. If you want to design a different default template, you can do so by saving any Set you create as default from the *Preferences>File Folder* tab under

"Save Current Set as Default". If you decide that you want your default setup to always start at 105 beats per minute (BPM), for example, you can save a new default reflecting the new custom parameter. You could also add several more audio or MIDI tracks to a Set then select "Save Current Set as Default". It's that easy.

Set is the name given to your work file in Live. Other programs use the terms Session, Song, Project, and so on. Looking at your new Set, familiarize yourself with how to use the Show/Hide Browser on the left-hand side of the main Live screen. Do this by *clicking* the arrow tab in the upper left just under the *Tap Tempo* switch. The Live Browser is where you will find all of the media and tools needed to work on your music, for now at least. Later you will use the Browser for a number of other tasks, including loading Live Sets and more. From the Browser you will find all of Live content, such as *instruments*, *effects*, *presets*, *sounds*, *clips*, *samples*, and more. That being said, your new best friend will be Live's Core Library that was installed with your version of Live 9. This is the content that has populated your Live Browser Categories and Places.

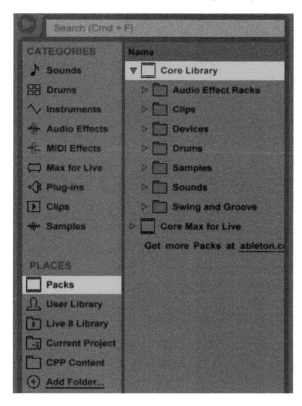

On the opposite side of the main Live screen is *Help View*. As mentioned before, you can show and hide this window from view from the *View Menu*, or by *dragging* the right edge of the main Live screen to the left using your mouse. From Help View you can quickly access all of Live's built-in tutorials (Lessons) and get extra help setting up your audio and MIDI hardware.

The last thing you need to be aware of is Live's two views for creating, producing, and performing music. Currently, you are looking at the Session View. If you hit the Tab key on your computer keyboard, the view will switch to the Arrangement View. These two views are independent working environments that reside within Live's main screen. Use the Session View to construct your musical ideas by importing and recording audio and MIDI into tracks. When your ideas are ready, you can record (perform) them to the Arrangement

Figure 3.2
Live Browser.

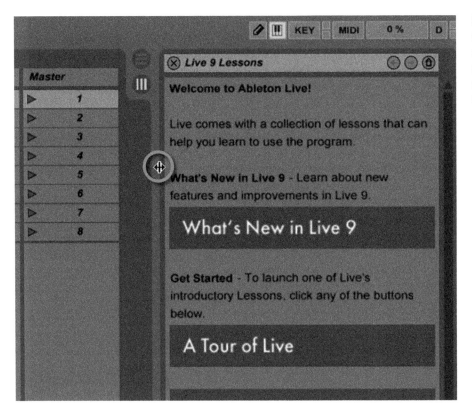

Figure 3.3
Click + drag the
right edge of the
main Live screen
to access Help
View.

View as a sequenced song (arrangement). This is the fundamental concept of Live, but the Arrangement View can also be used for traditional importing and recording audio and MIDI too. For that reason, let's first get you up to speed on how to work in, and with, both views.

With the basics covered, it's time to start working in Live 9! If this moves too fast, you can launch to the next Scene where concepts are broken down in more detail.

3.3 Audio in Live

The first thing you will need to do is create a "New Live Set" from the File Menu ([⌘+ N] Mac/[ctrl + N] PC). Once your Set is open, you'll need an audio source to generate sound. You can import pre-existing audio files or record your own sound source from an external audio input. Let's start out simple by importing samples to an audio track. Before importing audio, you have to consider which view to work in: Session View or the Arrangement View. Your Set should be currently open in Session View. Use the Tab key on your computer keyboard to toggle between views quickly. As mentioned above, one of the most common approaches is to import or record audio into Session View *Clip Slots* first, then later record a performance of these clips into the Arrangement View, thus creating a musical arrangement. Although this is the intended design of Live, by no means do you have to work that way. For this reason, we'll stick with importing and recording directly into each view for now and save the performance aspects for later: ▶ 3.6 ▶ Scene 4 ▶ 6.7 . Keep in mind while working through each of the following demonstrations that there are a number of ways to execute tasks in Live.

3.3.1 Importing Audio

To import an audio file to the Session View ▥, choose one of the following (if you need extra audio files to get started with, you can download some example samples from the companion website):

(a) *Drag* an audio clip directly from the Clips category in the Browser into an audio track *Clip Slot* or to the *Drop Area* in the middle of the Session View. *Double-clicking* on a clip will import it directly into a new audio track. You can also *drag* an audio file from the Samples category if you prefer.

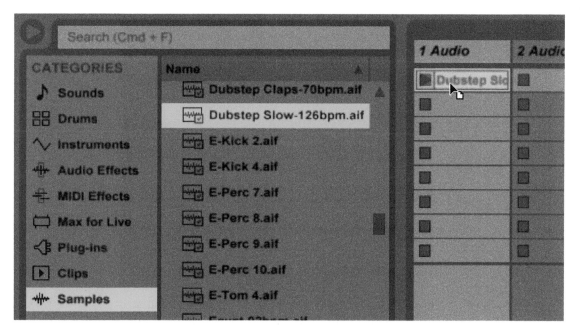

Figure 3.4 Session View: (Import) Drag and drop audio clip directly from the Browser into an audio track Clip Slot.

To import audio into the Arrangement View ▤, follow the same procedures as listed above. The main difference is that you cannot *double-click* a clip in the Browser to automatically import it. You must *drag* it into the *Track Display* area along the timeline in the Arrangement View.

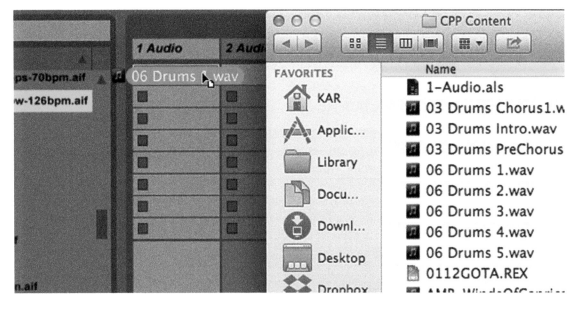

Figure 3.5 Arrangement View: (Import) Drag and drop audio clips directly from the Browser into an audio track's Track Display.

Notice that the Browser can be opened and accessed in both the Arrangement and Session Views. It's part of Live's main screen. If not visible, simply *click* on the Show/Hide Browser button in the top left-hand corner of the Live window.

(b) *Drag* an audio file directly from a Finder/Explorer Window into an audio track Clip Slot or to the Drop Area in the middle of the Session View.

For the Arrangement View, *drag* directly from the Finder/Explorer Window to the Track Display or Drop Area and place your audio file at the left most edge of the Track Display—the beginning of your Set's timeline.

Once you have imported an audio file into a Session View Clip Slot, you can *click* on its *Clip Launch button* to activate playback. After you have launched the clip and heard it loop a few times, *click* a *Clip Stop button* located in any empty Clip Slot in your audio track, then *press* the Stop button on the Control Bar Transport. The Stop button stops global playback in your Set since the

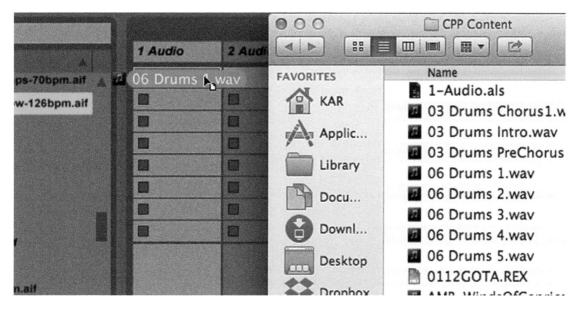

Figure 3.6 Session View: Drag and drop files directly from a Finder/Explorer Window into a Clip Slot.

Figure 3.7 Arrangement View: Drag and drop files directly from a Finder/Explorer Window into the Track Display or Drop Area.

Arrangement will always begin and keep running once a clip is launched in the Session View.

To listen back to a clip that has been imported into the Arrangement View, *double-click* the Stop button on the Control Bar Transport to set playback at the beginning of the Arrangement, ensuring the proper playback position. If there is an orange-lit button (*Back to Arrangement button*) in the upper right corner of the *Beat Time Ruler*, *click* that too, so it is no longer illuminated, and then *click* the Play button on the Control Bar. The Back to Arrangement button indicates that the current playback state differs from the stored/captured arrangement. Turn it off in order to revert playback priority to the Arrangement View tracks and clips.

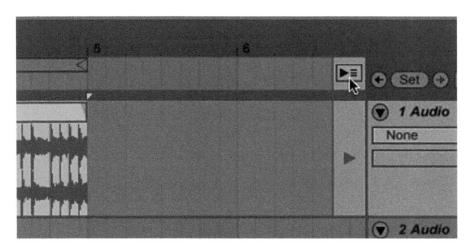

Figure 3.8
General Session View playback controls.

Figure 3.9
Make sure the Back to Arrangement button is not illuminated orange. Click it to turn it off in order to revert playback to the stored arrangement.

After your clip plays back, *click* the Stop button to stop playback.

Figure 3.10
Arrangement View Playback controls.

3.3.2 Recording Audio

You can record an audio input source directly into Live's Session View or Arrangement View. It's completely up to you. Each has its own benefits. No matter which you choose, you must always arm (record-enable) an audio track from its *Arm Session/Arrangement Recording* button located directly on the track. Now, before recording an external audio source, verify that your audio and record preferences are configured properly in Live's preferences: ▶ 2.3.4 . Once you have set them up, follow the next steps in succession:

1. Set the I/O section to be shown in view (View menu or I•O button—to the right of the Master track).

Figure 3.11
Set Audio Input Channel.

2. Using an audio track in the Session View, set "Audio From" to "Ext. In" (if not already set), and then set the *Input Channel chooser* to your desired audio input channel. Let's use input "1" since we are recording a mono source.

Later when you start working in the Arrangement View, you may need to set the I/O section to be shown in view so that you can access it on the Arrangement View track itself. Otherwise it's usually hidden from view. Note that the I/O section and other related display attributes are unique to each view (Session/Arrangement) and therefore can be shown/hidden independently for each view.

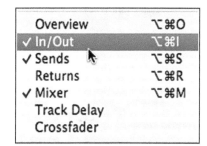

Figure 3.12
Select In/Out from the View Menu.

Figure 3.13
Arrangement View In/Out section mirrors the Session View In/Out.

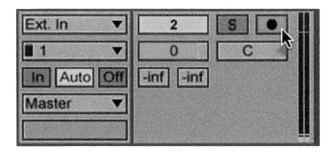

Figure 3.14
Arm Arrangement Recording (record-enable).

Figure 3.15
Arm Session Recording (record-enable).

In addition, be aware that there are no vertical track volume faders in the Arrangement View; instead, there are horizontal track volume sliders. Vertical faders belong to the Session View Mixer only. It is not possible to view the Session View Mixer and the Arrangement View simultaneously!

3. *Monitor* should be set to "Auto" and "Audio To" set to "Master" (Master track), then *arm* your audio track for recording by *clicking* on the track's Record button.

4. Now that your track is armed you can begin recording. If you wish to use Live's *Recording Count-In,* simply *click* on the Metronome dropdown in the Control Bar and choose a setting. This can also be set as a default setting in the *Preferences>Record* tab. The count-in will be both audible and viewable in the Control Bar when activated. Note that this is not a click track, only a pre-roll to recording. For a click track, activate the Metronome itself.

   (a) *Session View:* be ready! *Click* the Record button in the first Clip Slot on your audio track. This button will appear as a grey circle until activated. It will be a red triangle during recording.

Figure 3.16
Record Button.

Figure 3.17
Record button launched and recording.

Once the clip Record button is *clicked*, count-in will begin (optional) and recording will commence.

(b) *Arrangement View: click* the *Arrangement Record button* on the Control Bar to initiate recording.

Figure 3.18
Arrangement View recording.

Once the Arrangement Record button is *clicked*, count-in will begin (optional), and then recording will commence. Change this behavior to a three-step process. *Shift + click* or *right-click* the Arrangement Record button. To stop recording, simply *press* the spacebar to stop or the Arrangement Record button again to disengage recording at any time.

If you need a click track during recording, activate the Metronome. Adjust the click track's volume by selecting the blue highlighted *Preview/Cue Volume* knob or slider located at the bottom of the Master track.

Figure 3.19
Metronome On/Off switch.

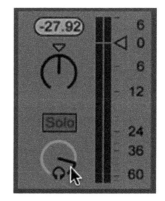

Figure 3.20
Adjust Metronome volume.

You can also loop record in the Arrangement View by selecting a specific range (timespan) in the Track Display and then recording multiple takes in *Loop Mode*: ▶ 8.2.2 . To do this, *click + drag* horizontally within Track Display following along the Beat Time Ruler to create a yellow highlighted selection, then select "Loop Selection" from the Edit Menu or *right click* to bring up a contextual menu ([⌘+L] Mac/[ctrl+L] PC). This will activate the Loop Switch, thus Loop Mode.

Figure 3.21
Activate the Loop Switch to loop a timespan selection in the Arrangement View.

Live will now record in an Arrangement Loop. Each record pass will be stored inside the Arrangement clip and will show up in the Clip View's Sample Editor as one continuous audio sample. To hear and playback your various takes, move the *Start Marker* to the point in the audio file where your desired take(s) occurred during loop recording. For an in-depth look at recording concepts and processes, refer to ▶ *Scene 8* .

3.4 MIDI in Live

The concept of MIDI in Live is no different than any other DAW's sequencer. MIDI is still just information without sound, which requires that you have some type of MIDI instrument to generate sound and MIDI data to tell that instrument what and how to create the sound. Therefore, you need to import a MIDI file or input your own MIDI note data. In order to input MIDI note data, you will either use the mouse, a MIDI keyboard controller, a trigger pad, or use the clever computer MIDI keyboard feature in Live. Whichever you use, an instrument must translate the data into sound. This can be a Live software instrument, third-party virtual instrument plug-in, or an external MIDI hardware synth. For now, let's stick to Live's built-in instrument devices. If you prefer to import a MIDI file or MIDI clip, then you can create an instrument after the file or clip has been imported. The order does not matter. Obviously, if you input your own data, then you will choose an instrument first, and then input notes to a clip when you're ready.

For the following walkthrough we'll use an Impulse instrument preset to playback MIDI. An Impulse preset will load up as an Instrument Rack. Your Impulse presets may vary from the example, depending upon which version of Live 9 you are using. To load an Impulse preset into your Set, first *click* on your MIDI track's title bar, then navigate to *Categories>Instruments>Impulse* in the Browser and *double click* on a preset's name or *drag* the preset from the Browser directly to a MIDI track or Drop Area. Choose a preset such as "Backbeat Room" or one that is well suited for standard acoustic MIDI drum kit loops.

Figure 3.22
Choose an Impulse preset from the Live Devices Browser.

3.4.1 Importing MIDI

There are two ways to import MIDI in Live: through MIDI clips (Live Clips) or MIDI files. Clips are exclusive to Live, whereas MIDI files are universal across all MIDI compatible software and related platforms. Amongst MIDI files, there are two types of files that you may encounter: multi-track (format 1) or single track (format 0). We will go over both since this dictates how many tracks will be needed and how MIDI signals are routed. Keep in mind that MIDI clips and MIDI files are different; but as far as Live MIDI Clips are concerned, they are formatted as single-track data.

MIDI clips and files can be imported via the Browser or directly from a Finder/Explorer window by *dragging* them directly to a MIDI track, the Drop Area, or when *double clicked* in the Browser. Depending upon the type of MIDI files you're working with (single or multi-track), Live will auto create tracks upon import to accommodate. For multi-track MIDI files, keep in mind that you will have to route any additional MIDI tracks' MIDI output channels to a software instrument in order to be translated into sound. That being said, don't forget that you must manually insert a software instrument to play back your MIDI files. By design, if you use Live Clip Presets, Live will automatically create the associated instrument upon import of the clip. This is very useful, especially if you are creating your own clip presets, which will behave in the same way. To learn more about these unique Live Clip features, launch to ▶ 15.2.3 .

Before you can start importing MIDI into your Live Set, download the CPP (Create, Produce, Perform) content from the companion website and install the legacy "Impulse" Pack, if you haven't already. This is also a good time to download the free Packs "Konkrete Breaks" and "Mad Beatz" from the Ableton website. Once downloaded, open Live and then *drag* the Pack onto the main Live screen. This will populate Live with a ton of MIDI clips and more: www.ableton.com/en/packs. Once you've downloaded the CPP Content, in Live *click* on "Add Folder ... " under Places in the Browser and then navigate to the CPP folder you just downloaded. Select the folder and then it will appear in Places ready for you to access. You can also access it directly from your Finder/Explorer window for this walkthrough.

To import a single-track MIDI file or MIDI clip to the Session View or Arrangement View, try the following:

(a) Navigate the CPP MIDI Loops folder and *drag* any MIDI Clip from it into the first Session View Clip Slot of your Impulse track or to its Track Display in the Arrangement View. This concept also applies to importing directly from a Finder/Explorer window.

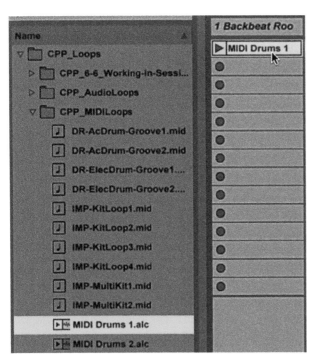

Figure 3.23
Drag any MIDI Clip from the Browser into the first Session View Clip Slot of your Impulse track.

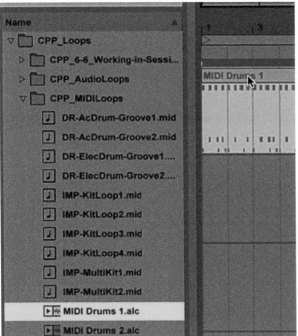

Figure 3.24
Drag any MIDI Clip from the Browser into the Impulse Track Display.

(b) Navigate through the same hierarchy, but this time choose a different MIDI clip and *double click* the clip to import it into the Session View, or *drag* it to the Drop Area in the Arrangement View. Either way, Live will automatically create a new MIDI track and the instrument associated to the clip!

Figure 3.25
Double-click a clip to import it into the Session View. This will automatically create a new MIDI track and the instrument associated to the clip.

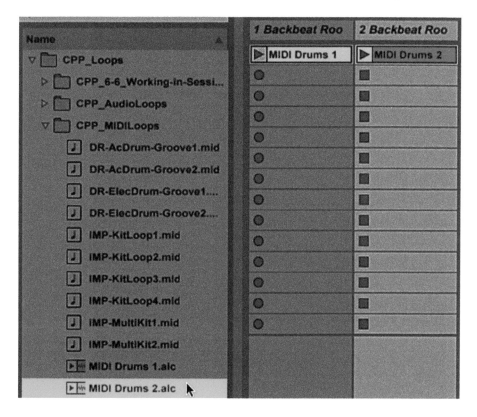

To import a multi-track MIDI file to the Session or Arrangement View, execute the following steps in succession:

1. Create a new Live Set (for simplicity).
2. Select one of the default MIDI tracks, then load up a "Drums" preset (Drum Rack) onto the track (*Categories>Drums*). For this example: "Kit-Core 808.adg".
3. Locate the MIDI file "**MT-KitLoop-1**" that you downloaded from the companion website in the folder you added to the Browser, then *double click*, *drag* it into the first unused MIDI track (not the "Kit-Core 808" track), or *drag* it into the Session View Drop Area. You could also *drag* it to the Drop Area in the Arrangement View. This will create the necessary MIDI tracks to accommodate

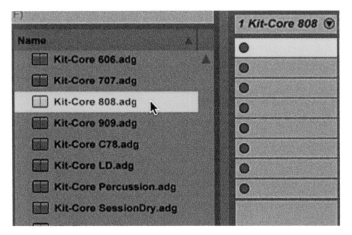

Figure 3.26
Select the default MIDI track in Session View and then double-click on a Drums preset to load it up.

the multi-track MIDI file: four in this case. Notice that MIDI files have different icons than MIDI clips. Use this same process for importing directly from a Finder/Explorer window.

Note: when importing to the Arrangement you may be prompted to import tempo and time signature data.

Figure 3.27
Double-click or drag MIDI file "MT-KitLoop-1" into the Session View Drop Area.

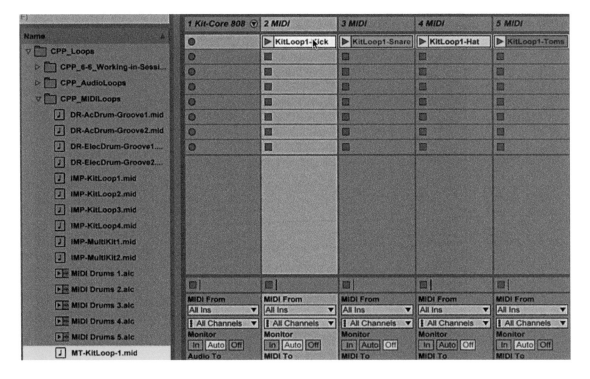

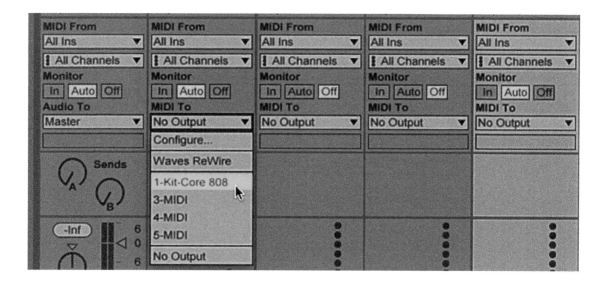

Figure 3.28
Route the MIDI To
chooser (Output
Type) to your
Drum Rack.

Figure 3.29
Route the MIDI To
Output Channel to
your Drum Rack.

4. Now you will need to route each MIDI track to the track that contains your Drum Rack. To do this, make sure that the In/Out section is shown in view, then select all of the new MIDI tracks by holding *shift* and *clicking* each one's title bar.

5. While still selected, use one of the multi-selected tracks to route the *MIDI To* chooser (Output Type) and *Output Channel* to your Drum Rack's name. The Output Channel chooser can also be routed to "Track In" to achieve the same result. It is possible to route each MIDI track to an individual drum sample in your Rack, but we'll keep it very simple and route to all channels for now.

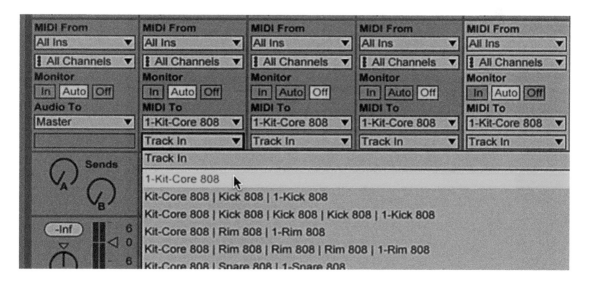

6. Last, set the "808" track to monitor input. *Click* "In" on the "Monitor" section or arm the track for recording.

Your Set should now look something like the example.

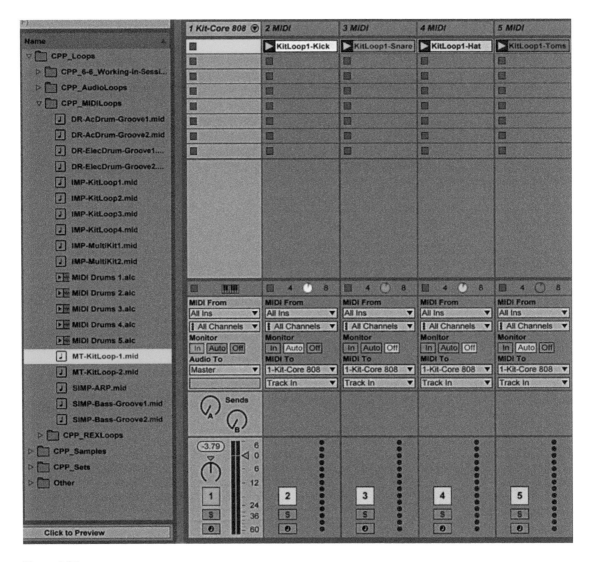

Figure 3.30
Example of multitrack MIDI routing.

Take a moment and launch the Scene—the entire row of clips—from the *Scene Launch button* on the far right of the screen, as well as trigger some of your Drum Rack's samples with your keyboard or controller. Remember to arm your Drum track in order to trigger it with a controller.

Figure 3.31
Scene.

3.4.2 Recording MIDI

To record MIDI directly into the Session or Arrangement View, follow these steps:

1. Starting with a new Set, load one of the same Impulse "Backbeat Room" drum kit from earlier into a MIDI track and set "MIDI From" to "All Ins" (if not already). You can also set this to your MIDI input device name instead.

2. Directly below "MIDI From", set the *MIDI Input Channel* chooser to "All Channels". Make sure the In/Out section is in view so that you can access the I/O section on the track.

3. *Monitor* should be set to "Auto", "Audio To" set to "Master", and then arm your MIDI track for recording by *clicking* on the track's Arm Session Record button.

Figure 3.32
MIDI routing in Session View.

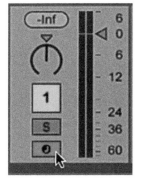

Figure 3.33
Session View.

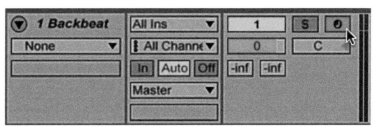

Figure 3.34
Arrangement View.

47

4. Now that your track is armed, you can begin recording:

    (a) *Session View: click* (Launch) the first Clip Record button in your Impulse track. This will initiate count-in (if enabled) and then recording will begin.

    (b) *Arrangement View: click* the Arrangement Record button on the Control Bar to begin recording. This will initiate the count-in (if enabled) and then recording will start.

Figure 3.35
Launch clip Record button.

Figure 3.36 Click the Arrangement Record button to begin recording MIDI directly into the Arrangement View.

Activating the *MIDI Arrangement Overdub* switch ⊞ (yellow when activated) will allow you to continually record MIDI notes to a MIDI clip, building it up with multiple notes or voices—for example, a drum kit—without erasing the existing notes recorded in previous takes or loop cycles except when they are the same note overlapping, hence "overdub". In this way Live is able to continually record your MIDI input, looping over and over until you tell it to stop recording, stop playback entirely, or disarm the Track. The overdub feature is great for inputting selective notes one at a time, pass after pass, or whenever you feel inspired to add more notes to your current MIDI recording. For more on looping, launch to ▶ 8.2 and for recording and loop record, launch to ▶ Scene 8 .

3.5 Essential Operations and Tasks

Now that you have had a chance to go over importing and recording audio and MIDI, let's take a moment to look at how to use a few of the more important and essential everyday operations and tasks for working in Live. Throughout the following group of exercises we'll use the same Set and media, so keep your work open as we go.

3.5.1 Looping your Loops

There is no doubt that at some point or another you will need to loop your MIDI or audio. The truth is that looping is the underlying principal of Live's Session View, and so you'll need to understand how to loop audio and MIDI clips. As far as the

Arrangement View is concerned, you may or may not have to loop audio and MIDI clips. This all depends upon how you create and produce your music. In the following section we will run through the quickest ways to execute looping operations with audio. As you read along, start to conceptualize how Live handles looping. You won't get it all right (that's not what this section is for), but when you get to ▶Scene14, you will be well on your way. To that end, think on this: clips loop in multiples ways. Clips are containers, therefore they loop as Session clips governed by their Clip View Loop Brace. They also loop as Arrangement clips governed by their physical clip length on the Track Display, yet dictated by the clip's Loop Brace. Got it?

Session Looping

Looping in the Session View is driven by the content of your clips. For example, it's possible to see a track in the session that never loops. Think of it as a one-shot recording intended to play through an entire song section without repeating. On the other hand, your rhythmic tracks are more likely to loop in two-, four-, or eight-bar phrases. Working with premade loops makes looping simple. When you *drop* one into a Clip Slot it will loop automatically. This goes for both MIDI and audio loops. If you recall from the audio *Preferences Tab*, short samples are automatically looped upon import unless you change that setting. Go ahead and try this out for yourself. Locate "**S3-AudioLoop-1**" and "**S3-MIDILoop-1**" from your CPP content folder via the Browser or directly from the Finder/Explorer window. *Drag* audio or MIDI Loop 1 into the Session View, then launch it and you'll see that it loops automatically. While the clip is looping, take a close look at the *Clip View* toward the bottom of the main Live screen and you'll see the clip's contents (waveform or MIDI notes) and playback looping in the *Sample Editor/MIDI Note Editor*. This view displays a clip's waveform or MIDI notes aligned to a grid—bars and beats. Also notice the *Loop Brace* that bookends the clip's contents, establishing the loop selection and cycle points of the loop.

Figure 3.37
Loop Brace in
the Sample
Display/Editor.

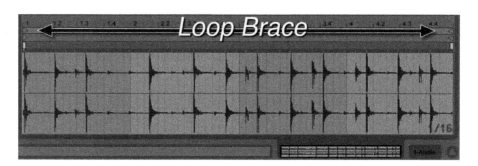

Move the Loop Brace *Loop Start* or *End* (bracket) to customize and loop a specific selection. You can also move the entire Loop Brace by selecting it (*clicking* and highlighting) or move the Loop Start/End individually. *Click + hold* and *drag* to reposition the Loop Brace as a whole or just the Start/End points.

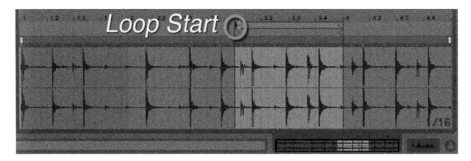

Figure 3.38
Loop Brace/Loop Start.

Taking this a step further, you can set a unique clip Start Point that differs from your Loop Start by moving the *Start Marker* on its own.

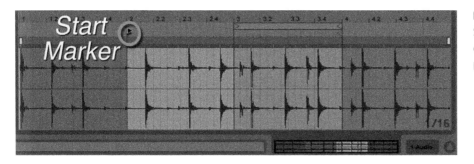

Figure 3.39
Start Marker determines where playback begins.

Arrangement Looping

Looping in the Arrangement is very similar to any traditional DAW or sequencer. A clip can be set to loop in playback by setting the Arrangement's Loop Brace to a specific timespan selection, then activating the Control Bar Loop Switch. You can also select a clip(s) and *press* ([⌘+L] Mac/[ctrl+L] PC) to set and activate the Loop switch all in one keystroke.

A clip can also be activated to loop internally—as mentioned—then physically looped for a range of time along the Beat Time Ruler by *dragging* the clip's edges to extend its length. The difference is that loop playback is for the Arrangement as a whole and a clip looping is about the clip itself within an arrangement.

Figure 3.40
Arrangement
Loop: Select a clip
or timespan with
the Arrangement
Loop Brace then
activate the Loop
switch.

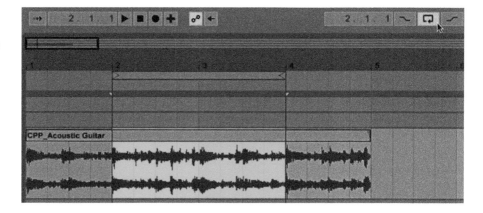

Figure 3.41
With a clip's Loop
switch activated it
can be looped
infinitely by
dragging its edge
to extend its
length.

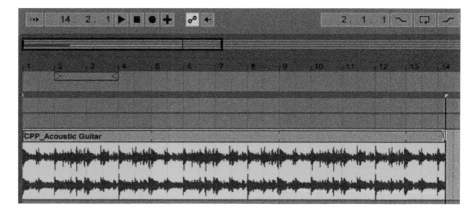

If the clip's *Loop switch* is deactivated in its Sample Box (Clip View), the clip cannot be extended beyond its set length. The alternative then is to simply duplicate the clip in the Arrangement View ([⌘+D] Mac/[ctrl+D] PC) as many times as necessary along the timeline.

Figure 3.42
Manually duplicate
a clip in the Track
Display to simulate
looping by
repeating.

A big difference with looping in Live's Arrangement View than other DAWs is that when a clip's Loop Brace/loop points are altered, the actual audible selection changes with it in the Arrangement. This is because a clip's Loop Brace determines what content is looped and the Start Marker dictates where in the waveform/MIDI playback begins. When a clip Loop Switch is deactivated, the Start/End markers dictate its playable length in the Arrangement View. For more information on this, read about takes in Clip ▶ 8.2.3 , which talks all about how Live handles Start Markers and the Loop Brace, etc., in relation to looping and takes.

3.5.2 Adding Effects

With all of the effects built into Live 9 and Suite, you'll definitely want to learn how to use them quickly and efficiently, keeping them ready at your fingertips. Audio effects can be applied to any audio track, which includes Returns and the Master track. They can also be connected to the output chain of any instrument. MIDI effects can be applied only to MIDI tracks and must be placed before instruments in the device chain. Simply *drag* the effect you want from the Browser to the appropriate track or *double click* it in the Browser. To put this into practice, let's add an audio effect to one of the audio loops located in the "**CPP_AudioLoops**" folder from the companion website. Feel free to use your own audio if you like. Let's start with Live's *Auto Filter*.

Audio Effects

1. Using a new audio track, *drag* any one of the audio loops into the first empty Clip Slot in the Session View.

2. Now launch the clip and make sure that it's looping (Loop Switch activated in the clip's Sample Box). *Double click* on your audio Clip to show the Clip View.

3. While the clip is running, *drag* the Auto Filter effect (not a preset) from the Audio Effects folder in the Browser to your audio track, or *double click* it to insert the effect on the selected track. Next, *click* on the *Hot-Swap* button in the upper right corner of Auto Filter 🔄 and select "Cut-O-Move L" from the Auto Filter presets folder in the content pane. Pretty cool, huh?!

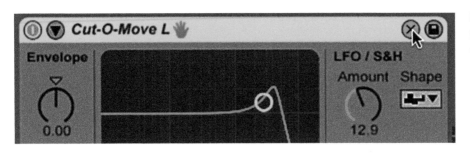

Figure 3.43
Hot-Swapping device presets.

4. Let the loop continue to play and *click* the Hot-Swap Preset button to turn it "On" again (if not already) so that you can audition other presets within the Auto Filter preset folder.

5. Listen to a few different effects and presets by *double clicking* each one. This will temporarily load them in place of the previous effect. During this process you will be so inspired that you'll want to stop and play around for a while. Before you get too distracted, let's add two more effects to create a chain.

6. Delete and reload Auto Filter. You can easily delete a device by selecting it from its title bar at the top of the device and choosing Delete in the Edit Menu or using the delete/backspace key on your computer keyboard.

7. Once you have reloaded it, navigate to Flanger in the Browser and *drag* the default Flanger effect to the right of Auto Filter in your *Track View* (you could also *double click*).

8. Select *EQ Eight* and add it to the right of *Flanger*. Leave Flanger as is, and move the EQ Eight's settings around with your mouse.

Figure 3.44
Customize frequency band settings with your mouse within the graphic display.

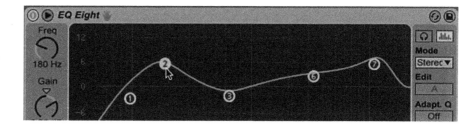

9. Now listen to your loop with and without the effects. To do this, deactivate and reactivate each effect on-the-fly by *clicking* the *Device Activator* button ⓘ in the upper left corner of each of the effects. This should help you to get familiar with the interface while observing the audible differences. You can also switch the order and placement of the audio effects in Track View by *clicking + holding*, then *dragging* the highlighted audio effect *left* or *right* to a new position alongside your existing two audio effects. A vertical yellow line will appear between the effects when you reach the correct Drop Area.

Figure 3.45
Three audio effects inserted on a single audio track.

MIDI Effects

For adding MIDI effects, let's use a Live software instrument and a MIDI clip "SIMP-ARP.mid" located in the "**CPP_MIDI Loops**" download. If you have a MIDI keyboard controller or want to use the computer MIDI keyboard, then you may choose not to use the CPP MIDI clip provided and play notes on your own instead. Since your instruments and presets will vary depending upon your version of Live 9, download the free "Solid Sounds" and "Samplification" Packs from the Ableton website. You'll find some nice instruments in there:

1. On a new MIDI track, load up the Instrument Rack preset "Hard Picked Dist Lead Guitar" (*Instruments>Instrument Rack>Guitar and Plucked*). This uses Live's Tension instrument! Feel free to use any other plucked-type staccato instrument for this exercise, such as a Simpler/Sampler preset.

Figure 3.46
Instrument Rack
(macros) in Track
View.

2. From the "Hard Picked" Macros section shown above, set "Release" to "23%".

3. Add an Arpeggiator preset "ClassicUpDown 8th" (*MIDI Effects>Arpeggiator> ClassicUpDown 8th*). Change the *Rate* to 1/16 and *Gate* to "200%" from the Arpeggiator interface.

4. Now *drag* in "SIMP-ARP.mid" into a Clip Slot, *launch* it, and take a listen. If you would rather play your own MIDI, then do so by playing long sustained chords. This way you will hear a nice arpeggiated effect on your MIDI instrument. The Arpeggiator is synced to the tempo of your Set, so feel free to adjust it to taste.

Figure 3.47
Device View
showing a MIDI
effect, Instrument
Rack (Macros), and
an audio effect
creating a complex
instrument.

5. Last, we'll add an audio effect to the output of our Instrument. Add the default Ping Pong Delay audio effect by *dragging* it to the "Hard Picked" MIDI track or to the end of the instrument/output chain in the Device View just after the "Hard Picked Dist Lead Guitar" preset (*Audio Effects>Ping Pong Delay*).

    Now *hold* chords out again and enjoy the complex rhythmic effect.

Hot Tip If you don't have a MIDI keyboard controller, then you can use the computer keyboard to supplement. This feature can be activated from the upper right corner of the main Live screen next to Key and MIDI:

Figure 3.48 Computer MIDI Keyboard On/Off Switch.

It will light up yellow when activated. Hold down multiple notes to hear the arpeggiated effects. The general layout is the row of "A, S, D, F, G, H, J, K, L" keys resulting as the white keys on a keyboard and the "W, E, T, Y, U, O" keys designating your black keys. This is a very useful tool for inputting MIDI notes on the road with your laptop, and it works for all MIDI instruments!

3.6 Performance to Arrangement

The final step of the creation and production process is to record a session performance into the Arrangement View. This is one of the features that makes Live special. It is so unique and creative that you will definitely want to spend some time reading about it in ▶ **Scene 4** Global Recording. For now, we'll just scratch the surface of this concept. For this exercise you can use either the "Hard Picked Dist Lead Guitar" instrument or the audio loop you used in the previous exercises.

In order to record your Session performance to the Arrangement View, you should follow these important steps to learn early on how to avoid confusion or launching of clips abruptly:

1. *Click* the Stop All Clips button in the Master track of the Session View to clear any clips that are paused. When paused, the launch button will be illuminated green while playback is stopped.

2. *Double click* the Stop button on the Control Bar transport to return your Arrangement to the beginning of the timeline (1.1.1 start time).

3. Now, if you're ready, *click* the Arrangement Record button on the Control Bar. After the count-in (optional), the Arrangement will begin recording and logging all of your actions made in Session View into the Arrangement View. *Launch* your audio or MIDI clip so that it plays while the Arrangement is recording. Let your clips loop for about 15 seconds or so, then *click* a Clip Stop button on the same track to stop it. (You could play and record Simpler in real-time if wanted.)

4. *Click* the spacebar or Stop button on the transport to stop the Arrangement recording/playback.

5. Now, *press* the Tab key to flip over to the Arrangement View to see your performance and then *click* the Back to Arrangement button.

6. *Double click the* Control Bar Stop Button to return the beginning of the timeline and then *press* play to listen back to your Arrangement.

3.7 Finishing Your Work

Once you get to the final phase of your work, or at least to the end of your workday, there are a few considerations and operations you must deal with before calling it quits. Saving and organizing your work are essential operations to completing a successful experience with Live. This includes exporting an arrangement or audio tracks and managing your files.

3.7.1 Exporting Your Audio

When you are finished working on your song, you will most likely want to share it or listen to it as a final two-track mixdown. To do this, you will have to export your audio out of Live to disk—automatically share it to SoundCloud too! Other DAWs may refer to this process as bouncing to disk. In Live, it is called exporting or rendering. The exporting process offers many options, such as rendering an entire song, track(s), clip(s), or video. Although this may sound like it's an easy process, it can get confusing very quickly. It all begins with what output source you want to render from and what content you want to render (Session clips or Arrangement clips). Although you will see that each view presents a different protocol for executing the rendering process, ultimately the view you are rendering from has no bearing on the actual rendered output. The rule is that whatever is supposed to play will be rendered regardless of the view that contains the content.

Arrangement

To export an entire arrangement or clips from the Arrangement View, select a timespan across the Track Display, and then select the *Export Audio/Video* command from the File Menu ([⌘+shift+R] Mac/[ctrl+shift+R] PC). This brings up an Export Audio/Video window that allows you to choose the desired output source, method for exporting your audio, and export settings (i.e., bit depth, sample rate, and so on). Choose an output source to render: Master track, Return track, individual track, or all tracks. The rendering of all tracks individually is also known as creating stems. Whichever you choose, just make sure that what you are listening to is generated from the Arrangement View and does not include clips from the Session View, unless that's what you want. Without getting too deep into the conceptual relationship between the Session View and Arrangement View, understand that the views are separate from one another, but they are capable of sharing playback when clips in both views are playing simultaneously, as long as they don't share the same track. Session clips always have priority. That being said, what you hear during playback is what will be exported, unless you choose a single track or all tracks as the Rendered track source in the export window. In such a case, the render outputs will correlate directly to what is to play from its track output. So when it comes to rendering only from Arrangement tracks, make sure that the Back to Arrangement button is not illuminated. This will ensure that you are exporting only the stored arrangement.

Session

Rendering a Session clip or Scene can be a bit trickier than in the Arrangement View. For this, you have to launch your clip(s) or Scene, then stop playback with the Control Bar Stop button or spacebar (pausing clips that are playing) so that their clip launch buttons remain green. This is in lieu of *clicking* their Clip Stop buttons. Think of them as if they are on standby. The clips you have been listening to (now on standby) will be what is rendered, meaning that what you would hear during playback is what will be rendered. Of course, all of this is contingent upon what output you choose to render from (i.e., Master, track, all tracks, etc.). Make sure to set the "Length [*bars.beats.sixteenths*]" in order to render the selection of audio you wish to save to disk.

3.7.2 Set versus Project

Creating, producing, and performing with Live begins with a Set. As you already know, this is the environment you have been working in all along. Sets are stored in Project Folders, which optionally store all of the media related to your Set(s). Sets can either refer to media located in various places such as the Core Library, custom folders on your computer, or from its own Project folder.

Figure 3.49
Project Folders store Set(s) and their related media.

This includes samples (imported, recorded, and processed), instruments, clips, presets, and other pertinent information. When you save a Set for the first time, it will automatically be saved in its own new Project Folder named to match the Set name. Live Sets are identified by the *.als* file extension. As you work, save as often as you like. This overwrites the current Set each time you save. You can also save your Set as a new name inside the same Project folder, or outside in a new location. The purpose of Project folders is to store the information for a particular body of work (i.e., a project with multiple drafts of the same Set). If you are saving it outside the folder, Live will again create a new Project folder matching the name of the new Set. If you save a Set with a new name outside the current Project folder structure that contains imported and or recorded audio, the samples will continue to reference the original project's samples folder. This is fine for now, but when you take your project to a different computer or delete the other project, you will be missing your samples. Not a good situation to be in. To avoid this, use Live's *Collect All and Save* feature accessed from the File Menu or the File Manager before taking your project on the road or archiving it. It is imperative that you manage your Projects and Sets!

Here's a quick example. Let's say that you have been working on a Set and you have saved and named it as usual. If you look inside its Project folder you will notice that if you have recorded audio, it is there in a samples folder, but if you imported audio, it is not. This is because Live is referencing the imported audio from its original file folder location. If you manipulate or process the audio, it will then show up in the Project folder. This is why you use the "Collect and Save" option before you take your project on the road. The general procedure is as follows: when you first create a new Live Set, choose "Save Live Set" form the File Menu. When you want to save as a different name within the Project folder, then choose "Save Live Set as … " and if you want to save it to a new location or as a separate back-up, choose "Save a Copy … "—which creates a new name and file directory—or "Save Live Set as … ". If you want any one of the Sets that you just saved to be self-contained, then open that Set up and choose "Collect and Save". It's up to you how you want to manage. When you do this, Live will prompt you to select the files you wish to copy into your Project folder, which includes copying Library content if you so choose. You can also access the more detailed save options from the File Manager.

One key difference with Live in regards to the saving process is in the Live Browser itself. You can simply *drag* a clip directly into folders in the Browser Places to save your clip as a Live Clip. You can *drag* a track(s) or an entire selection of clips to Places, thus saving them as a Live Set (.als). Choose how you wish to copy your samples and assorted files for this feature in the "Collect Files on Export" chooser in the Live Preferences *Library>Browser Behavior* Tab or when prompted. This is a very productive and strategic way to keep your sets organized and ready to load at a moment's notice while you are running Live. Once you develop a system of dedicated folders for your projects, this method becomes an efficient way to save your Sets and all-in-one Clip Presets. For more on customizing and saving presets with the Live Browser, launch to ▶ **15.2.4** .

As easy as it may seem, to close your Live Set, select "Close Live Set" from the file menu or use ([⌘+W] Mac/[ctrl+W] PC). This will close your Set and reopen the default Set. If you choose the close window button in the upper corner of the main Live screen, the Live program will quit as opposed to just the Set.

3.7.3 Managing Files

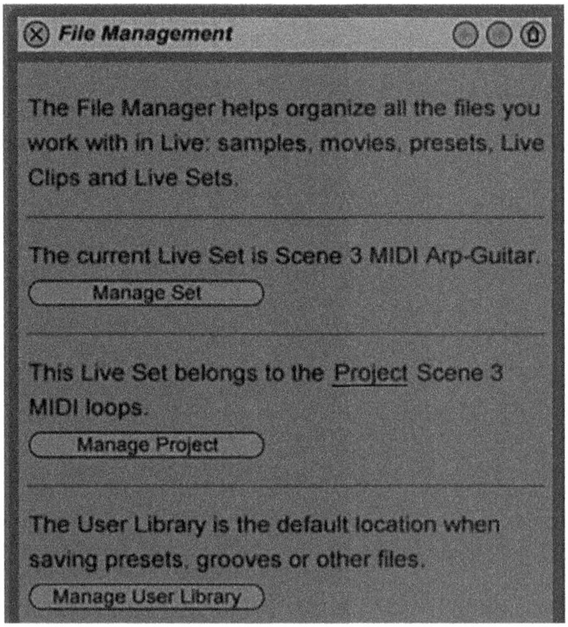

The File Manager helps to manage and administrate all of the files contained in your Set, Project, and in the User Library. This is necessary since the content in your Sets can be referenced from various locations rather than from the actual Project folder as mentioned. The File Manager provides a way to view all of your media, locate any missing files, and copy the external files—those being referenced from elsewhere—to your current Set, among other things. Managing your files is very important. On the one hand, you want your projects to be self-contained and complete when you transfer them to another studio or computer, but you don't want to copy your entire Core Library into a single Project while working at your computer. That would be redundant and would use up a lot of hard drive space. To ensure that situations such as these don't arise, use the File Manager to manage your Sets and Projects, and Collect and Save. If a sample is missing, or unable to be located when you open a Set/Project, Live will warn you. Use the File Manager to search and rescue your files—all part of the Live Browser system.

Figure 3.50
The File Manager, when in view, is located to the right of the main Live screen.

To use the File Manager, go to the View Menu and select "File Manager". It will appear on the right hand side of the main Live screen where the Help View would have been located. From there, choose what to manage. There are three options: *Manage Set* is specific to the current Set; *Manage Project* catalogs all of the Sets and information included within the Project Folder; and *Manage User Library* indexes and lists important information related to the User Library.

Searching for Missing Files

At some point in your Live career you will run into a scenario where one of your Sets loses track of some important files that may have been moved or renamed, etc. Things like this do happen. How do you know when there are missing files? Live will notify you. You'll see the Status Bar at the bottom left of the main Live screen highlighted *orange*, displaying "Media files are missing. Please click here to learn more." Great, now what's the solution? Live has a built-in search feature that is accessed through the File Manager.

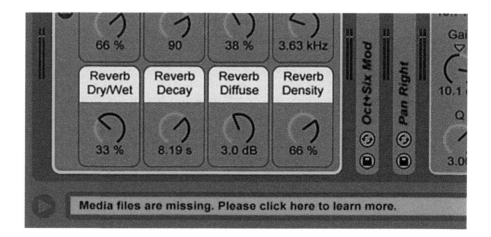

Figure 3.51
Status Bar Media
Files Missing!

So there is no need to panic. *Clicking* on the orange prompt will immediately open up the File Management Menu. Here you have a few options to help you find your missing file(s). If you have a general idea of where they were last located, you can tell Live where to search by choosing a folder directory and activating the *Search Folder* button. If you don't have a clue, then let Live do all the work. Either way, select *Go*.

Look closely at the top of the File Management Menu and you'll see an exact number of files missing from your current Set. This count references the listing of files located further down in the menu, which Live will look for.

You also have the choice to manually find and repair your files by *clicking* the *Find File* button. This will open up the Browser where you can start a search similar to Hot-Swapping. More information on this can be found by selecting the "Tell me more" prompt at the top of the menu next to the Missing Media File count. More often than not, Live will successfully locate your missing files. Whichever process you choose, it can take a while depending upon the size of your hard drive and various locations of your files in question. Be patient; nobody likes missing or corrupt files, but Live's File Manager will help ease the pain. Once the candidates are located, select "OK" and then Save from the bottom of the menu.

Creating a Pack

All this talk about downloading Packs, but what about creating your own? Don't overlook that the File Manager is not just for correcting missing file associations, it's also a very powerful tool for collecting and saving your media into one location, especially with its ability to save your entire Set as a Pack. This option is found

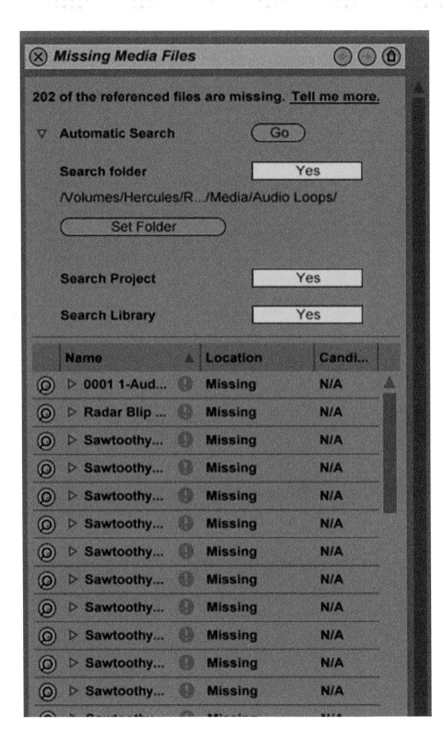

Figure 3.52
The File Manager
Menu: Missing
Media Files.

within the Manage Project area within the File Manager. *Click* on "Manage Project" and select the triangle unfold button named "Packing" where you can select "Create Pack". In doing so, name it and choose a location to save it. This is used for copying and compressing the entire contents of your project into a single .alp file easy for archiving and sharing. Live will use a specific lossless compression process to minimize your Set to a smaller size. Not only does this make it easier to transfer, but it is also a safe way to archive your precious data.

Global Recording:
Arrangements "On-the-Fly"

4.1 Introduction

As a musician, composer, or performer, Live provides you with the tools to literally launch your music in a forward direction: forward and constant motion like a band of musicians jamming on stage, striving for that perfect flow. This is what sets Live apart from other Digital Audio Workstations (DAWs); but by the same token, with so many unique ways to control your music, it can quickly become overwhelming when you are first starting out. In this Scene you will learn how to elevate your music production to the next level, making it more exciting and even better than it was before. The focus here is on Live's global recording concept and how to utilize those features to create, produce, and perform music.

Remember to keep an eye out for launch points to Scenes ▶ **Scene 1** and Clips ▶ **1.3.1** as you read. These will help to connect related concepts for expanding your knowledge and focus in on your specific area of concentration.

4.2 The Global Recording Concept

No matter which sequencing software you have used prior to Live, one thing is for sure, they all share the same basic design concept: *linear recording*. This means the program follows a horizontally aligned timeline that guides your music (audio/MIDI) in a strict sequence over time. However, Live goes beyond the linear timeline in *non-linear recording*. Non-linear means not in a straight line—rather, out of sequence or free of, with no strict or confining order. Think of it as a coloring book without lines or a clock without numbers. Everything is based on relative timing. So, what does this have to do with the global record concept? Well, the global record concept is all about giving you the composer, producer, and performer a bridge from relative time (Session) to linear time (Arrangement)—in other words, capturing musical improvisations and performances into a recorded sequenced arrangement.

> **Info Box**
>
> The **Arrangement Record Button** is located on the Control Bar Transport. When the Record Button is activated, it records all of your actions into the Arrangement.
>
> Use this to capture a real-time performance, improvisation, or to create an arrangement of your Session clips.
>
> ▶ **3.6**
> ▶ **5.7.1**

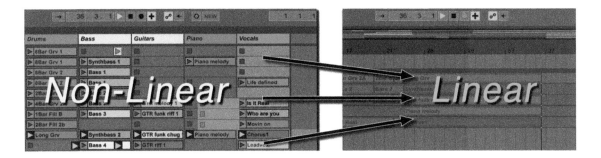

This concept can be mind-blowing, but it is a powerful, inventive, and a logical way of realizing music. So, how does this all work? Well, let's start by taking a look at one of the most important features in Live: the *Arrangement Record button*.

The Arrangement Record button is located at the top of the main Live screen between the Stop and MIDI Arrangement Overdub buttons on the Control Bar Transport. At first glance this button ![button] looks like a standard transport record button, but it's the concept behind it that unlocks a whole new way of approaching and arranging your music productions. It gives you the ability to create linear musical arrangements out of non-linear musical sketches and improvised musical ideas through a performance in real-time. Not only does it provide the power to create these arrangements free of a linear timeline, but it also allows you to perform non-linear improvisations in real-time—on stage or in the studio! Don't worry; in no way does Live abandon linearity; it just adds the ability to create music on-the-fly free and unconfined.

Figure 4.1
Capturing a Session View performance into an sequenced arrangement in the Arrangement View.

Figure 4.2
The Arrangement Record button.

4.3 Music On-the-Fly

As a musician, pretend that you are holding a large handful of written phrases, index cards labeled A to Z, if you will. Now, if you were asked to place these phrases contained on the cards into an arrangement, you'd probably begin laying them down on a table one by one in alphabetical order, or a relative, but logical sequence. After placing them neatly and reading them back, you might decide to pick some up and reorder them a bit to improve their overall arrangement. That's fine, but imagine if you could simply toss these card phrases onto a table one at a time, or handful at a time in any order instead of neatly placing them. Then imagine that they are read back to you instantly in the order they land and perfectly aligned in view without overlapping. Now you can read them again in their new arrangement just as they fell to see what worked. Ok … not a bad arrangement; in fact, you might find it quite inspiring the way card, let's say C, transitioned into card A this time. Possibly a transition you would never have thought of, but now it would be really cool to run that sequence and transition twice and then end with card D altogether, repeating four times. Interesting and inspiring, but the arranging doesn't stop there. You can pick them up and toss these parts again on the table in the new order just described and off you go! Whether you realize it or not, you have just created a new arrangement in real-time without shuffling, swapping, or moving any card along a fixed sequence in any way. You simply threw the cards on the table in an order you felt could sound good as a whole. With the musical phrases right in front of you, you essentially created an arrangement all in real-time, launching any card you wanted at any time. In perfect sync too! This analogy is no different than working in Live's Session View. Just substitute the index cards of musical phrases for audio and MIDI clips.

4.4 User Interfacing: Two Parallel Worlds

Live consists of two main views that make up its overall interface. This is critical! These views, Session (vertical) and Arrangement (horizontal), share the same tracks/channel strips, instruments, plug-ins, effects, and tempo, etc., but function independently in regards to musical performance and arrangement playback. Although they do work in tandem in relation to signal flow via the track signal flow and architecture (audio and MIDI input/output), the Arrangement and the Session Views are completely different in functionality—in other words, until the Arrangement Record button bridges the two together, granting you the ability to launch audio and MIDI parts from one view into the other as a linear arrangement in real-time. This is how Live's global record concept is used for making music on-the-fly. It is the ability to launch your audio and MIDI parts in real-time from the Session directly into the Arrangement, even while experimenting with effects, loops, and automation. With that unique interaction between the Session and Arrangement Views bridged by the Arrangement Record button, the sky is the limit. Taking it one step further, once you have captured a musical arrangement from the Session View, you can edit your tracks with a fine-tooth comb, down to the barest detail within the Arrangement View. To get a better handle on this, let's examine both views. The goal is to ensure that you grasp this parallel concept from the very beginning since it is an entirely unique way to record and sequence arrangements.

> **Hot Tip** *To increase the overall display size of your main Live screen, go to Live's Preferences Look/Feel tab and scale the Zoom Display to increase or decrease your workflow environment. This applies all views in Live.*

4.4.1 Arrangement View

Working along a linear timeline is great for sequencing, editing, copying, and pasting MIDI and audio. It's been done that way for years. In fact, the most common way to experience music is in a linear progression. This is because music is experienced through a set beginning and end that encapsulate a sequence of events that follow a strict pattern or timeline. The only exceptions are forms of improvisations, or jam sessions where things take a few unpredictable twists and turns before coming to an end.

The Arrangement View can function in two ways: first, as a traditional sequencer, providing a view where you can record, input, import, and edit audio and MIDI events in tracks along a linear timeline; and, second, as a canvas for recording whole multi-track musical performances on in real-time. It's totally up to you and your project's needs. The main point is that the Arrangement View in Live gives you the ability to create an entire musical arrangement in real-time as a predesigned or improvisational performance rather than just piecing it together one region (clip) or track at a time. Sure, you could *click* and *drag* tracks or parts in across the timeline and paste them together, but one of the best things about using Live is that you don't have to. Live provides the freedom to choose between two ways of working: a launch pad style of creating a production, or a more traditional linear way of recording and editing. There is no right or wrong way to work; the methods work great together as well as wonderfully on their own.

The general layout of a completed arrangement should look quite familiar in this Arrangement View example (Figure 4.3).

You can see that the Arrangement View displays all of the usual information found in a traditional DAW such as tracks, MIDI events, audio waveforms, regions, automation, etc.

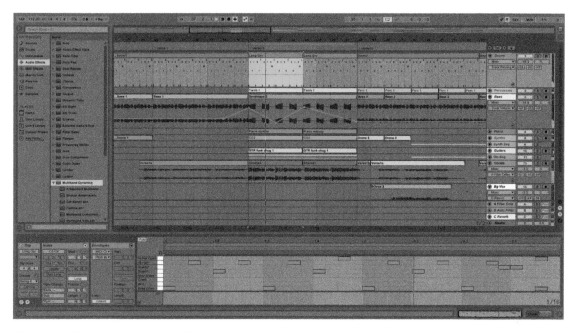

Figure 4.3 The Arrangement View displaying an arrangement.

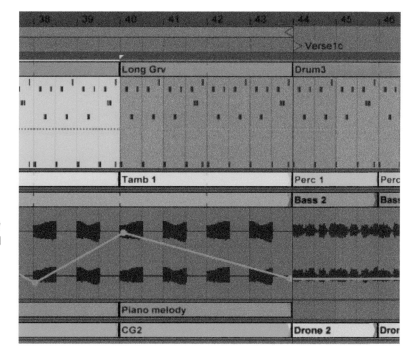

Figure 4.4
Arrangement View displays traditional linear/horizontal tracks that can store MIDI notes, audio waveforms, automation and much more.

This particular example is a direct result of launching clips in the Session View as a performance and capturing them into the Arrangement View by using the Arrangement Record button. Now that you know what it looks like as a finished product, understand that an arrangement is captured and logged (recorded) in real-time as a performance is generated from the Session View. This could either be a completely predesigned arrangement that is simply launched as a whole or an improvisation of clips performed on-the-fly from the Session View. From here, the arrangement can be altered however you see fit. For an in-depth breakdown of the Arrangement View details, launch to ▶ Scene 6 . Of course, clips can be recorded directly into the Arrangement without ever existing as a clip in the Session View ▶ 8.4 .

4.4.2 Session View

Live's Session View acts as a musical sketch and launch pad, allowing you to try out new ideas and improvise freely, as mentioned earlier. Each Clip Slot in the Session View grid can hold a predefined audio recording (audio file/sample), MIDI file, or can be recorded directly into on-the-fly. Audio/MIDI can also be easily imported through the Live Browser or Finder/Explorer window. Once clips exist in the Session, they can be launched for playback in any order and at any time you desire.

Info Box

The **Session View** serves as a musical sketch and launch pad for trying out new ideas and improvising freely. Each Clip Slot in the Session View can hold any type of musical idea. ▶ Scene 5

Hot Tip Ideas can be recorded on the fly or dragged in and played in any order and at any time. Clip performances can also be dragged and dropped between the Session and Arrangement View. ▶ 5.7

Let's dive deeper into the Session View and look at how clips and Scenes make music flow. This will help you to understand the different ways in which you can arrange music on-the-fly with the global record concept. For a closer look at the Session View and how it works, launch to ▶ Scene 5 .

As shown, Live's main screen is designed to accommodate both Arrangement and Session Views along with sidebars for the Live Browser and additional detailed overviews. Unlike the Arrangement View, the Session View has a totally original look and feel that is different than your old DAW. The only similarities are the mixer channel strips with volume unit (VU) meters, faders, and the record, mute, and solo buttons at the bottom of each track. (Figure 4.5.)

4.4.3 Clips

Live displays both audio and MIDI as a self-contained block of information or container called a *clip*—audio clip and MIDI clip. Each one of these clips

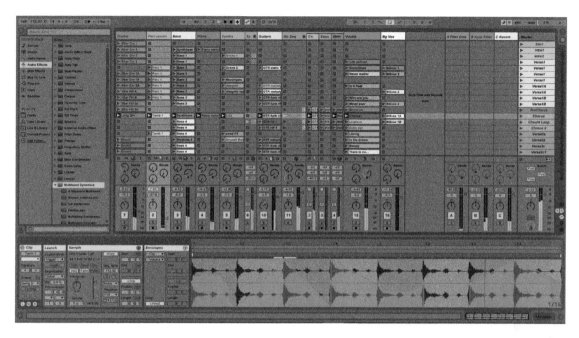

Figure 4.5 Session View, Live Browser, and Device View (Detail View).

holds unique information, not just audio and MIDI events; but we'll get to that later.

All clips have a built-in activity button called a *Clip Launch Button*. It is a triangular button located at the left of the clip. When it is pressed, the clip is activated and starts to play back.

There are many ways to describe the clip launch function—"firing off a clip", "playing a clip", or "launching a clip". Every track can contain any number of clips and has a unique *Clip Stop Button*. These buttons are located in empty clip slots except when the track is armed for recording (record-enabled). There is also a Clip Stop button in every track column just above the track Mixer section in the *Track Status Display* area. Having a stop button on each track is an integral part of the Live's real-time power! They will stop any running clip that is located in the same track as the Clip Stop button itself. Clips can also be stopped with the *Stop All Clips Button* on the Master Track. A common misunderstanding of Live is to use the *Space Bar* or *Control Bar Stop button* to stop playback of all the clips and to deactivate their playback. Although they do stop playback, they actually only pause the clips. You will notice that the Clip Launch buttons are still green (active), meaning that

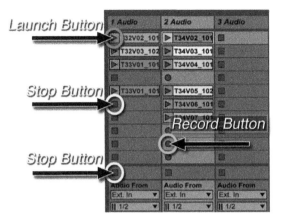

Launch Button

Stop Button

Record Button

Stop Button

Figure 4.6 Stop All Clips Button.

Figure 4.7 Control Bar Stop button. This is the same as the computer keyboard spacebar.

the clips are in paused and ready to restart upon when playback starts again. There are uses for this, specifically exporting/rendering your audio. Launch to ▶ 8.5.1 for more.

While launching clips you will come across one restriction: *only one clip can be active at any given time within a track*. So, what does that mean? It means you'll use Scenes. What are Scenes? Let's read on.

4.4.4 Scenes

The Session is built on a grid system made up of rows and columns (Session Grid). Each horizontal row is a Scene and vertical columns are tracks. Since tracks can only play one clip at a time, clips are often spread out across multiple tracks as rows known as Scenes. As mentioned earlier, clips are the basic musical building blocks of Live. These musical ideas— melodies, beats, and bass lines, etc.—can be constructed into larger Scene structures. In this scenario, Scenes form the musical structure of your songs or compositions in the Session View. Scenes are labeled, organized, and launched from the Master Track column. Putting Scenes in the context of global recording arrangements, they are generally used to launch entire musical sections of a

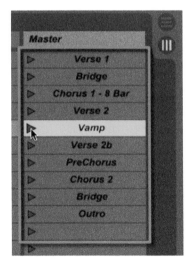

Figure 4.8
Scenes organize and launch entire horizontal rows of track clips all at once.

performance such as a verse, chorus, bridge, or outro, etc. In this way you are able to activate an entire set of clips simultaneously as opposed to each individual clip one by one. Of course, a Scene can be used in many different ways as determined by you, the user. There will be much more on how to use Scenes creatively in the sections to follow. Additionally, you can launch to ▶ **Scene 5** to see how Scenes function within the Session View and to ▶ **Scene 9** for an in-depth look at their advanced functions and musical concepts.

4.4.5 Session View as a Mixer

Even though the Session View and Arrangement View operate independently, their Mixer Sections are linked. This includes signal routing, I/O, Sends, and all of the Mixer's functionality. The Arrangement and Session View share all of the same controls, parameters, and information. The difference is that the Arrangement is laid out in a horizontal view and the Sends and Pan controls have been moved to the Track Display via the *Control Chooser* instead of knobs as in the Session View. In other words, the Session View Mixer Section with its traditional vertical faders is ideal for mixing your music regardless of which view sound is generated from! You can work on your mix from whichever view you prefer, but most users will gravitate to the vertical track mixing of the Session View and rely on the Arrangement View for fine-tuning and automation of a mix. For more on Track/Mixer Automation, launch to Clip ▶ **10.2** .

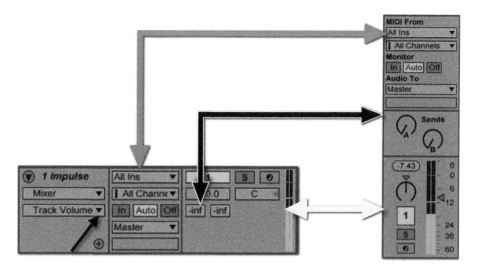

Figure 4.9 Tracks and Audio/MIDI signal flow is shared between the Session View and Arrangement View. This includes I/O, Sends, Returns, and Mixer Section.

4.5 The Linear Approach

By now you have heard the term "linear" mentioned several times throughout this book. There is a good reason for that. Until the advent of the DAW, we were unable to randomly move or process data outside of real-time. In the same way, sheet music is written and subsequently performed linearly. In fact, we need timelines and the ability to string musical ideas together so they can generate a consistent and repeatable sequence—a good reason to not throw the concept out the window. Sequencing is the heart of music programming and for that reason Live still embraces the linear arrangement. There are two ways to work in a linear fashion in Live: first, build and sequence your arrangements solely in the Arrangement View; or, second, build and perform a precise arrangement in the Session View.

4.5.1 Create

As a composer, you can use the Arrangement View to write out your musical ideas just like you always have. A great program inspires the composer and Live is definitely built to kindle and empower your imagination, inspiring unlimited creative possibilities. The ability to record audio and MIDI events is the key building block for the digital composer; therefore, ideas can be composed modularly in the either view. This includes simply adding prerecorded audio and MIDI clips. From a composer's perspective, think of the two views as working together. The Session is your musical sketch, and the Arrangement is your finished work. Of course, nothing in a DAW is permanent. You can edit and make changes as much as you like. Another way to think of the concept is that the Arrangement is a multi-track recorder that is used to record your session sketches (clips) as they are played in real-time for when you are ready to print them as a final sequence, song, arrangement, etc. This direct link between the Session and Arrangement View is very significant for composing. Let's take a closer look at specific ways to utilize this relationship. First and foremost, keep in mind that there is no one right way to work in Live. You must determine what works best for you! Here are two specific approaches.

Capturing a Session Performance as an Arrangement

The first approach is to begin by composing in the Session View and then recording a performance of the Session clips as an arrangement into the Arrangement View. This is the way in which the global recording process was intended. You will begin sketching and refining your ideas in the Session View, then recording your ideas as a logical sequence, arranged as a performance in the Arrangement View. For techniques on composing in the Session View, launch to ▶ **5.6** .

Let's look at the Session and Arrangement through the eyes of the Arrangement Record button.

The process begins with importing or recording clips into track Clip Slots. Working track by track, you build musical phrases, riffs, licks, or whole musical sections. Of course, this really depends upon how you compose your music. As you develop your ideas, you have the option to stack them up vertically in each track—clip by clip—or assemble them into Scenes (song sections)—row by row—stacking those Scenes into a sequence. Again, this will depend upon how you prefer to

Figure 4.10

Active real-time performance in the Session View.

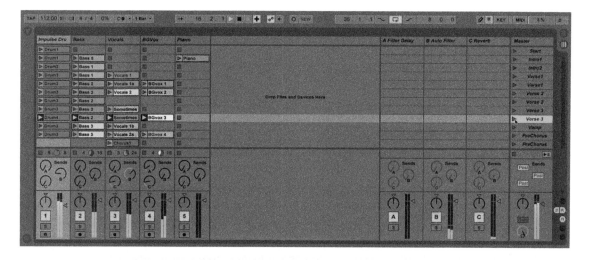

Figure 4.11 Capturing a Session performance into the Arrangement View on the fly.

compose. Once you have constructed your ideas or entire sequence, it's time to capture (record) them as an arrangement. Here you have two choices, but first you must *press* the Arrangement Record button. Then you may either begin launching individual clips, as you deem appropriate, or launch your constructed Scenes as you see fit. Taking this a step further, you can also assign and automate *Follow Actions* that will execute designated clip launch behaviors automatically ▶ 7.4.5 . Either way, once you stop recording, flip over to the Arrangement View to see your new arrangement. From there, feel free to edit and tweak any elements of your arrangement directly in the Arrangement View.

Working in Both Views

The second approach to global recording is more of a hybrid process that involves working back and forth between both views. When composing directly into Session View you may feel that loops, beats, and rhythmic patterns are naturally a perfect fit for recording into Clip Slots, but that you prefer to write out lead parts or record vocal tracks along a linear timeline in Arrangement clips. The rationale is that rhythmic loops tend to be short in nature and melodic phrases tend to be longer motifs or themes and often not looped. In this case, one way to work is to lay out your rhythmic patterns and foundation first in the Session View. Then, record your keyboard, guitar, or vocal layers in the Arrangement View using the Session View to playback the rhythm section or backing tracks as a guide while recording the lead elements directly into the Arrangement View. Once you've finished recording your lead and harmonic elements, you could then capture the backing tracks into the Arrangement View or move either the lead or harmonic elements or rhythmic elements manually into the other view using the *drag and drop from view-to-view* functionality. Simply select the material you desire to move and *click, hold + drag* it over the specific *View Selector* or *press* the Tab key and then place the material wherever you want. When *dragging and dropping* into the Arrangement View you will need to duplicate and or edit clip lengths along the timeline (loop them) in order to fill out song sections that you wish to loop for multiple bars in the arrangement. In this way you are constantly building your arrangement back and forth between views using the strengths of each to piece together an arrangement. Through a combination of session recording, arrangement recording, and editing, you can create and produce your musical ideas into full arrangements. To learn about Arrangement View concepts, launch to ▶ *Scene 6* .

4.5.2 Produce

As a producer, you need the ability to edit and rearrange tracks in order to manipulate or remix a song. You will find that Live's linear Arrangement View allows

Info Box

Only one clip can play within a single track at a time. Once a clip is activated, it willl automatically mute and deactivate any other clip on that same track. This concept also applies to Session vs. Arrangement tracks. Only one track can play at a time. The Session always has priority.

Use global **Quantization** timing resolutions to enhance the function of these rules.

you to do this, as well as work and produce your music in the same way you always have. To that end, with any other DAW, the song edit you create is the end of the line. In Live, you can bring new life to your productions with the opportunity to evolve your work by trying out (auditioning) in real-time different clips while your linear arrangement plays back. Simply launch a clip or clips in the Session View while the arrangement is playing and the Session clip/track will take over playback priority, muting out the equivalent arrangement track(s).

If that sounds interesting, how about this ... through the Arrangement Record feature it's possible to record your MIDI clips into the arrangement and then overdub them as desired, or record audio clips and then punch record over them, making Live a powerful producer's tool. Similar to the concept described in the section above, you can produce your edit or mix in the Arrangement View first—importing new material, editing audio stems, etc.— and then bring it back into the Session View as a whole or sectionalized via the *drag and drop* process. This is a fantastic feature that adds flexibility to any sequenced arrangement, especially when you are working/remixing from pre-existing audio stems. Imagine taking an entire studio multi-track as it was recorded and sequenced as an arrangement, then cutting out pieces, grooves, riffs, and putting them into the Session View where you can launch them into a new order or over the original sequence. When copying clips from view-to-view, you will have to get used to how the Session View interprets the Arrangement clip sequences as it relates to launching and re-sequencing them for playback in the Session View.

Take a look at the following walkthrough on how to *drag and drop* between views. This can be executed from either view—session to arrangement or arrangement to session. For this exercise we will go from the arrangement to the session:

1. *Select* all clips (or whichever ones you want) in the Arrangement View by holding down the *shift* key and *clicking* on each clip. To select all at once *press* ([⌘+A] Mac/[ctrl+A] PC).

2. *Click + hold* on the selected clips and then *drag* over to the Session or Arrangement View Selector in the upper right of the screen. You could also *press* the Tab key while *clicking + holding* a clip(s). This flips the current view to the Arrangement View (the opposing view).

3. While still *holding/hovering* the clips over the Session View, line them up with their respective tracks, then *drop* them (let go of the mouse button) into the Session View tracks.

In the example you will see how the Arrangement View transferred into the Session View when its clips were *dragged* and *dropped*.

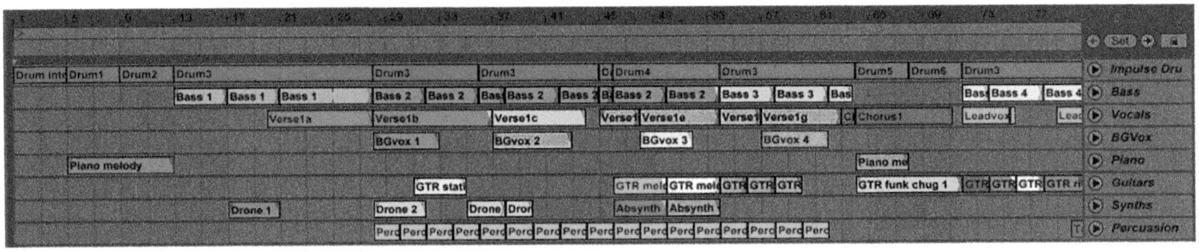

Figure 4.12 Arrangement in Arrangement View.

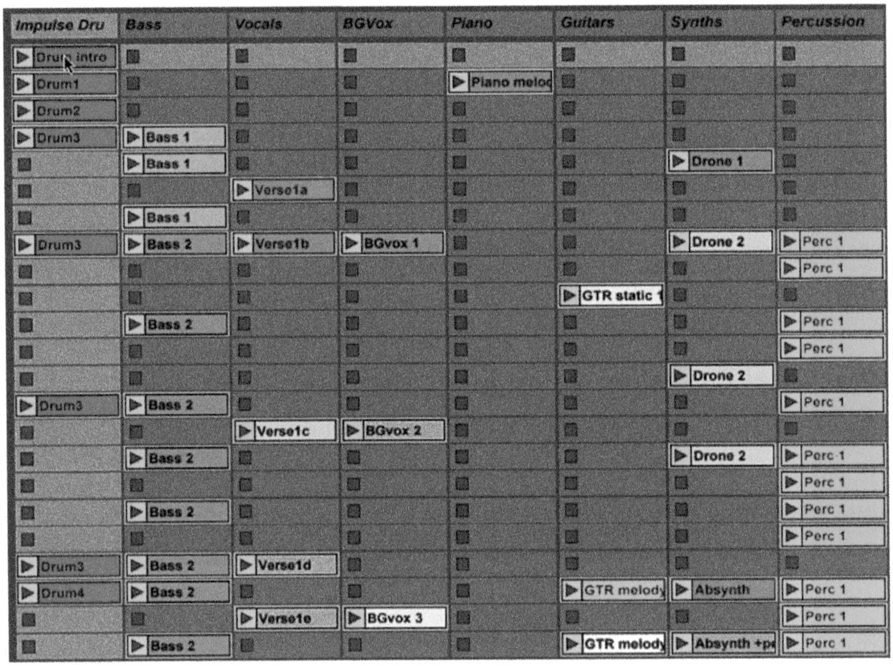

Figure 4.13 Drag and drop an arrangement into the Session View from the Arrangement View.

You will see that the Session View now contains a vertical representation of the Arrangement clips. Notice that the Session clips are relative to where the Arrangement clips were located along the Beat Time Ruler (timeline). This is a very literal interpretation of the Arrangement clips. Each clip is laid out based on the individual clip lengths as originally shown in the Arrangement Track Display. This means that no matter how long a clip looped for, it will only take up one Clip Slot in the Session View, leaving the clip slots below it empty. The actual clip itself will contain the exact amount of bars and beats that it played for in the Arrangement View. Since the clips are positioned in a relative relationship, launching individual clips and Scenes will have to be handled manually in order to recreate the exact arrangement. That's a little obsessive. The point really is not to necessarily recreate the arrangement *per se*, but to use this feature to capture clips from the Session View in the Arrangement View and then bring them back into the Session View just the way they were performed. Maybe it was an impromptu performance or a deliberate sequence of clips. Reverse this process and you can build grooves and modular musical ideas in the arrangement, then bring them into the Session View. No matter the reasoning, you can move clips in this manner in either direction from the Session to Arrangement View, or vice versa. The great thing about all of this is it can be done in real-time without stopping playback.

Remixing

When it comes down to remixing, the possibilities are endless. If your song has already been assembled in the session, then it's easy to try your hand at capturing multiple performances or passes of the song. Each time you might use different bass lines with different beat loops. Instead of launching entire Scenes, maybe it would sound cool for the rhythmical clips of one Scene row to playback with the vocal clips of another Scene row. No matter how you formulate your remix, you have unlimited flexibility. Feel free to work in either view or both to produce your remixes. *Drag* and *drop* your session to the arrangement or vice versa. In the arrangement you can manipulate, edit, and loop in a linear fashion, rebuilding the song from the ground up. You can also use the hybrid approach by overdubbing ideas from the session to the arrangement.

Beat Making

Making beats is an art form requiring patience, skill, and imagination: all character traits of a great producer. Coming from a linear approach, you can produce beats just like you always have using the Arrangement View. This hasn't changed. Create and sequence along the timeline as you go. The most common method is to program beats with a drum machine approach by looping playback and adding a layer during each record pass in *MIDI Arrangement Overdub* mode. Program beats with a MIDI controller or insert them in *Draw Mode* ▶ 7.7.4 . Creating beats in the Session View is just about the same process as far as programming MIDI goes, but note that MIDI is always overdubbed and merged during recording ▶ 8.2 . The real difference is in how you go about linking ideas, loops, and phrases together. As far as producing a beat, the process is still a linear concept, but you are working with vertical tracks and Scenes as opposed to horizontal tracks along a timeline. In the Session View, there is no timeline. Instead, you will create clips that loop and Scenes to launch simultaneous clips—your loops. in this case. When you wish to make forward progress or move to a new groove (next song section or new loop), launch the next clip(s) or Scene. As you can tell, herein lies the departure from linearity. The point to understand is that producing a beat loop in the Session View is a linear process; linking loops and song segments is not.

4.5.3 Perform

Enough about making music in your project studio, it's time to get out and perform! With your Live Set in hand, Session clips laid out into Scenes, and a little inspiration, you can perform and even improvise your music from the stage—never missing a beat. Live is literally a playable instrument, but let's save the serious improvisations for the next section. For now, we'll focus on using Live as an on-stage sequencer and musical trigger/synth device. Of course, you can use Live to play any number of virtual instruments in a performance situation, but Live is a lot more than that. The true linear performance aspect is in the ability to open a Set and playback a predesigned song from the arrangement while performing with it as a band or solo performer. Another idea would be to playback your arrangement while performing additional parts and clips in real-time from the Session View. Last, you might launch your entire song from the Session View generating a predesigned arrangement in real-time while making a few game time decisions about the performance. You could even globally record your session performance onstage while performing! Since the last idea borderlines on non-linearity, we will move on; but before we do, understand that in the real world you will always walk the line between the linear and non-linear aspects of Live. That's the point of it all!

4.6 The Non-linear Approach

When have you ever had the opportunity to take your linear composition and remix it on-the-fly, swapping bass lines and beats without hesitation or stopping playback? How about composing and improvising on-the-fly? Now you do! We have already hinted at ways to perform with Live in a non-linear fashion and are sure you are already dreaming up a million ways to do so. On the other hand, you might be thinking: how can I take advantage of this as a singer songwriter or producer? Before we get ahead of ourselves, let's break it down into our three categories: create, produce, and perform.

4.6.1 Create

When looking at the global recording concept from a non-linear perspective, things can start to get confusing, especially since Live gives you the ability to create without ever having to stop playback. For that reason, we'll go into more detail.

One of the easiest ways to work in Live is to use the Session View to experiment and sketch out modular ideas. There is no timeline; therefore, you can just keep running the clips as long as you want. For example, you could launch a loop and let it keep running. While it's playing, you can experiment with other musical elements such as a bass line, more rhythm, or a melody. This could be an audio loop or a musical line you play in yourself. Either way, you can test them out together in any way you like. *Drag* in other loops or ideas from the browser and just start building music as you go. If you like what you hear, you can toggle the Arrangement Record button and capture your idea as an arrangement, or at least as a fixed performance. Once you have your performance idea recorded, you can then bring the Arrangement clips into the session to create a Scene while clips are still running and looping. In the same vein, if you already have an arrangement, but want to change it or create new ideas to add to it, just use Arrangement Record.

Figure 4.14
Back to Arrangement button.

Play your arrangement and, when ready, toggle Arrangement Record, *launch* your clip(s), and lay down your new takes. When you're done, *click* the Arrangement Record button again, then *click* the *Back to Arrangement button* and the arrangement will take over playback from the session. Similarly, you can experiment with alternate clips instead of recording them and at any point return back to the arrangement by *clicking* the Back to Arrangement button.

4.6.2 Produce

Live's non-linear principles give you the ability to literally mold and shape any song on-the-fly. It's all about the way in which Live lets you *drop* in tracks, clips, loops, and effects without stopping playback. You can even *Hot-Swap* instruments and effects in real-time ▶ **2.4.2** . The idea is to let your musical ideas and talents flow without interruption.

Remixing

From a producer's point of view, you will find Live remixing friendly, especially when the linear shackles and chains are off. It is amazing how quickly you can begin to remix a song on-the-fly by simply starting playback and launching clips wherever you might want to them to go. Even more intriguing, think about a song you want to remix. In an ideal situation you'll have the multi-track session with each element or layer of the mix isolated or as stems. Normally this would be on a timeline locked to a linear grid. Now think of the Session View as the multi-track session, free of a timeline. There you will be able to listen to the song sections in any order you wish. Launch the hook whenever you like. Swap out the accompaniment or beat groove with a new one or one from another song. It's that simple and that powerful.

4.6.3 Perform

When it comes to Live's non-linear capabilities, it's all about performing and improvising. It is very easy to use Live as an instrument: yes, a musical instrument, which you really cannot say about any other DAW out there. Sure, you can play back sequences or use virtual instruments with other DAWs, but with Live your music is free flowing and always in your control. For example, while playing a song from the Arrangement View, you can use the mouse pointer to move around and re-launch playback in real-time without stopping the current playback position. Based on *Quantization Menu* settings, you can vamp and loop sections as long as you like . There are simply no rules governing what you can do. On top of that, with global Quantization, you never miss a beat! I'm sure you're thinking: this all sounds great, but how does this apply to global recording or, more specifically, the Arrangement Record feature?

Figure 4.15
Quantization Menu.

Performing and improvising onstage in real-time sounds almost impossible with a DAW, doesn't it? Well, with Live, whether working as a DJ or performing with a band, you can launch audio and MIDI clips while mixing and effecting your music in real-time. Live has all the tools to tweak, automate, and sweep through real-time effects on-the-fly. It all starts out with pre-existing Session clips. This can be a complete song, basic foundation of a song, or a bunch of loops—hopefully organized. Simply start playback and launch and manipulate audio, MIDI, and effects however and whenever you like. Interact and improvise with your music while it's playing. You can dial up instruments, effects, and *drop* clips of all sorts. Ok, this is great, but it really doesn't incorporate global recording. The truth is, you will not perform with the Arrangement Record button *per se*, but rather employ it to predesign elements of your Set for use and interaction onstage. What you *can* do is ... while

Info Box

Global Quantization establishes a discrete rhythmic value in which all clips adhere to when they are launched. This allows clips to playback free of timing errors.

Clip-Level Quantization allows you to override the global settings for individual clip customization. To learn more about Clip Quantization, launch to ▶ 7.4.1

all of this impromptu stuff is going on, you can be recording your performance through Arrangement Record. This means that when all is said and done, you have your entire performance logged into the arrangement for use later! Thinking out of the box, you can also come up with creative ways to incorporate global recording into your performance, such as recording an impromptu section, then pasting it ahead or pulling it into the session as a Scene, etc. In the end, you will have a sequenced performance.

Session View Concepts

5.1 Real-time Launching Base

The Session View serves as both a musical sketchpad and a traditional track mixer. The term sketchpad comes from the fact that it was designed as a real-time launching base for all of your musical ideas, musical motives, beats, riffs, and much more. The Session View is free of a fixed timeline, opening up the possibility for experimentation, improvisation, and real-time development of musical ideas without ever stopping the creative workflow—in this case, literally Live's playback. This makes it an ideal place to sketch out your ideas before you know how they should evolve, how long they should remain, and where they should go in your arrangement or production. The workflow can be as simple as this: create an instrument, build a phrase with clips, layer on more tracks, and then *launch* various clips while reacting to the musical moment. The idea is to let sounds and grooves inspire how you create. In this way, the Session View makes Live a musical instrument, in the studio and onstage. This what makes Live so different and potentially intimidating.

Here's an analogy. Creating in the Session View is like cooking, but with the ability to experiment with flavors and ingredients without the possibility of over-seasoning or burning your food. In Live, you can mix it up, add, subtract, and let it simmer for as long as you want. Take your time while clips are playing to choose your next musical idea. Cooking up a musical feast in Live is quite different than other Digital Audio Workstations (DAWs), so let's take a look under the hood to see what makes the Session View so unique and how you can take your music in new directions. It all boils down to understanding Live's usefulness and real-time flexibility. With a clear understanding, you should discover numerous new creative possibilities, especially when it comes to launching new musical ideas into your own productions. This will foster a whole new perspective on your personal workflow. With that said, the Session View has infinite potential, so let's start with some of the more conventional ways to utilize it.

5.2 Layout

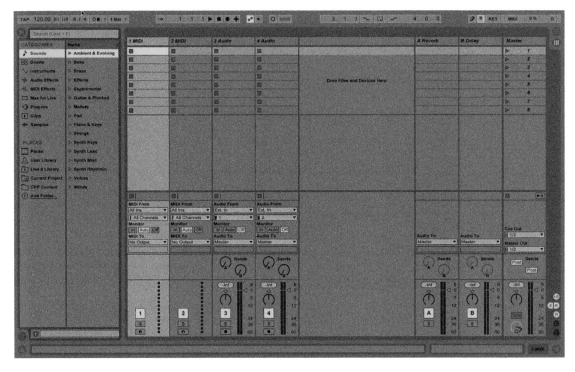

When you first launch Live, it should open up the Session View. For many of you, this will be like a case of *déjà vu*. The view will feel oddly familiar, but not exactly what you are used to. That is because you are looking at what could be considered a vertical track mixer, but with a ton of unfamiliar rectangular slots stacked above the traditional channel strips. What are these slots? They're where my plug-in inserts usually go! Those are *Clip Slots*, but don't worry, we'll get to all that. In the meantime, looking at the overall layout, the Session View is made up of tracks—*Audio, MIDI, Returns, Master*—and *Clip/Device Drop Areas*. Included within the layout are *Clip Slots, Mixer, Drop Areas* and *Scenes*. That's a lot of new stuff, so let's break down each element, starting with the basics.

Figure 5.1
Live 9 Session View.

5.2.1 Tracks

There are five track types in Live: *Audio, MIDI, Return, Group*, and *Master*. Each of these will appear as columns in the Session View. The basic default Set for Live will open up with four tracks (two MIDI, two Audio). Audio and MIDI tracks will

Figure 5.2

Four basic track types: Audio, MIDI, Returns, Master.

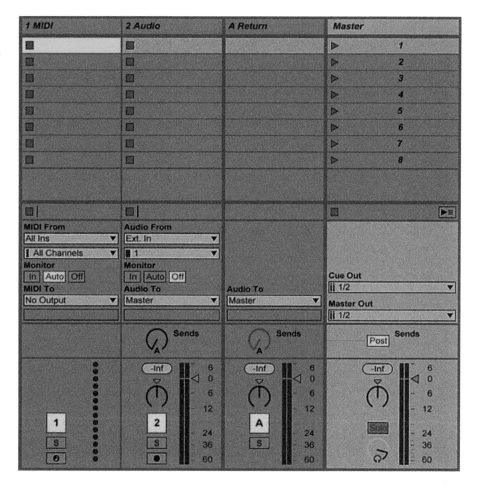

always be viewed and organized left to right, starting at the left-hand side of the Session next to the Live Browser. The Returns and Master remain on the right side of the Session, with the Master track always furthest to the right. By default, each track is appropriately labeled, Audio/MIDI by number and type, Returns by letters, and Master as is. It should be noted that the first two Returns have been pre-named: A Reverb, B Delay.

All of these track types function as you might expect. Audio tracks handle audio files, audio recording, and are used to playback audio clips. MIDI tracks handle MIDI files, events, information (note data), and are used to play, trigger, and perform MIDI instruments (software/virtual and external) and MIDI clips. A Return track is essentially an aux track that handles audio signals sent (bussed) and received from track Sends. Live allows for up to 12 Return tracks. They are generally used for hosting send effects (Audio Effect devices), such as reverb or delay, but also have many other uses, such as for submixing. A submix is the combination of multiple audio signals split off from their specific track's main channel output and sent via their Send knobs into a single aux track—in Live's case, a Return. The submix is then the result of these various audio signals mixed together at the Returns output. In this way, you can use the Return track to control the output of multiple tracks using a single channel fader. This is common for handling multiple backing vocals or a multi-channel drum kit and so on. To learn about submixing with Return tracks, launch to Clip ▶ 5.8.1 . The Master track serves as the default track where the overall outputs of all tracks are summed as a mix. It's the main output and generally from where you listen back to Live's playback. Group tracks are a special track type for managing and playing multiple tracks and clips all at once as a single unit with a summed audio output (submix). In fact, this is Live's unique way of creating submixes. More on this later, or launch to Clip ▶ 11.3.1 .

Tracks are physically divided into two halves from top to bottom. The top half contains the *Track Title Bar* followed by Clip Slots. This is where you can *drag* and *drop* audio and MIDI files (clips) in order to build your musical ideas in the Session View.

The bottom half of a track column contains the *Track Status Display*, *In/Out* section, *Sends*, the *Mixer Section*, and the *Crossfader Section*. Each component can be hidden from view to maximize screen space as described in the next section.

The Track Meter section can also be resized horizontally to increase or decrease its width. As an added bonus, you can customize track colors. *Right click* or *ctrl + click* on a track Title Bar and choose any color you like.

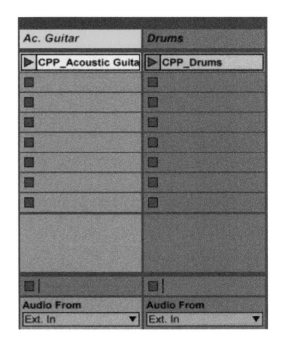

Figure 5.3
Track Title and Clip Slots.

| Hot Tip | Expand a track's Peak Level and RMS meter: click + drag the dividing line below the Sends in an upward direction. |

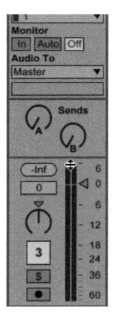

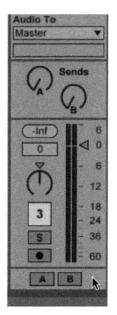

Figure 5.4
Track In/Out, Sends, and Mixer Section.

Figure 5.5
Expanded Track Signal Level Output Meter.

Figure 5.6
Track Crossfader Section.

5.2.2 Session View Mixer

The Session View Mixer does exactly what it says: it mixes the audio signals of all of your tracks using traditional vertical faders. The mix you create here is the same mix that will be heard in the Arrangement View, meaning that all mixer parameters (faders, pan, sends, etc.) are linked between views. Remember, they share the same signal path; it is the actual clips (contents of the tracks) that are different. In other words, the Session View and Arrangement View share the same audio and MIDI input and output signals, but handle different clips and sequencing functionality. Notice that the Session View Mixer is not viewable in the Arrangement View!

The general layout of the Session View Mixer consists of multiple components the can be shown or hidden from view. Customize your setup in the View Menu or from the yellow buttons to the bottom right of the Master track.

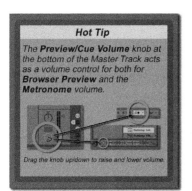

Hot Tip

The **Preview/Cue Volume** knob at the bottom of the Master Track acts as a volume control for both for **Browser Preview** and the **Metronome** volume.

Drag the knob up/down to raise and lower volume.

The **Mixer Section** consists of: *Pan, Volume fader, Track Activator button* (deactivated = off/bypass/mute), *Solo/Cue button,* and the *Arm Session Recording* switch (also referred to as record-enable or track record button). Return tracks do not have Arm buttons. Also notice that the Master track does not have a Track Activator button. Instead it has a unique *Solo/Cue* button and *Preview/Cue Volume* knob (appears blue and with little headphones). Just above this knob is the Solo/Cue button. When this is set to "Solo", track solo buttons behave as expected. When set to "Cue", soloing a track re-routes its track output temporarily to the Cue Out. This is designed for "cueing" up tracks privately in headphones DJ style. The Cue feature is only available when Cue Output is set to a different audio output than the Master Out. To view Cue Out, *click* on the I/O button towards the bottom right of the Master track.

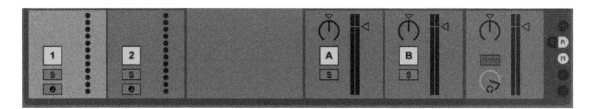

The Mixer can be resized to your custom needs. As mentioned, *dragging* the dividing line at the top of the Mixer expands the height of the Mixer Section and subsequently the *Peak Level Meters.* This adds tick level marks, a *Track Volume* numeric scalar field, and *Peak Level* indicator buttons. Increasing a track's width in this state will add a decibel scale alongside the meter's tick marks.

Figure 5.7
Session View
Mixer.

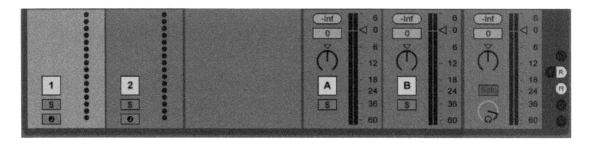

The **Sends Section** consists of the traditional bus send knobs labeled A, B, C ... and so on through L. The Master track does not have send knobs; rather, it has *Pre/Post Toggle (pre-fader/post-fader)* for setting the way in which Return tracks tap into each track's audio signal.

Figure 5.8
Expanded Session
View Mixer.

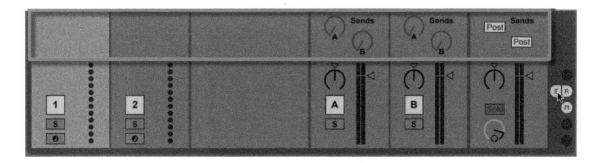

Figure 5.9
Session View
Sends Section.

The **In/Out Section** (I/O) consists of: *Input Type, Input Channel, Monitor Section, Output Type, and Output Channel*. Returns and Master track only have output options.

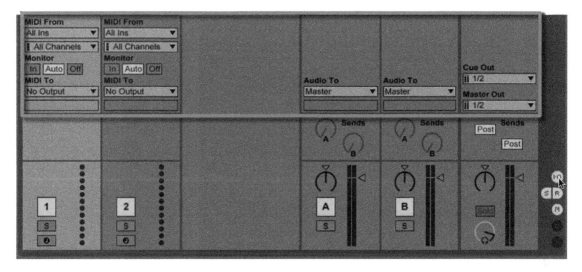

Figure 5.10
Session In/Out
Section.

Notice that Cue Out is available only when the In/Out section is in view. Remember to set this to a secondary output—if supported by your audio interface (i.e., 3/4) in order to privately cue track outputs that are in cue mode.

5.2.3 Drop Areas

The Clip/Device Drop Area is defined as the empty space between the audio/MIDI tracks and the Returns and Master track above the Mixer Section. You can create entirely new MIDI or audio tracks by *dragging* and *dropping* files (MIDI or audio) and devices (instruments/effects). Whichever you *drop*, Live will automatically create the appropriate track type.

The Mixer Drop Area is the empty space to the right of the Mixer section below the *Clip/Device Drop Area*. Here, too, you can create new tracks by simply *dragging* and *dropping* devices. Notice you can't drop files here.

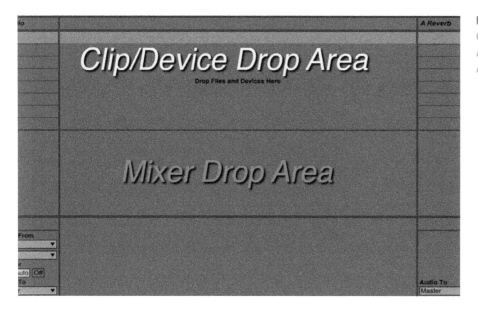

Figure 5.11
Clip/Device Drop Area, Mixer Drop Area.

5.3 Clips

Session clips are the basic music building blocks of Live. They store your audio and MIDI-based musical materials like a container—similar in nature to regions or blocks in other DAWs. These clips can be expanded, recorded, and arranged into a basic musical idea or an entire song. Each clip can be customized and manipulated with its own unique set of parameters and information. They are used to build and grow larger musical structures, eventually becoming the musical elements to create songs, scores, sound design, jingles, remixes, DJ sets, live stage performances, and interactive performance installations, just to name a few.

5.3.1 Clip Slots/Session Grid

When opening a new Live Set you will see many empty Clip Slots. These are the rectangular boxes, stacked one on top of the other, in the track columns. As mentioned, session tracks are divided in two halves, the top half being dedicated to Clip Slots as part of the Session Grid. This is where your audio and MIDI clips will live.

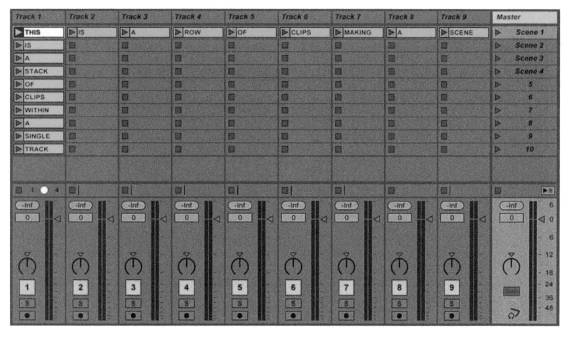

Figure 5.12 The Session Grid: Clip Slots aligned in vertical track columns and horizontal Scene Rows.

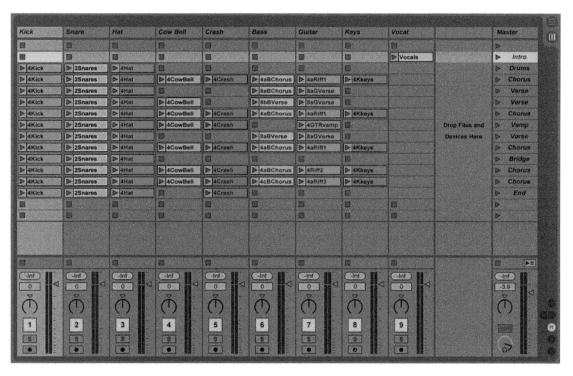

Figure 5.13 Session clips and Scenes.

The Session Grid is formed by the array of Clip Slots laid out as vertical columns and horizontal rows (Scenes ▶ 9.1 | ▶ Scene 9). The alignment of the Session Grid outlines the potential organization and structure of clips as they can be laid out in Session View Clip Slots. When a Clip Slot is empty, there will be a small gray square in the left corner of the empty Clip Slot called *Clip Stop Buttons*. When a clip is loaded into a Clips Slot, the clip will automatically assume a color and fill in the Clip Slot, replacing the Clip Stop button with a triangular *Clip Launch Button*. You can choose a default color settings in Live's preferences. When a track is armed the Clip Stop buttons in that track will turn into round *Record Buttons*.

Now that you have a general idea about what you are looking at, let's open up a Set with some clips and demonstrate how to launch, record, and edit clips.

5.3.2 Launching Clips

For this section, use the Live Set called "**CPP_5–3-2_GlobalQuantize-Off**" located on the companion website. This way you can try out the concepts in real-time as you read. Download it now and open the Set.

Now that you are looking at a Set with filled Clip Slots, notice the Clip Launch button to the left of each clip. This is how a clip is activated for playback—hence, launching clips.

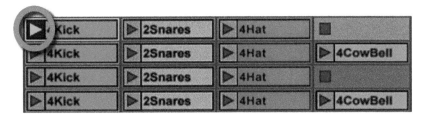

Figure 5.14 Clip Launch button activates a clips playback.

To launch a clip, *click* on its Launch button with (or select a clip with) the mouse and then *press* the return/enter key on your computer keyboard. To stop a clip, *click* on a Clip Stop button located above or below the clip in any one of the empty Clip Slots on the same track. You can also use the spacebar to stop and start playback of the entire Set globally, which will pause all launched clips upon stopping and restart those clips upon playing. More on that coming up later.

Quickly navigate from Clip Slot to Clip Slot using the computer keyboard's left/right, up/down arrow keys. This is a great practice to develop early on in your exposure to Live's Session View. A selected clip will be highlighted light blue around its border and the Launch button. When a clip is playing, its Launch button will be illuminated green. Clips can also be launched via custom computer keys and/or a MIDI controller using Live's *Key Map* and *MIDI Map* function. Launch to ▶ Scene17 for more mapping and control.

Rules of the clip: two important rules to remember!

1. Clips can be launched at any time in whatever order you choose, but only one clip per track can play at the same time. Each track can contain multiple clips from top to bottom, but only a single clip can be activated at once per track. In other words, you can play as many clips as you want simultaneously as long as they are not on the same track.

2. The order in which clips are laid out in the session does not dictate or determine their playback order. Only you can do that. This goes back to the non-linear concept behind Live's Session View.

Stop here! Take some time now to review these governing principles of Live. Practice with the example Set you already have open. Here you have multiple clips with different colors and titles. Begin by launching clip 1 "Groove 1" in Clip Slot 1; then, after a moment, launch "Groove 2" and so on ...

Hey, wait a second! My clips are not lining up with each other and things sound off. What gives?!?!

Don't worry; this was deliberate and can all be corrected with a *click* of a button. Live has provided a fantastic way to avoid timing errors when launching clips: global Quantization. On the Control Bar you will find a dropdown chooser called the *Quantization Menu*. This can be set to various rhythmic values (resolutions) that globally determine when clips will begin playback after their Launch button is activated.

In this example you can see that "1 Bar" has been selected. This means that when your Set is already playing, clip launches will commence playback at the beginning of the next bar. For example, if you launch a clip while the Session is in playback, Live will wait until the next bar—based on the quantize value—to launch or re-launch the clip no matter when you *click* the Launch button. With Quantization, Live will force the clip to launch at the appropriate time based on the resolution of the Quantization Menu, whether it is on the 8th, 16th, or 32nd note. Quantization settings always round to the nearest rhythmic value based on the setting you choose. Conveniently, while a clip is waiting for the correct launch point, its launch button will flash green.

Figure 5.15
Quantization Menu set to 1 Bar.

Remember, to stop a playing clip, *click* either a Clip Stop button in that track column (generally located in the next Clip Slot just below the currently activated clip) or in the *Track Status Display* located above that particular track's I/O section. Stopping clips in this way will only affect the clip on that particular track. All other clips will continue running, as will the arrangement playback. This is because the arrangement always runs in the background, in parallel with the session, so that it can potentially log (globally record) all of your actions. For more on this, launch to ▶ 4.4 or ▶ Scene 4.

When the arrangement is running, the Play button on the Control Bar will be solid green. Its current or running position is indicated in the *Arrangement Position* field just to the left of the Play Button. This is a way of keeping track of where you are in the arrangement timeline while in Session View. Remember that Session clips are independent of that timeline and of the arrangement. Session clips are not bound by linear order or time. If you want to stop all running clips at once in the Session View, *click* on the *Stop All Clips Button* located on the Master track just above the Master track's VU meters. Do not confuse the Stop All Clips button with the Control Bar Stop button! Using the Control Bar or computer keyboard spacebar to stop playback stops the entire Set, arrangement included, and active clips will remain paused. Thus, they will continue to show the green Launch button inside the clip. When you *press* play/spacebar or launch a clip after this, all clips that were playing automatically begin playing again. *Pressing* the Stop All Clips button on the Master will deactivate *all* playing clips. Since all other DAWs start and stop playback from the transport or spacebar, this can sometimes lead to confusion. Of course, there are many more ways to take control of clip launching and playback using a MIDI controller or your computer keyboard. Launch to ▶ Scene17 for more details on mapping your clips.

5.3.3 Basic Editing

For the most part, moving clips around among Clip Slots and executing the common tasks—selecting, cut, copy, paste, duplicate—are conceptually similar to most other programs. The real difference lies in the actual editing of waveforms and MIDI events, which takes on a whole new approach. First of all, a clip's content cannot physically be edited from a Clip Slot. This happens in the Clip View. Second, audio clip waveforms cannot actually be physically cut or separated in the Clip View— rather, only cropped by a selection within (truncated on each end) and manipulated by selecting where in the waveform a clip will start and end playback. For more traditional waveform editing you must use the Arrangement View. Detailed information on Clip View is discussed in ▶ Scene 7. For now, just be aware that

Clip View will appear across the bottom of the main Live screen when clips are selected—same goes for clips in the Arrangement View. Let's stay focused on the basic editing functions of the Session View.

Use your mouse or computer keyboard arrow keys to navigate and select clips in the Session View Grid. Select multiple clips using *shift + click*. Selecting multiple clips allows you to make multi-parameter or global edits on the selected clips, such as copying, pasting, and more general changes to their properties.

Here are some common key commands for editing functions. You will also see these in Live menus and supporting documentation. Launch to ▶ Web for more key commands.

Modifiers:

- Command = ?
- Control = ? (ctrl)
- Shift = **Place a "shift up arrow" here **
- Opt(ion)/Alt = ?
- Contextual Menu = right-click or ctrl + click

Commands:

- Cut = [⌘+X] Mac/[ctrl+X] PC
- Copy = [⌘+C] Mac/[ctrl+C] PC
- Paste = [⌘+V] Mac/[ctrl+V] PC
- Delete = delete or backspace key

5.4 Tracks versus Scenes

We have already established that tracks are vertical in the Session View as opposed to horizontal track lanes (Track Display), as in the Arrangement View, or in any other DAW for that matter. We have also established that only one clip can play at a time from the same track. In other words, an unlimited number of clips can exist in one vertical track, but never can more than one clip in the same track play at the same time. Following this principle, clips from the same instrument or audio source that are intended to be played in an alternating fashion would generally be placed on the same track since they never need to play simultaneously. For example, each rhythmic variation of a drum kit track or drum loop would be placed in a new clip on the same track and organized and stacked in a vertical progressive

Figure 5.16
Related clips on one track.

direction. Since each section or part of a song generally uses one drum kit throughout, the rhythmic patterns and beats will vary from time to time, section to section; therefore, each variation will have its own clip. These clips are then launched in an alternating fashion to create different beat pattern variations. You can see this in the example picture that clips are all from an *Impulse* Drum kit.

Ok, but what if you want to play multiple clips at the same time in order to layer the drum kit—or any source—into a more complex instrument or rhythmic pattern? In that case you would spread the clips across multiple tracks so they can play simultaneously. Of course, if it's MIDI you're talking about, then you will need to route each additional MIDI track's MIDI output to share the same MIDI instrument source so that it is generating the sound for all of the tracks you want to play at the same time ▶ 15.10.1 . With that said, how do you launch them all at the same time so that they initiate playback simultaneously? A very good question! As you might recall from earlier, this is the job of Scenes, the horizontal rows in the Session Grid. Take another look at an example.

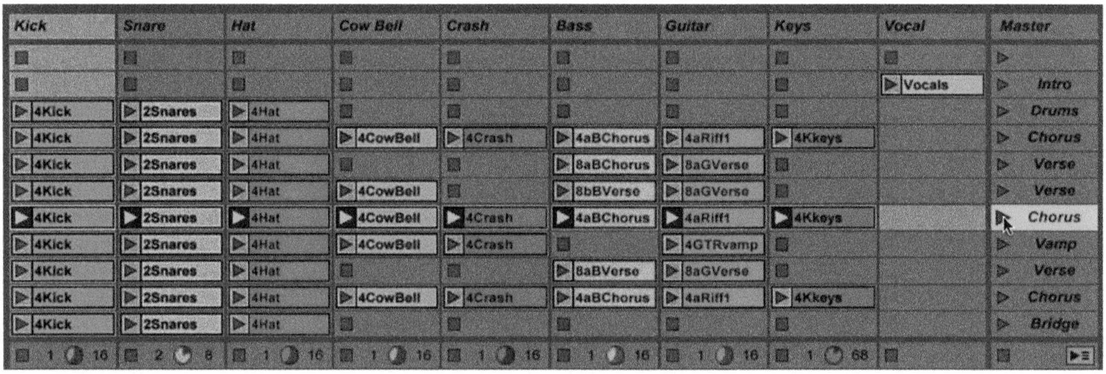

When aligned in rows, clips are usually intended to be played together. Scenes make launching multiple clips at the same time possible. They are used most commonly for one of two reasons: first, for launching several clips all at once, layered together as one; or, second for launching a row of clips that make up a song's specific part or section, such as a hook or bridge. Scenes can also be used in other inventive ways; but of the two mentioned, the latter will be the most common. The reason for this is that once you get the hang of using Scenes, you will want to expand your skills and begin using *Group Tracks* for launching groups of clips and leave the song sections launches to Scenes. To learn more about Group Tracks, launch to ▶*Scene11* .

Figure 5.17
Scenes.

As you can see, Scenes and their Launch buttons are located on the Master track. There you will also establish a method for labeling and listing Scenes. In this example, Scenes follow traditional song form nomenclature, which happens to be a descending sequence from top to bottom. Their physical ordering does not determine the order you have to play them in. This is just a logical way of organizing a session considering our brains usually desire some sort of continuity. To rename a Scene, select it and choose the "Rename" command from the Edit Menu (*right-click* or *ctrl + click*) on the name. The keyboard shortcut is ([⌘+R] Mac/[ctrl+R] PC). In addition, all of the standard editing functions—cut, copy, paste, etc.—are available from the Scene contextual menu. *Right-click* or *ctrl + click* on a Scene name. You can also find all of these commands from Live's menus, and, of course, traditional keyboard shortcuts also apply. One very clever feature under this menu is "Capture and Insert Scene". Launch to ▶ 9.3 to check this out.

5.4.1 Launching Scenes

Scenes are launched from the Master track as located on the right of the Session View. To launch playback of all clips in a single row, do one of the following:

1. *Click* on the *Scene Launch button* to the left of the Scene Name.
2. Select the Scene and *press* the return/enter key.
3. Use a pre-assigned MIDI or Key Map ▶Scene17.

Launching Scenes follows the same Quantization rules as launching clips. This gives you the ability to launch Scenes effortlessly and error free. Launch them at any time, in any order. Feel free to stop or launch any individual clips while a Scene is running. When a Scene is launched it activates each Clips Slot in the row, triggering a launch or stop command. If you want a specific track clip to continue playing unaffected when launching a new Scene, you can remove the Clip Stop button(s) below the clip that you wish to remain running as needed. This can be done from the "Add/Remove Stop button" located in the Edit Menu ([⌘+E] Mac/[ctrl+E] PC). Take a look at the example. The "4aBChorus" and "4aRiff1" clips continue to play the Chorus Scene while the "4Hat" and "4CowBell" move on to the Bridge Scene row upon the launching of the Bridge Scene. Notice the missing Clip Stop buttons on the Bass and Riff track. The "4Crash" and "4Keys" clips stopped upon launch of the new Scene because their Clip Stop buttons were triggered with the new Scene Launch.

Keep in mind that you can always use the Stop All Clips button to stop Scenes and all clips—even those without Stop Buttons—while keeping the arrangement playback running. This allows you to activate a new Scene again in sync with your tempo, and is especially useful when you have the *Metronome* active. All of this can be executed in real-time without ever stopping the flow. This is an integral part of Live's mission: non-stop workflow.

Figure 5.18
Clip Stop Button Removed.

5.4.2 Select on Launch

As you become more advanced with launching clips and Scenes, you may want to start customizing the way in which Live responds to your launch actions, especially how it handles selecting clips or Scenes after either has been launched. Such changes can be made under the *Live>Preferences> Record/Warp/Launch*.

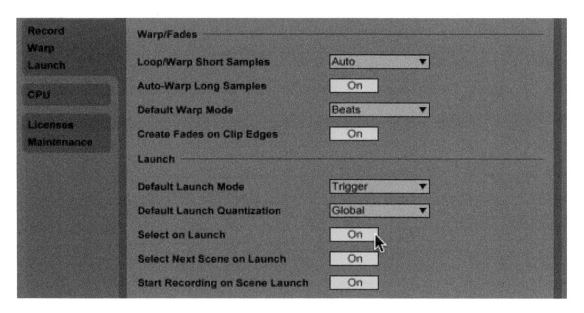

Figure 5.19 Select on Launch determines whether or not a clip/Scene will be selected when it's physically activated from its Launch button.

Take notice of the third option under the *Launch* section called "Select on Launch". If this is set to "On", a clip or Scene will become selected when it is launched, meaning that the session grid is focused on that particular Scene's clip. This also causes Clip View to change focus from the previous clip onto the newly launched clip, displaying the clip's properties and contents. When launching from Scenes, Clip View will display the clip from whichever track column had been selected. If you prefer to maintain focus on the previous clip as it was prior to launching a Scene or clip, then set Select on Launch to "Off". This preference works well if you need to concentrate and focus on a particular clip's contents for editing or parameters while activating and playing other clips and Scenes in real-time.

To learn about Clip View, launch to ▶ **Scene 7**. Under the same preference tab you can also customize how Live responds to Scene launches. This option is called "Select Next Scene on Launch". To learn about Scene launch settings, refer to ▶ **9.2**.

5.5 Track Status Display

Located at the bottom of each track's Clip Slot section right next to the Clip Stop button you will find the *Track Status Display*.

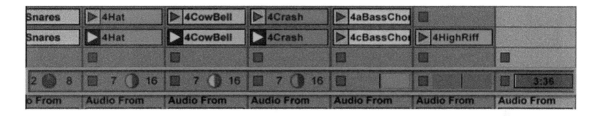

Figure 5.20
Track Status
Display.

This keeps track of how long a clip has been playing and how many times it has looped. When a clip is playing back from the Session View you will see a little clock-like circle (pie graph) along with two sets of numbers. When a clip is playing back from the Arrangement View, the Track Status will display a mini version of that Arrangement clip. In the example, the Crash, Hat, and Tamb tracks are looping and have been launched from the Session View. The last track on the right with removed Clip Stop buttons is a non-looping or one-shot clip that has also been launched from the session. The rest of the clips—Kick, Snares, CowBell—are playing back from the Arrangement View. The pie graphs indicate that the clip is looping. The number on the left indicates how many times the clip has looped—three times in this case—and the number on the right indicates how many beats make up the loop's length—16 beats. The non-looping clips' tracks status is represented by a progress bar (timer) showing how much time (minutes: seconds) is remaining before the clip comes to an end. When recording Session clips, the Track Status displays a count of how many bars and beats have passed during the recording.

Be aware that the Track Status Display is also used to display information related to input monitoring ▶ 8.1 .

5.6 Working in Session View

Now that you have an idea what clips and Scenes are and how to launch them, it's time to import and work with your own. Let's keep it simple and begin with a "New Live Set". Open a new Set from the File Menu ([⌘+N] Mac/[ctrl+N] PC) and make sure that it is empty (no clips). You should now be looking at the Session View and have two audio tracks and two MIDI tracks in view. Let's focus on the audio track for now.

For the following exercises we'll use the drum audio loops found in "**CPP_5–6_ Working-in-SessionView**" located on the companion website. You may download these now or use your own drum audio loops, just be sure to choose four that sound similar to each and have the same tempo (BPM).

5.6.1 Audio Clips

Audio files and Live Clips can be located literally anywhere in your computer filing system. Hopefully you have them somewhat organized! Although you can locate these files and import them into your Set using your computer's Finder/Explorer window, using Live's browser system is the best and most efficient way. The advantage is that the Live Browser keeps you organized and always ready to work with content on-the-fly since they are integrated into Live itself. That being said, let's use the Live Browser to access and *drag* in loops for the following exercises. For a refresher on how to use content through the Browser, launch to ▶ 2.4 .

Example A: Single Track and Clips

1. In a new Live Set, delete one MIDI track and one audio track. You should now have two tracks in all.

2. Add your folder containing your drum audio loops to Places in the Browser.

3. *Click + drag* "**S5-DrumLoop-1**" to the first Clip Slot in track 2 "2 Audio". Live will quickly analyze the file. You will see a progress bar in place of the clip

during analysis. When this is finished, your Set will have automatically adopted the tempo of the audio loop. Make sure it is set to 120 BPM.

4. *Click* on the Clip Launch button for DrumLoop-1. You should hear it playback and its Clip View should appear at the bottom of the main Live screen. We'll discuss this in more detail later, but for now notice the loop's waveform as it appears in the *Sample Editor* ▶ **7.2.1**.

5. *Click + drag* "**S5-DrumLoop-2**" to the empty slot below "DrumLoop-1" on the same track "2 Audio". Notice that DrumLoop-1 continues to playback uninterrupted.

6. When you are ready, launch "Drum Loop-2". Notice that no matter when you *click* its Launch button it launches in tempo on beat in sync, taking over playback accordingly—thank you Quantization Settings! This Set is set to "1 Bar" quantization as indicated in the Quantization Menu. Practice switching back and forth between the two drum loops to get the hang of launching clips. Try using the keyboard arrow keys and return/enter key to launch clips as well.

7. *Click* the Clip Stop button in the empty Clip Slot below "DrumLoop-2". This stops the clip, but the Arrangement Position will continue to run. You can also use the arrow keys to navigate to an empty Clip Slot and *press* return/enter to stop a clip. To stop the arrangement, *click* the Stop button on the Control Bar or *press* the spacebar.

Okay, now that you have the hang of launching clips from the same track, let's switch things up and use multiple tracks.

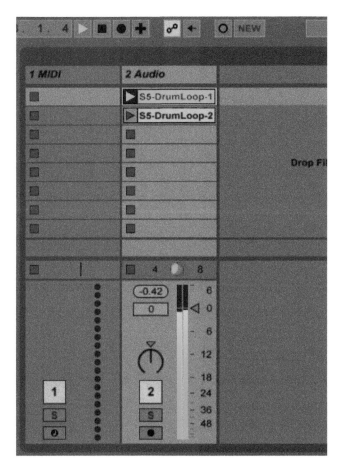

Figure 5.21
Launching clips on the same track with 1 Bar Quantization.

Example B: Multiple Tracks and Clips

1. Select "DrumLoop-2" from its Clip Slot and *press* the delete/backspace key on your computer keyboard to remove it.

2. Go back to your folder in the Browser and *double click* on "DrumLoop-2". This will create a new audio track: "3 Audio".

3. Now move the MIDI track's column position. *Click + hold* track "1 MIDI" from its Title Bar, then *drag* it so that it resides after track 2 and 3. It should now be called "3 MIDI".

4. Launch "DrumLoop-1" and let it play for a few cycles, then launch "DrumLoop-2". Feel free to adjust the volume of either loop from the mixer section to balance their output. Keep an eye out for overloading (clipping).

5. Stop "DrumLoop-1" using the Clip Stop button below it. You should now only hear "DrumLoop-2". Practice stopping and launching each clip so they play on their own and play together. Once you get the hang of this we'll add another loop to the mix on-the-fly.

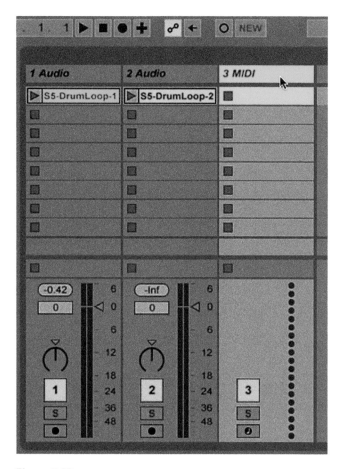

Figure 5.22
Drag a track to a new location from its Title Bar.

6. With "DrumLoop-1" and "2" running, *double click* on "DrumLoop-3" from the Browser. "DrumLoop-3" should now appear on a new audio track. Move this track column so that it sits right next to the other two audio tracks as we did for the MIDI track.

7. Launch "DrumLoop-3". All three loops should now be playing simultaneously. Check the mix and then add "DrumLoop-4". Experiment playing them in different combinations. When you're done, stop all clips from the Master track, then stop the arrangement.

Figure 5.23
Launching
multiple clips
across tracks.

Figure 5.23
Launching
multiple clips
across tracks.

Example C: Scenes

It's time to use some basic editing functions and launch clips with Scenes. There are many ways to copy/paste clips—Edit Menu, keyboard shortcuts, arrow keys, mousing, and contextual menus—so you will have to decide what works best for you. If you are using a laptop track pad, then keyboard shortcuts with arrow keys will probably work best. If you have a mouse, then you might opt for mouse-based editing. Whenever possible, you should use keyboard shortcuts:

1. With all of your clips still in place from Exercise B (four tracks and four loops), select "DrumLoop-1" and *copy drag* [opt/alt+drag] or duplicate ([⌘+D Mac/[ctrl+D] PC) it to the next Clip Slot below (Clip Slot 2) placing a duplicate there.

You can also use the arrow keys and copy/paste keyboard commands:

> **Copy** ([⌘+C] Mac/[ctrl+C] PC)
>
> **Paste** ([⌘+V] Mac/[ctrl+V] PC).

2. Repeat this process, pasting "Loop-1" in Clip Slot 4 and 6, skipping 3 and 5.

3. Select "DrumLoop-2" and copy it into Clip Slots 3, 4, and 5 on its track (Slot 2 will remain empty).

4. Select "DrumLoop-3" and copy it into Clip Slots 2, 4, and 5 on its track.

5. Select "DrumLoop-4" and copy it into Clip Slots 3, 5, and 6 on its track.

6. Now that the clips are in place, launch each Scene successively and listen to the playback.

Figure 5.24 Session clips and Scenes.

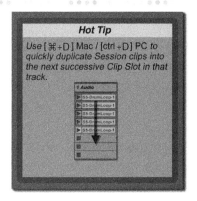

Hot Tip

Use [⌘+D] Mac / [ctrl +D] PC to quickly duplicate Session clips into the next successive Clip Slot in that track.

After a moment or two, begin experimenting with launching Scenes in a random order. After you get the hang of it, start launching individual clips and observe the behavior of the session. Alternate between custom clip combinations then back to Scene launching and so forth. Feel free to reorder the clips within each Scene or create new Scenes. While all of your clips are running, experiment with the Mixer Section, deactivating, soloing, panning, and balancing tracks. When you're done, stop your clips and playback.

5.6.2 MIDI Clips

As a producer you will find yourself wearing many hats, some of which will include making beats. Live makes it very easy to make your own beats without having to purchase additional virtual instrument plug-ins. In this section we will focus on using an *Instrument Rack* preset to trigger drum samples via MIDI. For a closer look at Impulse, launch to Clip ▶ 15.3.1 .

Let's begin with a new Live Set. For the following exercise we will use the MIDI loops from "**CPP_5–6_Working-in-SessionView**". You will also want do download the Packs mentioned in Clip ▶ 3.4.1 from the Ableton website if you haven't already. Be sure to add it to Places. Once you have a new Set open, follow these steps:

1. From the Live Browser, go to *Categories>Instruments* and unfold "Impulse" to view its presets.

2. Select any Impulse preset and *drag* it directly to a MIDI track in the Session View or *double click* the preset name to insert the instrument and create a new MIDI track.

3. Locate "**S5-IMP-KitLoop1**" from your folder in Places, then *click + drag* KitLoop1 to the first Clip Slot in your MIDI track. Don't forget to set a Quantization value.

4. *Click* the Clip Launch button for" KitLoop1" to hear it playback. Notice that your Impulse drum kit instrument appears in the Device View (Detail View) at the bottom of the Live's main screen. If all you see are knobs and are feeling adventurous, then *click* the last round button (Show/Hide Devices) to the left of the knobs to unfold the device and take a closer look at the instrument. Listen for a moment, then *click* the Clip Stop button on your track, and then *click* the Stop button on the Control Bar transport to stop the arrangement playback.

5. Add "**S5-IMP-KitLoop2**" to the next Clips Slot beneath "KitLoop1".

6. Launch "KitLoop1" again, then after a few loop cycles launch "KitLoop2".

7. While "KitLoop2" is still running, *drop* in "**S5-IMP-KitLoop3**" from your folder into the next available Clip Slot and then launch it.

8. After a moment or two, start launching back and forth between the three MIDI loops; when ready, *press* a Clip Stop button on the same MIDI track, then *click* Stop on the Control Bar transport.

MIDI and Audio clips both function and follow the same concepts in regards to launching, editing, and creating Scenes. As you work more and more with Live you will come to realize that Audio and MIDI flow seamlessly, almost to the point that they are one in the same, or at least that is the goal.

5.6.3 Crossfader Section

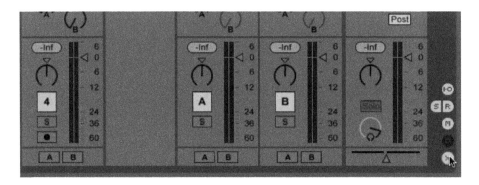

Figure 5.25 Crossfader Section.

When in view the A/B *Crossfade Assign Switches* are located along the very bottom of each audio track and the *Crossfader* below the Master track. This area is called the *Crossfader Section*. When you open Live for the first time it won't be displayed, but it can be shown from the View Menu or the *Show/Hide Crossfader Section button* at the bottom right corner of the main Live screen visible with the X symbol. The Crossfade Assign Switches (A/B) allow you to physically assign a track's output to the Crossfader. By activating one of the switches, the track's output level is routed to the Crossfader, which controls all of its assigned outputs in the overall mix. In this way, the level of any track assigned to a Crossfade Switch is affected by the position of the Crossfader from left to right. In the center is an equal balance. As you slide it right or left, the Crossfader's assigned track outputs will fade and crossfade in or out as the Crossfader's location changes from left to right. All the way to the left mutes all B track's output and all the way to the right mutes all A track's output. *Right click* or *ctrl + click* to access Live's seven different crossfade curves available for the Crossfader.

A Quick Example

Let's say you have one set of tracks assigned to A, generating "Mix 1" and another set of tracks assigned to B, generating a different mix, "Mix 2". Using Crossfader you could alternate between each "Mix" by sliding the fader from side to side. This fades in the B clips' track output, while the A tracks fade out, thus crossfading between "Mix 1's" and "Mix 2's" track output as your Master output transfers from monitoring "Mix 1" to "Mix 2". Obviously, the mixes will share an equal portion of the Master output the closer the Crossfader gets to its middle position (equal balance between the A tracks and the B tracks).

The easiest way to practice with the Crossfader is to set up two audio tracks. Track 1 is assigned to switch A and Track 2 is assigned to switch B. Use your mouse to move the crossfader to the left and right while listening to the results. You'll notice the audio levels rise and fall on each track depending upon the position of the Crossfader. After working with this basic method you can assign crossfade switch buttons to multiple tracks. For example, use switch A for all of your vocals and instruments and switch B for all of your drums and bass tracks. Using the Crossfader with this selection of A and B switches creates some interesting results. Last, go ahead and activate the Arrangement Record button to record your Crossfader moves directly into the Arrangement View.

5.7 Sessions into Arrangements

Live's Session View is a great place to develop and sketch out your musical ideas in real-time. From there, you can literally record *all* of your ideas, improvisations, and songs sections from the Session View into the Arrangement View in real-time. When the Arrangement Record button is engaged and a clip or Scene is launched, all of your actions are captured and logged into Arrangement clips or as automation along the *Beat Time Ruler* (linear timeline) of the Arrangement View. This includes the launching of clips, Scenes, and changes made to the Mixer, device parameters, and much more. To review this exciting and powerful concept, launch to ▶ **Scene 4** Global Recording. For more on Track/Mixer Automation, launch to ▶**Scene10**.

5.7.1 Capturing a Session Performance

To record Session clips into the Arrangement View, you will use the *Arrangement Record button* located in the Control Bar Transport.

Figure 5.26
Arrangement
Record button.

Click the Arrangement Record button to engage and begin global recording, then launch your Session clips or Scenes in any order you like while it's running. This behavior can be customized. *Right click* the Arrangement Record button to disable/enable "Start Transport with Record Button" or use *shift-click to disable*. When disabled you must *click* record, then play. The moment you fire off your first clip or Scene, Live will be logging your actions into the arrangement. Press the Tab key or *View Selectors* to view the arrangement as it's recorded in real-time. *Press* the Tab key or View Selectors again to go back to the Session View and continue launching your clips and/or Scenes. *Click* the Arrangement Record button again to stop recording. The arrangement will continue to run. If you want to stop recording and playback simultaneously, *click* the Stop button on the transport or *press* the spacebar. Now you will have an arrangement of your exact performance awaiting you in the Arrangement View. This will be shown as clips laid out in a linear fashion along the timeline just as you launched them from the Session View.

Figure 5.27
View Selectors.

5.7.2 Playing Back a Performance

By toggling over to the Arrangement View you will see your performance just as you played it, but shown as Arrangement clips working from left to right in a linear horizontal fashion and, of course, without Launch buttons. Now that you have a captured arrangement, *click* the Back to Arrangement button and then play your arrangement back using the transport Play button or spacebar. Playback is now generated from the Arrangement clips rather than the Session clips. At any time you can flip back to the Session View to launch or record more clips.

Back to Arrangement

As you begin to experiment with launching and recording Session clips, you must pay attention to the rules that govern the relationship between the two views! Session tracks and Arrangement tracks manage and pass the same audio signals (signal flow); therefore, Session clips and Arrangement clips on the same track cannot play back simultaneously. Whenever a Session clip is launched, it, along with its Session track, takes over playback priority from the Arrangement track. When this situation arises, you will notice in both the Session View and Arrangement View that the Back to Arrangement button will turn orange, indicating that the current playback differs from the stored (captured) arrangement. This indicates that some or no material is not playing back directly from the arrangement and that one or several clips are playing back from within the Session View.

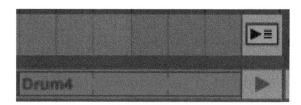

Figure 5.28 Back to Arrangement button is illuminated indicating that a clip(s) or parameter has been altered and are playing back from the Session View.

If you toggle over to the Arrangement View you will see that some or all tracks are grayed out—a transparent look to them.

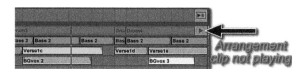

Figure 5.29 Arrangement Track/clip not playing.

To make Live play or revert back to the recorded arrangement in its entirety, *click* on the Back to Arrangement button located on the Master Track in the Session View or the upper right of the Arrangement View in the *Scrub Area*. To revert only to a specific track, then *click* the Back to Arrangement area on the right of the specific track. Live will flip back to the arrangement playback on-the-fly in real-time so you can keep working. Understand that simply stopping a clip in the Session View does not cause the Arrangement clips/tracks to resume playback or revert to their stored state. The truth of the matter is that it's sometimes easy to forget that you have inadvertently launched a session while working back and forth between the Session and Arrangement Views. The Back to Arrangement will let you know when this happens. Keep in mind that a similar effect also occurs when an automated parameter is physically manipulated in the Session View and will be indicated when the *Re-Enable Automation button* is illuminated orange ▶ **10.1.1** .

5.8 Musical Concepts

It's safe to say that once you are comfortable in Live, you will spend a whole lot of time creating, producing, and performing in the Session View. Not only will you enjoy its unique non-linear approach to creating music with clips, but it also has the familiar Track Mixer with channel strips, VUs, and Returns that will allow you to mix on-the-fly. These are all necessary tools for working with audio in any DAW. From basic mixing concepts to advanced audio production, these tools will allow you to perfect your music within the Session View. From there, the goal is to get you working in both views in a hybrid manner, toggling back and forth between the Session and Arrangement View to maximize their strengths for your workflow.

5.8.1 Produce: Submixing with Return Tracks

The fact that the Session View has a traditional Track Mixer with channel strips is not only a comforting familiarity, but also means that there are a number of useful ways to manage and manipulate audio playback—signal flow—for producing your music. One interesting way to manipulate audio inside Live is to route track Sends to Return tracks for creating "submixes". This is useful for routing a group of tracks to one output where they can be affected and mixed with the same effect and balanced together before going to the Master track. In this scenario, you would use the track send knob to control the level of the track. This allows flexibility in regards to submixing and creating new blends of audio from multiple tracks. Let's walk through this:

1. Create three tracks with different audio clips in each track in the first Clip Slot (Scene 1). You should now have a Scene row consisting of three separate audio track clips.

2. Bring the Return tracks into view from the View Menu or its Show/Hide button click on "Return B's" Title Bar and then insert a new Return track from the Create menu.

3. Bring the I/O into view (Show/Hide button) and then for each track, turn Send knob C all the way to the right.

4. Set the *Output Chooser* for each track to "Sends Only".

5. Launch your Scene 1, then add or subtract a precise amount of audio signal to be routed to your Return track.

As an alternative, use Group Tracks to achieve similar results. Launch to ▶Scene11 .

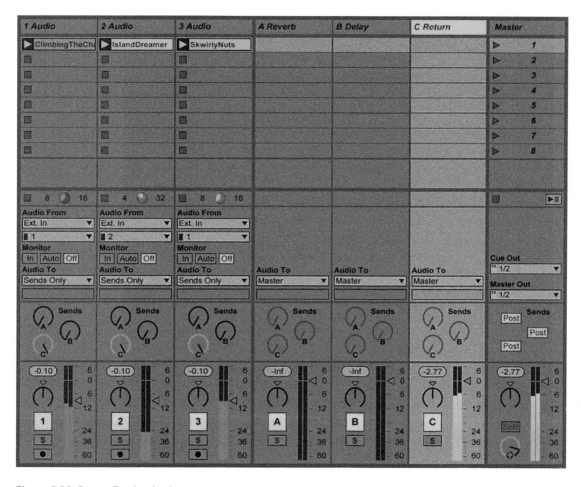

Figure 5.30 Return Track submix.

6.1 Musical Timeline

The main purpose of the Arrangement View is to capture the musical performance of Session clips and provide a linear representation of that performance to be edited, mixed, and exported as a final production. It functions as a musical timeline or linear timeline where clips are sequenced into fixed arrangements. This is where Live globally records and stores all of your actions, movements, and commands. You can record clips, clip launches, clip manipulation, mixer automation, and device automation. It is also a place for editing and altering clips and clip sequences.

6.2 Layout

To view an arrangement, toggle to the Arrangement View by *pressing* the Tab key on your computer keyboard, or toggling the View Selector on the upper right of the main Live screen ▤. The Arrangement View layout is basically like any other Digital Audio Workstation (DAW). It consists of horizontal tracks and lanes that host clips (regions) and manage audio signals for virtual instruments, external audio input sources, and plug-ins, Returns, and a Master Track.

Track names and parameters are located to the far right of the *Track Display*—often called track lanes in other DAWs. The triangle to the left of the Track Name is the *Unfold Track button*. *Click* this to view track contents such as MIDI events, waveforms, breakpoint envelopes (automation), and to make selections within a Track Display. To the right of the Track Name you will find the In/Out Section and the Mixer Section, when they are in view. The In/Out Section contains track input and output choosers for routing audio and MIDI in Live. The Mixer Section contains three buttons and additional input fields and sliders. The numbered yellow button is called the *Track Activator*. It allows you to bypass/mute a track. Next to that is the *Solo/Cue button*. This solos a track, isolating it during playback. To learn about the Cue feature, launch to ▶ **5.2.2** . The last button on the right of the Mixer

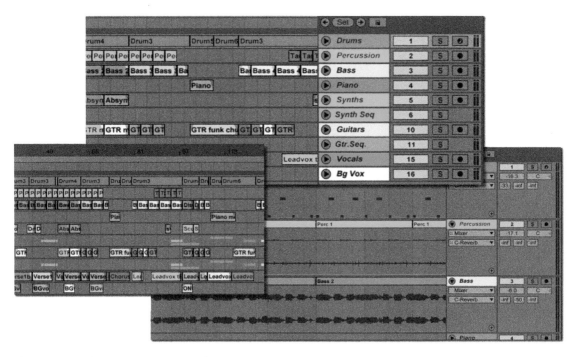

Figure 6.1 Live's Linear Arrangement View.

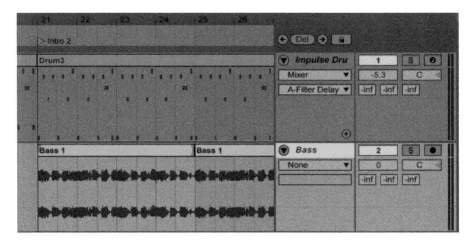

Figure 6.2 Arrangement View layout with MIDI and audio track.

Section is the *Arm Arrangement Recording*. The Mixer Section is where volume, pan, sends, automation lanes, and envelopes are displayed and adjusted. If you do not wish to have these parameters in view, then you can hide them. Check or uncheck Mixer from the View Menu or from the Show/Hide button at the lower right of the main Live screen:  *Show/Hide Mixer*. Also notice the other view options available from this menu (Figure 6.3).

Figure 6.3
Show/Hide Mixer button.

6.3 Navigating

Arrangements are fixed along two linear timelines in the Arrangement View: the *Beat Time Ruler* measured in beats and bars, based on the designated time signature, and the *Time Ruler* measured in hours: minutes: seconds. The Time Ruler can also be set to display other standard Film/TV time code formats (SMPTE), often important when working with video in Live (Options Menu) ▶ **Web** . Navigation in Live is quite unique and differs significantly from other DAWs function. This is most likely due the required real-time flexibilities of the arrangement, or maybe one programmer's preference. Either way, the Arrangement View consists of many parts, so let's start by looking at how to navigate along the timeline and take control of your music.

Hot Tip
Expand the parameters of the Time Ruler by choosing a Frame Rate (SMPTE timecode [hr:min:sec:frame]) to be displayed as the **Time Ruler Format**. *This is ideal for working and syncing with picture (video) and for scoring music to Film/TV and other image-based multimedia* ▶ **Web**

6.3.1 Scroll and Zoom

Zooming and scrolling through the Arrangement View and displays is very different than most of the common zooming features in other audio production applications. As mentioned above, the Arrangement View consists of two rulers for measuring time and location: Beat Time Ruler and Time Ruler. Navigation primarily takes place via these two rulers or from the *Overview*. The overview is a horizontal display located directly below the Control Bar. It is subdivided into the *Clip Overview/Zooming Hot Spot*, a box-shaped outline indicating the area currently in view, which can be helpful for navigating the arrangement.

Figure 6.4 Overview of the Arrangement View. Also used for navigation and zooming.

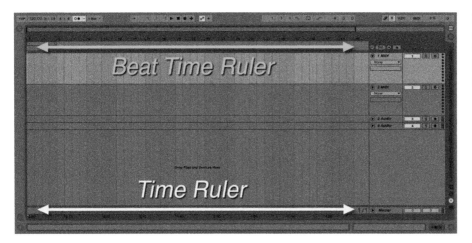

Figure 6.5
Arrangement View: Beat Time Ruler and Time Ruler.

The Beat Time Ruler is located just below the Overview in the Arrangement View. The Time Ruler runs parallel to the Beat Time Ruler and is located at the bottom of the track display. Placing your mouse over any one of these areas will cause the mouse pointer to switch to a navigation tool.

A magnifying glass will appear when your mouse pointer is placed over the Overview or Beat Time Ruler. With the magnifying glass, *click + drag* horizontally (right/left) to scroll and vertically (up and down) to zoom. A single *click* anywhere along the Overview will refocus the display on that specific area. To zoom in closer on your arrangement, *click + drag* the magnifier downward towards the bottom of the display in the general area you want to focus on. Repeat this gesture to zoom in to the finest resolution possible. Keep an eye on the bars and beats displayed on the Beat Time Ruler to get a sense of where you are along the timeline. As you zoom in, more bar numbers will be displayed indicating your exact location. *Double clicking* anywhere along the Beat Time Ruler will zoom back all the way out to fit the entire arrangement as long as there is no selection made in the Track Display. If there is a selection made, *double clicking* will zoom in to show the selection. When zoomed out, clips can become mashed together, making it difficult to identify the exact location or identities of the clips. Zooming around the Arrangement View will take practice!

Scrolling through the arrangement is very quick and easy. As mentioned, you can use the Overview, Beat Time Ruler, or Time Ruler. Simply *click + drag* along the timeline of either. It should be pointed out that the Overview does not display the arrangement true to scale. This means that it does not physically take up the same screen space as the track display does. Think of it as a bird's eye view or

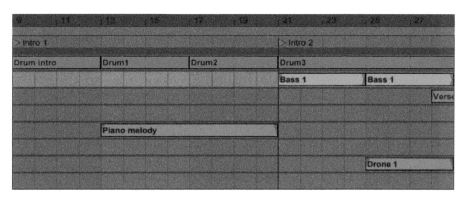

Figure 6.6 Moderate zoom level of Arrangement View track clips.

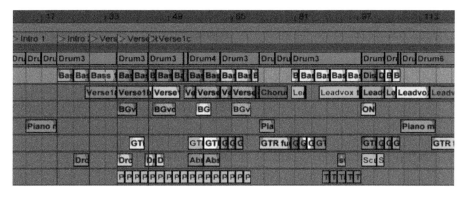

Figure 6.7 Tight zoom level showing Arrangement clips mashed together.

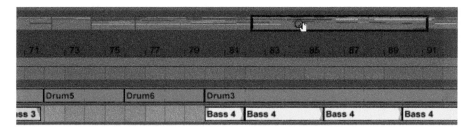

Figure 6.8 Zooming Hot Spot is the black outlined square within the Overview. Use this to focus on a specific area of an arrangement.

approximation—like viewing a holiday parade from a sky cam or blimp. In other words, be careful where you *click* along the Overview or you might end up very far from your desired location. Take notice of the Zooming Hot Spot to help narrow in on your location.

Depending upon how you like to work, you can also have the display auto scroll during playback with *Follow* activated, or stay focused on the current visible song position. Follow is located on the control bar transport section. *Click* the Follow button to activate scrolling with playback (yellow = on).

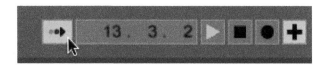

Figure 6.9
Follow On/Off
Button.

As mentioned, you can scroll through the arrangement via the Time Ruler. A hand-shaped scroll tool will appear when the mouse pointer is placed over the Time Ruler. The hand tool is for scrolling only. You may also access the hand scroll tool anywhere in the Track Display by holding ([⌘+opt] Mac/[ctrl+alt] PC) + *dragging*. In this way you can also allow for vertical scrolling. This is probably the most efficient method for navigating through the arrangement since you can begin scrolling from anywhere within the Track Display.

6.3.2 Transport

The main playback controls for the Arrangement View are located in the Transport section of the Control Bar. From left to right you will find: *Arrangement Position*, *Play*, *Stop*, *Arrangement Record*, and *MIDI Arrangement Overdub*.

Figure 6.10
Transport Controls
and Arrangement
Position display
field.

The Arrangement Position display field indicates the current playback start location of the Arrangement View when playback is stopped. While playback is running, it displays the current arrangement position by following playback. The exact position is displayed as: [*bars. beats. sixteenths.*]. From this display field you can manually *drag* or *type* number values to position playback at any location. This will move the *Arrangement Insert Marker* to your desired playback start location. You can also achieve the same result by *clicking* your mouse pointer in the Track Display on or around the location you wish to start playback. After an insertion is made, *press* play.

The Play, Stop, and Arrangement Record buttons are fairly self-explanatory. They function just like any other DAW. One quick note is that *double clicking* the stop button will return the Insert Marker to the beginning of the set "1.1.1". As an alternative to using the play button or the spacebar, you can start playback anywhere along the timeline by *clicking* in the *Scrub Area*. This area is located between the Beat Time Ruler and the Track Display. It is also identified by the speaker icon that appears at the end of the mouse pointer when hovering your mouse over the specific Scrub Area. *Click* anywhere in the Scrub Area to jump to and start playback. This can be done while the arrangement is stopped or already in playback. If in playback, Live will jump to the new position based on the Quantization setting. To take advantage of this feature, make sure that "Permanent Scrub Areas" is "On" in the *Preferences>Look/Feel Tab*. If disabled, you must *hold* the Shift key while *clicking* in this area to affect playback.

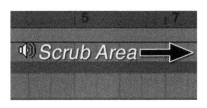

Figure 6.11
The Scrub Area is for starting/jumping playback to a clicked location or to scrub" audio.

6.3.3 Locators

At any point along the timeline you can add a *Locator* to mark sections, start points, hits, or anything else you might want to identify. Some DAWs refer to these as "markers" or "memory locations". They make your arrangement very clear and easy to navigate. Locators can be added during playback or when the arrangement is stopped. They appear along the Scrub Area as a triangle connected to a vertical line that extends vertically across all tracks.

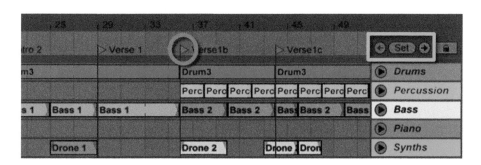

Figure 6.12 Locators are identified by the triangles and connected to a vertical line that extends the entire track display. Locator controls are directly above the Track names next to the Scrub Area.

To add a Locator, select "Add Locator" from the Create Menu at the top of the main Live screen. It will prompt you to create a name for the Locator. If you wish to change it later, select "Rename" from the Edit Menu ([⌘+R] Mac/[ctrl+R] PC). You can also add a locator from the *Set/Delete Locator* button located on the far right of the Scrub Area above the track names. This button provides the fastest way to create and delete locators. Once a locator has been selected, the *Set button* changes to "Del" for delete. The arrows to either side of the *Set button* are for navigating to and from locators (next/previous locator).

When adding locators during playback or recording, they will be inserted according to the Quantization setting. When not in playback, they are inserted wherever the Insert Marker is or where your selection begins. Locators make navigating and launching sections of an arrangement as simple as a mouse *click*. A single *click* on a locator will move the Insert Marker or playback to that locator. *Double clicking* will start playback immediately when the arrangement is not in playback. Jumping or launching from markers during playback or recording is also subject to the global Quantization. Locators can be triggered via MIDI or computer keys per custom assignments. Remote controls are assigned through Live's *MIDI/Key Map Modes* ▶Scene17. Combined with global Quantization, Locators and the Arrangement View can become non-linear!

To move locators, *drag* them manually or use the arrow keys on the keyboard. They move along the grid based on the *Snap to Grid (Marker Snap)* settings. Snap to Grid and other grid settings are accessed from the Options Menu or Track Display contextual menu (*right click or ctrl + click*). Both menus offer a choice of grid values for snapping and insertions made while working in the arrangement. The current Marker Snap setting, or the spacing between grid lines, is displayed in the lower right

Figure 6.13
Marker Snap (current spacing between grid lines).

corner of the Track Display. There are two setting types: *Adaptive Grid* and *Fixed Grid*. These values are based on beat values/subdivisions and are used to show the number of beats (grid lines) per bar. For example, "1/2" means half note subdivision with one grid line and "1/4" equals quarter note division with three grid lines, etc.

The Adaptive Grid concept is governed by zoom levels and labeled as sizes such as "narrow" or "wide". As you zoom, the grid will adjust the resolution of the grid lines relative to the zoom level. For example, with a wide grid you may only snap at every two bars, but as you zoom in it may snap at every two beats. The Fixed Grid is independent of zoom. It remains set to the value you have chosen. To speed up your workflow, use the keyboard shortcuts for setting and adjusting the Marker Snap settings ([⌘+1–5] Mac/[ctrl+1–5] PC).

6.4 Working in the Arrangement View

How you ultimately work in Live is up to you, and working in the Arrangement View is a totally acceptable way for composing new music or working on an existing production. Live is more than a stage-performing tool; it's a full-blown DAW! You can create, produce, and perform in it too, although the arrangement is most commonly used to edit clips, manage a sequenced or performed arrangement, and a place for drawing and editing detailed automation. These are all part of creating and producing the final product, a finished and well-produced piece of music. That being said, now is a good time to use one of Live 9's Lessons. From *Help View*, load "What's New in Live 9" or "A Tour of Live". This will be useful for putting concepts to practice as we discuss them, especially since they are accompanied with Live Sets.

6.4.1 Launching

There are four basic ways to play back an arrangement and clips:

1. *Click* the *Play button* from the Transport on the Control Bar.
2. *Press* the Space Bar.
3. *Double click* a *Locator*.
4. *Click* anywhere along the *Scrub Area* (*when preference enabled*).

If you have made a selection within the Track Display, playback will commence from that point when you *press* the space bar. When launching from a locator during playback or record, playback will jump to the locator position on the next global quantization value.

There is a very important feature that links the relationship of the Arrangement View with the Session View that has been continually emphasized throughout the book: the Back to Arrangement button.

Figure 6.14
Back to Arrangement button.

It's very common for Session clips to sneak into playback of an arrangement by accident because anytime you launch Session clips, the related Arrangement tracks and clips are bypassed. Session clips and Session automation always have priority over Arrangement clips, tracks, and their track automation. In this case, the affected Arrangement track lanes will be grayed out in the Arrangement View and the Back to Arrangement button will illuminate orange. This indicates that some or all Arrangement tracks are bypassed and not playing back the stored arrangement. If so, *click* the Back to Arrangement button to revert everything back to the recorded arrangement contents. This can also be executed on a per track basis. To the far-most right of each track lane that is bypassed there will be a triangular button. *Click* it to revert that track back to its arrangement content only. That being said, it's a good practice to periodically verify that you are listening to the playback of your stored arrangement. More on Back to Arrangement in Clip ▶ **5.7.2** .

6.4.2 Looping

There are two distinct types of looping that you will use when working in the Arrangement View: looping playback and looping clips. Generally, with looping playback you will make a timespan selection that you can listen to over and over again. You'll find this is useful when creating, editing, or mixing music as you can repeatedly listen to your selection and also for loop recording. That's not to say that there would never be a reason to loop playback on stage. Looping clips, on the other hand, is an integral part of the arrangement and clip workflow design. In this way, loop-enabled clips are *dragged* out (extended) to repeat many times to function as part of a musical element of your arrangement. Drum loops are a prime example of this type of looping. For now we will focus on looping playback selections of the arrangement for tasks such as editing, mixing, and recording. For an in-depth look at looped clips, launch to ▶ **3.5.1** and ▶ **7.5.9** . For looping in the arrangement, launch to ▶ **14.5** .

To create a loop selection, *click + drag* to highlight a selection of clips in the Track Display. Once selected, choose *Loop Selection* from the Edit Menu ([⌘+L] Mac/[ctrl+L] PC). Notice that this will also activate the *Loop Switch* located on the Control Bar to the On position. When this switch is "Off", the selection will not loop in playback. The loop length is indicated in the number field box to the right of the Loop Switch represented in [*bars. beats. sixteenths.*].

The loop selection is indicated by the Loop Brace, a gray box with triangle bookends, and can be moved by *dragging* that box horizontally along the Beat Time Ruler. It can also be shortened or lengthened as desired. The concept here is really no different than any other DAW. If you navigate away from your loop selection to edit something else, you can easily reselect it by choosing *Select Loop* from the Edit Menu or just *click* the Loop Brace. In case you need to loop an entire arrangement, select all ([⌘+A] Mac/[ctrl+A] PC) and choose Loop Selection as described. Loop Selection command is also accessible from a contextual menu. *Right click* or *ctrl +click* after highlighting the desired timespan. There you will also find additional features applicable to your arrangement.

Figure 6.15
Selection in the
Arrangement View
set to loop during
playback.

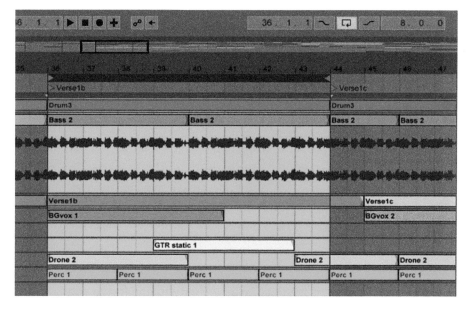

6.4.3 Selecting

When it comes time to edit and manipulate your clips in the Arrangement View, you will want to feel comfortable making selections with both the mouse pointer and using keyboard shortcuts—shortcuts are invaluable! The latter will come with time and practice. In the meantime, you can use menu commands and contextual menus (*right click* or *ctrl + click*). For a list of Mac/PC keyboard shortcuts, launch to the ▶ **Web** .

A single *click* anywhere in the Track Display will move the *Arrangement Insert Marker* to that position. Selections are indicated with blue highlighting. *Click + drag* anywhere in a Track Display to make a selection. To select an entire clip, *click* on the colored title bar at the top of the clip. You will then see two triangular arrows appear on both sides of the clip. You can also *click* on the loop brace to make time range selections even when the Arrangement Loop Switch in not "On". This will highlight all clips contained within the Loop Brace's boundaries—start to end point. Once you make a selection in the arrangement, playback will always begin at the beginning of the selection until changed.

Unless a track is unfolded, you will not be able to see the highlighted selection of a clip. To unfold a track, *click* the *Unfold Track button*, a triangular arrow next to the track name by the Mixer Section. When unfolded, you can make specific selections within a clip's boundaries in the Track Display. *Click + drag* within the clip waveform/MIDI area (below the title bar) to select a specific timespan or isolate an event. You may also adjust the height of tracks to unfold them and or to resize them to your liking by *dragging* the dividing line between track names and unfold buttons. Making an insertion or selection is governed by the Marker Snap settings. This means that the Insertion Marker will align to the nearest visible grid line. Feel free to adjust the snap settings and width of the grid from the Options Menu or by *right click* or *ctrl + click* in the Arrangement View.

6.4.4 Editing

When it comes to arranging, you will find yourself moving, resizing, splitting, copying, pasting, and consolidating clips quite often. These are just a few of the features incorporated as standard editing functions necessary to edit your arrangements.

Moving Clips

To move clips to different tracks or to a different location in the arrangement, simply *click + drag* from the clip's title bar to the desired location.

Figure 6.16
Moving clips around the Arrangement View Track Display.

Re-sizing/Trimming

Place your mouse pointer over the left or right edge of a clip. Once the mouse pointer changes to a bracket, *click + drag* the edge to your desired location.

Figure 6.17
Resizing/Trimming a clip.

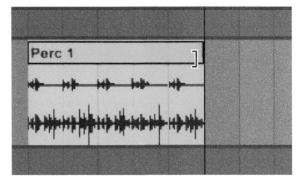

Splitting

To split (separate or divide) a clip into multiple clips, place the Insert Marker at the point where you want to split and select "Split" from the Edit Menu ([⌘+E] Mac/[ctrl+E] PC). This will split the clip in two. You can also split a clip by making a selection, therefore isolating a portion. Unfold the track, and then make your selection in the waveform or MIDI display. After you split the clip, it will turn into three clips. Remember, insertion is based on the snap settings. Alternatively, Live will automatically split clips for you when you *drag* a selection within clip to another location. This speeds up your workflow for certain operations, such as splitting out and copying selections of isolated percussive hits or beats. Once selected, you can *drag* the isolated beat directly into a sample-based instrument such as Impulse ▶ 15.3.1 or a Drum Rack ▶Scene16. Launch to these sections for more.

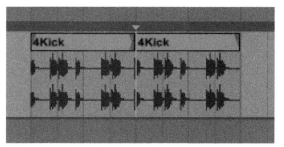

Figure 6.18 Split a clip from an insertion location.   **Figure 6.19** Split a clip by making a selection. The result is three clips.

Consolidating

In Live, consolidating means to join multiple clips together as one. To consolidate clips, select your desired clips or timespan across clips and select "Consolidate" from the Edit Menu ([⌘+J] Mac/[ctrl+J] PC). This will join the clips together, creating a new audio or MIDI clip. If you want to include silence before or after the new clip, simply extend your timespan selection to include silence pre- or post-clip.

Cut, Copy, Paste

These commands follow the standard word processing functions and use the same keyboard shortcuts. To cut a clip in the arrangement, make your selection and select "Cut" from the Edit Menu ([⌘+X] Mac/[ctrl+X] PC). To copy, make your selection and select "Copy" from the Edit Menu ([⌘+C] Mac/[Ctrl+C] PC). To paste, place the Insertion Marker or make a selection in the desired location and select

"Paste" from the Edit Menu ([⌘+V] Mac/[ctrl+V] PC). Keep in mind that you will need to unfold a track when choosing a discrete selection smaller than the entire Arrangement clip.

Duplicate

The *Duplicate* function creates an exact copy of a clip's selection and places it immediately after the selection's end point so that they are butted up against each other. In this way a selection can be repeated along the timeline as a discrete copy. You can select a portion of a clip(s), entire clip(s), or a selection that extends before or after a clip(s) edge. To duplicate, make your selection and select "Duplicate" from the Edit Menu ([⌘+D] Mac/[ctrl+D] PC).

Cut/Copy/Paste/Duplicate Time Commands

While navigating through the Edit Menu, you will notice the *Cut/Copy/Paste/ Duplicate Time Commands*. These commands affect the overall time of the arrangement, each adding or removing a time from it. This means that by adding or removing time, all tracks and clips will be shifted in time or removed from the arrangement to accommodate the command, therefore shortening or lengthening the overall duration of the song. To execute one of these time commands all you must do is make a selection on one track, and then initiate the command. The rest of the tracks will be displaced automatically.

For example, when duplicating time, make your timespan selection on any track you wish. It doesn't matter which because you're duplicating a timespan, not just a single clip. After you've made your selection, select "Duplicate Time" from the Edit Menu. Now your selection will have been duplicated immediately after the original selection, pushing all other tracks' clips to the right—later along the timeline. If you had used the traditional duplicate command, the duplicate would only be of the selected clip(s) and overwriting any clip(s) in the way, while maintaining the current length of the song or arrangement. Take a look at how Duplicate Time works!

Paste Time works the same, except you have the freedom to paste at any location along the timeline. Make sure that you select the same first track—top to bottom— in the Paste Time destination as you selected when you copied the original selection. This means that the paste destination should be selected on the same physical track you wish to paste your track clips in order to avoid pasting clips to the wrong tracks, as you will see in the example.

Obviously, there are times when you may wish to paste clips to different tracks in this manner. Just make sure you do it on purpose, not inadvertently.

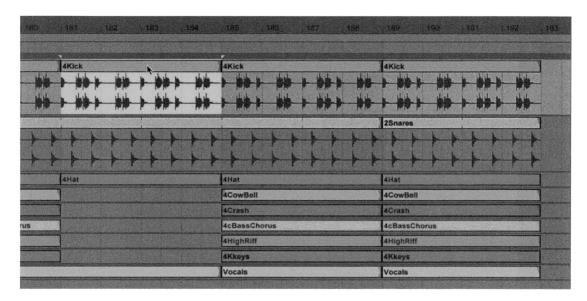

Figure 6.20
To Duplicate Time, make a selection in the track display then select the command from the Edit Menu.

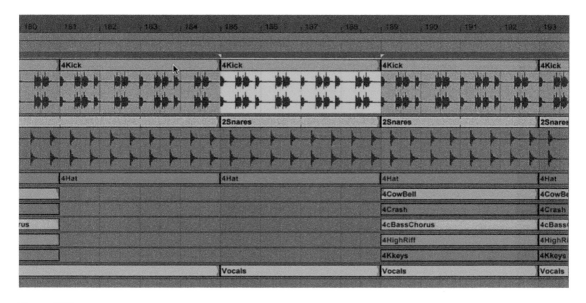

Figure 6.21
Notice that the duplicate selection shuffles all other track clips to the right to make room for the new track clips. This adds time to the length of the Arrangement relative to the selection duplicated.

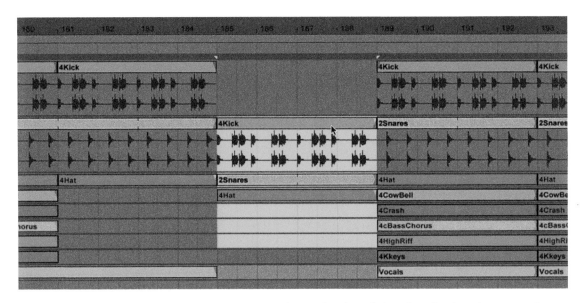

Figure 6.22 A different paste track destination was inadvertently selected; therefore clips have been pasted to the wrong tracks, offset by one track level.

6.4.5 Fades and Crossfades

Fades and Crossfades are an essential element of editing audio in any DAW. Fades are used at the edges of audio clips, and crossfades are used to join two adjacent audio clips. Their main purpose is to eliminate pops and clicks created by edits or cuts made in an audio waveform where the cuts don't exactly cut at a zero crossing point—where no signal is present—or blend and smooth out amplitude levels that are significantly different between clips. Manipulating their volume eliminates these imperfections, which is why they are called volume fades. It is generally a good idea to use fades and crossfades when editing audio content even if you think it already sounds perfect. For this reason Live gives you the choice to create fades by default. There are multiple types of fades and crossfade shapes that determine how quickly a signal is faded in or out. As you become more experienced with using fades and crossfades, you will learn that different shapes of fades (curves) work better than others for every situation. For now, just focus on what sounds good and gets the job done. You can worry about all the technical audio engineering details on another day.

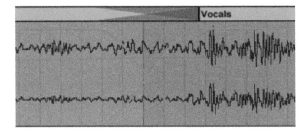

Figure 6.23
Unfolded track showing a crossfade in the Track Display.

Fades and Crossfades are viewed in the Track Display attached to a clip. You will also see reference to a fade in a clip's title bar and the result in its waveform where signal has been reduced. To view them in a track's Track Display, "Fades" must be selected from the *Fades/Device Chooser* menu below the track's name.

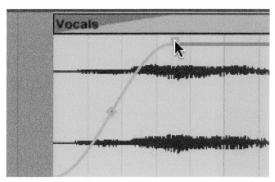

Figure 6.24
(above left)
To view fades in a track's Track Display, select "Fades" from the track control chooser.

Figure 6.25
(above right)
Fade in.

When a fade is applied, the title bar shows a wedge shape indicating that a fade or crossfade is present. The waveform display shows an accurate editable display of the fade location and shape. The waveform itself reflects the volume fade in the display by redrawing itself to match the curve. Obviously, the track must be unfolded in order to view this, but you must also select "Fades" from the Fades/Device chooser below the track name. Fades are manipulated from the *Fade In/Out Handle* (length) and the *Fade Curve Handle* (slope shape). *Click + drag* these to the position you desire in order to change how quickly or intensely a fade In/Out happens and how much of the audio signal is affected by its adjacent files when they are blended together.

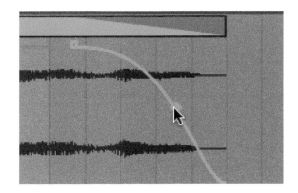

Figure 6.26
Fade out handle.

From the clip contextual menu (*right click* or *ctrl + click*) you can select *Show Fades, Reset Fades*, and *Create Fades/Crossfade*. Show Fades allows you to skip the chooser assignment step. Reset Fades reverts a fade or crossfade back to its default setting. Create Fades allows you to make a selection spanning from a clip's start or ending edge, or across two clips that are to be cross-faded. After you make a selection, the Create Fade command will do just that: create a fade or crossfade. By default—unless set differently in the *Record/Warp/Launch* Preferences—fades on clips edges and adjacent clips are automatically created with a 4 millisecond fade length. If the *Create Fade on Clip Edges* option is "Off" in the preferences tab, then you can delete fades by selecting the fade handle and *pressing* the delete/backspace key on the computer keyboard. If it is "On", then delete will only reset the fade to its original position. Keep in mind that fades and crossfades are independent of track volumes. Volume fades are clip based.

6.5 Arranging Concepts

By now it should be pretty obvious that the Arrangement View is very useful for working in a linear fashion. Everything you need to arrange, rearrange, edit, and master is readily available. You can create an entire arrangement from scratch or additional music elements without limitations. No matter how you like to work, Live can function just like any other DAW in this context—not to mention all of the effects and mixing capabilities! We'll save that for another Scene, but for now let's look at a few of the "create, produce, and perform" aspects of the Arrangement View. Keep in mind that any one of these concepts may share similarities with the other or fall into multiple categories.

6.5.1 Create

Creating in the Arrangement View is fairly straightforward. You can record or import audio and MIDI directly into the tracks just as you would with any other DAW. From a technical standpoint, building arrangements could not be easier—it's the creative muse or inspiration that is so elusive; just set up your desired track types, microphone, or load up instruments and off you go. Once your song elements have been laid into place, you can begin to edit and tweak each Arrangement clip as you see fit. This includes using effects and other manipulations to creatively affect your audio and MIDI events. Live allows you to work, build, and arrange multi-track arrangements (sequenced Sets) full of MIDI and audio without limitation. So, if you prefer to work in the Arrangement View, there is nothing to stop you. As a film or TV composer, you will have to view your video clips in the Arrangement View, which is important for scoring music to picture. Launch to the ▶ **Web** for information on using video with Live. The Arrangement View is a great way to transition from your old DAW to Live without the burden of learning an entirely new way of composing. Yes, it's true that the Arrangement View serves a greater purpose by design, but it will also make you feel right at home while you gradually discover the power of the Session View. For now, there is no reason to disrupt your workflow. Feel confident that you can continue working the way you always have during this period of transition. When you've completed the transition to Live, the arrangement will be a place more for working on edits and mixes rather than creation, but that's not to say you won't create in the arrangement now and again.

6.5.2 Produce

Remixing a song, programming beats, overdubbing, or simply punch recording is all part of producing in the Arrangement View. As mentioned, you can move clips around, duplicate them, and create a variety of loops—grooves, bass lines, rhythmic beds, and hooks. With all of Live's features and devices, you can start to remix songs to your heart's content. This is important to understand. Live is a linear production tool. That's the beauty of the Arrangement View. Copy, paste, cut, or lengthen large song sections—verses, hooks, themes, motives, etc.—in unique ways. Automate effects and mixer parameters, then copy and paste those too. Finally, master your entire production in the Arrangement View, then bounce a final mix.

6.5.3 Perform

An arrangement is a ready-made piece of music, a sequenced song that you can take with you onstage. This gives you the option to sing or play a guitar along with the arrangement. Think of Live as your band, or if you're a DJ, it's everything—your tracks, effects, mixer, and so on. Obviously, in the same way, you can perform with one of Live's virtual instruments simultaneously with your arrangement. Going a step further, you could also *ReWire* Live with another DAW. Use them together to perform with your arrangement in real-time or as another sequencer element. No matter how you conceive the relationship, ReWire can open up a whole world of hybrid performance situations. For more on ReWire, launch to ▶*Scene18*.

Those of you who are familiar with Live might be thinking that performing with an arrangement is more limiting than performing from the Session View. That really depends upon how you go about performing. Not everyone is a DJ or improviser, but there is a little performer inside all of us! That is why the arrangement allows for composers and producers to not only use the Arrangement View as a multi-track sequencer and for tracking audio, but also makes it possible for performing. From playing software instruments in real-time along with pre-programmed sequences or by manipulating pre-programmed sequences in real-time, the Arrangement View can do it all and without a hiccup. By design, an arrangement can be manipulated during playback, allowing you to re-launch playback from anywhere on the timeline using Locators and loop sections/selections on-the-fly using the Arrangement Loop feature. You can also record and manipulate automation in real-time during playback and while recording. No matter who you are, the Arrangement View can be a valuable performance tool suitable for DJs, performers, and composers.

Clips

7.1 Musical Building Blocks

Clips are the musical building blocks of Live. Each clip represents a musical event—for example, a melodic idea or phrase, percussive rhythm or beat loop, grooving bass line, transitional effect, or an entire music track or song. Therefore, clips are used to form the musical framework of songs, scores, hooks, remixes, DJ sets, live stage show performances, interactive music installations, and much more. Whether laid out in the Arrangement or Session Views, clips store a vast amount of information, along with audio and MIDI content. Each clip has its own unique set of editable properties and parameters (clip data) that dictate how its enclosed content is executed. This makes a clip a dynamic customizable container of musical information and performance data for use not only in one specific Set, but also across all of your Sets and Projects as customized presets. These are called Live Clips. *Drag* a clip to the Browser to create a dynamic Live Clip. Launch to ▶ 15.2.4 and see how clips retain device and performance data and how they can be stored and utilized via the Browser or from within another Set. Let's take a closer look at clip properties.

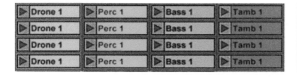

Figure 7.1
Session clips and
Arrangement clips.

7.2 Clip View

All clips consist of audio or MIDI content and carry playback information and properties. Such details are viewable from the *Clip View* located at the bottom of the main Live screen. Access Clip View by *clicking* or *double clicking* on a clip. The action is dependent upon what is already in view. This area of Live's main screen

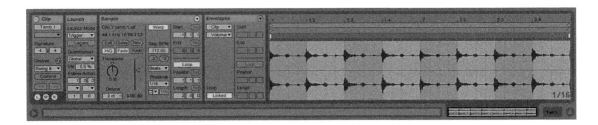

is known as Detail View since it dynamically displays various details depending upon what is selected in Live. Once a clip is selected, its contents and properties will appear in Clip View. Here you will see a number of clip *Property Boxes*. Notice that the color of the selected clip matches the color of the Property Boxes. Clip View displays everything you need to know about a clip, including all of its contents, properties, settings, and how it's launched and plays back. Most importantly, MIDI programming, editing, waveform manipulation, and automation are all executed from within Clip View. Accessing audio and MIDI content in this way is very different from other Digital Audio Workstations (DAWs) and requires some getting used to, but it suits Live's workflow very well. Remember, the concept of Live is one of accessibility and non-linearity, and Clip View provides an immediate focus on clip details and a means for them to be manipulated. This is especially important for Session clips, which require the flexibility of real-time manipulation on-the-fly.

Figure 7.2
Clip View
(Detail View).

7.2.1 Sample Display (Editor) and MIDI Note Editor

You have probably already noticed the *Sample Display/Note Editor*. It's kind of hard to miss when Clip View is open.

Figure 7.3
Sample Display/
Editor.

Figure 7.4
MIDI Note Editor.

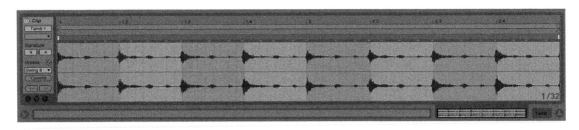

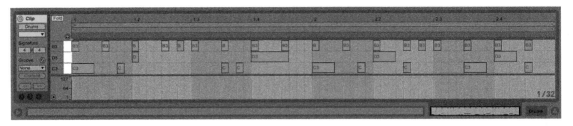

This display shows an overview of a clip's contents depending upon the type of clip selected—audio or MIDI—and will always be shown in Clip View for instant access to waveforms or MIDI events. It also serves as the graphical editor for *waveforms, Transients, Warp Markers, MIDI Notes, Velocities,* and *Clip Envelopes.* It should be pointed out that physical editing of sample waveforms in the Sample Editor is fairly limited. There is no cut/scissor tool *per se*, but you can, in fact, crop samples from a loop selection.

> **Hot Tip** If you really need to make detailed audio edits, you can access a third-party audio editor from the Clip View contextual menu. Right click or ctrl + click in the Sample Editor and select "Manage Sample File". This will bring up a new window on the right side of the main Live screen. From there you can click on the "Edit" button for the sample you want to edit and then Live will launch your external sample editing software as assigned in Live's preferences.

Use your mouse pointer to make selections, manipulations, and zoom content within the Sample Display/Note Editor. This will vary a bit depending upon whether you are working within the Sample Editor or MIDI Note Editor. Sample Editor: with the *Permanent Scrub Areas* preference turned "On" (Look/Feel preferences tab) the mouse pointer will show up as either an arrow pointer, magnifying glass, or speaker icon in the Sample Editor depending upon where in the display you are navigating. From anywhere along the bottom half of the display you can switch quickly from the speaker (scrub) icon to the magnifier tool. *Press + hold* the shift key to do this. With Permanent Scrub Areas turned "Off", the scrub icon will not appear, defaulting to the magnifying glass instead. When working in the MIDI Note Editor, the mouse pointer will alternate between an arrow, trim tool (bracket), or velocity scalar (⌘ Mac/[ctrl] PC + *drag*).

Figure 7.5
Follow Button.

The Sample Display/Note Editor also shows where the current playback position is located while a clip is running. Use the *Follow Switch* in conjunction with playback to allow the editor/display to scroll with playback as the sample unfolds along the Time Ruler. Activate Follow Mode from the Control Bar. This is especially helpful when zoomed in tight on a clip's contents.

As you can see in the example, the Sample Display/Note Editor also consists of navigation and playback elements that affect the function of a clip in the Session

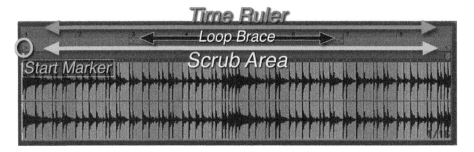

Figure 7.6 Time Ruler [bars.beats.sixteenths], Loop Brace, Scrub Area, Start Marker.

and Arrangement View. Let's break down each section and take a close look at what they do.

Time Ruler

At the top of the Sample Display/Note Editor is the Time Ruler measured in *bars: beats: sixteenths*. Use this area as a time-based reference and to zoom in/out on content. A Magnifying Glass icon will appear at the mouse pointer placed over the Time Ruler lane. *Click + drag* up or down to zoom. As you zoom in, finer beat time resolution will display. Keep an eye on the [*bars.beats.sixteenths*] labels to identify your location.[0]

Loop Brace

Below the Time Ruler is the Loop Brace, which indicates where looping begins and ends within a clip. Looping will occur only when the Loop Switch in the Sample Box (audio) or Notes Box (MIDI) is activated (yellow = On). At each end of the Loop Brace is the *Loop Start/End*. In the example, looping has been activated as shown by the slightly darker gray Loop Brace (as opposed to a lighter shade when not looped). To make a quick loop selection, highlight a region of audio or MIDI with your mouse pointer, then type ([⌘+L] Mac/[ctrl+L] PC).

Scrub Area

Just below the Loop Brace lane is the Scrub Area, available whenever *Permanent Scrub Areas* is enabled. When you hover your mouse over this area, the pointer will change into a speaker icon. *Clicking* along the Scrub Area will cause playback to jump or launch to that position relative to the Quantization settings.

Start/End Marker

Along the Scrub Area you will see the *Start* and *End Markers*. They are easy to overlook due to their small size and light gray color, but they are always located just above the Sample Display/Note Editor and below the Loop Brace lane. The Start Markers indicate where a clip's content will begin playback upon launch. The End Marker will only take effect when the clip's Loop Switch is deactivated ▶ 7.5.13 . *Hold* down the (⌘ Mac/[ctrl] PC) key while hovering over the Scrub Area and *click* anywhere along this area to instantly set the Start Marker at a new location along the Time Ruler. To place the End Marker, *hold* down the ([⌘+Shift] Mac/[ctrl+Shift] PC) key and do the same. You'll find this is easier than *clicking* on a Marker and *dragging* it to a new location. Note that the same process and key commands work for setting the start and end of the Loop Brace.

7.2.2 Clip Overview and Device View Selector

Hot Tip

*Increase the size of the **Scrub Area** in the Sample Display/Note Editor to include the entire Time Ruler area. Hold the **shift** key to convert your mouse into a speaker icon while navigating over this area.*

Hold shift then click anywhere along the enhanced Scrub Area to jump playback to the clicked location.

There are a few additional ways to navigate Clip View. If you haven't noticed already, there are two tabs always located at the bottom right of the main Live screen. On the left is *Clip Overview* and on the right is the *Device View Selector*. These are used to quickly toggle between Clip View and Device View. Since both are part of the Detail View, it's important to know how to navigate between them and which one you are looking at. To clarify, Device View shows the devices that are inserted on a track, such as instruments or effects, as opposed to clip contents as shown in Clip View. Use the Clip Overview and Device Selector to toggle between a selected track's Clip View and Device View, as seen in the example. *Click* directly on either tab or use the keyboard shortcut, *shift Tab*, to toggle between each view. Additionally, the *Clip Overview* tab also serves as a *Zooming Hot Spot* for zooming the Editor (*click + drag* up/down).

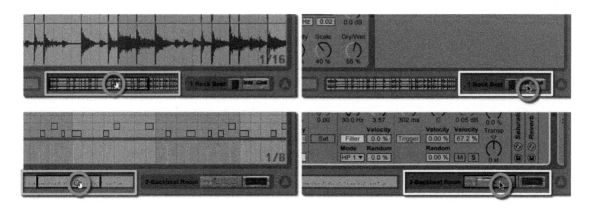

Figure 7.7   Clip Overview.   Device View Selector.

Each selector displays a condensed overview of its related contents, a clip wave-form/MIDI notes and track devices, for the selected clip/track. Clip Overview is very useful for navigating the Sample Display/Note Editor when you are zoomed in tight on specific events since you will not able to see the entire clip contents all at once otherwise. In situations like this, activate Follow Mode and watch the Display/Editor scroll during playback.

7.2.3 Audio versus MIDI Clips

Before moving on to dissect the various Clip View property boxes, you should know that there are a few differences between the properties displayed for audio versus MIDI clips. One major difference is that the *Sample Box* is only displayed for audio clips and the *Notes Box* is only displayed for MIDI clips. Further differences will be pointed out as we dig deeper into clip properties. Now, take a moment to look at the following example of Clip View for audio and MIDI clips to familiarize yourself with some of the noticeable differences, especially between the Sample and Notes Box properties.

Figure 7.8   Audio clip.   MIDI clip.

7.3 Clip Box

Moving from left to right in Clip View, we begin with the *Clip Box*. All properties here are the same for both audio and MIDI clips.

7.3.1 Clip Properties

In the upper left corner of the Clip Box is the *Clip Activator Switch*. This is used to deactivate (bypass) a clip, muting it from playback. This can also be accessed via a clip's contextual menu (*right click* or *ctrl + click*), or by pressing "0", which can be a very useful shortcut. A deactivated clip will appear colorless. In Arrangement View it was also labeled "Clip Deactivated".

Figure 7.9
Clip Properties.

Name, Color, and *Time Signature* properties do not affect playback; rather, they are for organization and reference only. To name or rename a clip from the Clip Box, type directly inside the Clip Name box. This will only affect the clip, not the file on disk. The Clip Color chooser allows you to assign any color to your clips and related title bars. Coloring clips is important for organizing your work. Clip Time Signature provides a visual reference of the clip's original time signature. Type or *drag* directly in the Signature box to adjust the signature label. Remember, this does not affect playback.

7.3.2 Groove Settings

These properties affect the playback of a clip and are used to alter and adjust its musical timing and feel of audio or MIDI information. Located under the Groove settings are *Clip Groove, Hot-Swap Groove, Commit Groove,* and *Nudge. Hot-Swap Groove* is used to quickly select, change, or audition grooves on-the-fly. The *Clip Groove* chooser is used to select grooves when they have been loaded in the *Groove Pool*. It is set to "None" (unavailable) until the clip is assigned to a groove or grooves become available in the pool. Commit Groove is used to permanently apply the assigned groove to the clip. When applied to MIDI clips, the notes within the Note Editor move forward or backward depending upon the timing and quantization strength. Note that this is a destructive process, meaning that when you commit a groove setting to a MIDI clip, its notes are permanently moved to fit the groove. When applied to audio clips, Warp Markers are added and adjusted to fit the groove (non-destructive). To learn about Groove, launch to ▶**Scene12**.

Just below the Commit Groove button is Nudge Backward and Nudge Forward. *Clicking* either of these buttons while a clip is running will skip playback forward or backward based on the rhythmic value set for Quantization. Use this for a stutter or skip effect in real-time. You can actually make the clip sound like it's skipping or stuck on a beat as if it were a scratched CD or record—remember those?

7.4 Launch Box

Launch Box properties are only displayed when in the Session View. They are dedicated strictly for Session clips and how they behave when launched. This box can be hidden from view by *clicking* the round yellow *Show/Hide Launch Box* button [L] just below the Clip Box on the lower left.

7.4.1 Clip Quantization

Live makes launching clips one after another a seamless process by applying quantization to clip launches—probably one of the most fun parts of Live. In addition to the global Quantization Menu that we have already discussed in Clip ▶ 5.3.2 each Session clip can be assigned to its own independent launch quantization setting via the *Clip Quantization* chooser. By default, Clip launches are assigned to global Quantization, but can be set independently of other resolutions. This gives you the freedom to launch clips as you see fit, perhaps based on a clip's feel or the beat flow of its contents. Maybe you want to delay the start or stop of a clip, while multi-tasking clip launches and other production or performance commands in real-time. Try out this concept by setting a handful of clips to various independent quantization values, then practice launching them in various sequences. Get creative and mix and match different quantize values in both the Clip Quantization chooser and the global Quantization Menu. It won't take long and you will quickly become an expert with clip launch properties in Session View.

Figure 7.10
Clip Quantization.

7.4.2 Launch Modes

Launch Modes determine how a clip responds when its Clip Launch button is activated (*pressed* or *clicked*). This is based on the concept that a button has two functional states (positions): *On* or *Off*. On is defined as *down* or *pressed* and Off is defined as *up* or *released*; therefore, the physical actions are: *click + hold* and *click + release* or *press + hold* and *press + release*. This applies to MIDI/Key Map-based launching as well: *MIDI key press/release*, etc. ▶*Scene17*. Of course, launch modes abide by a clip's quantization setting located in the Launch Box from the Quantization chooser. The best way to become familiar with the modes is to practice using them. You can then determine which works best for you in different scenarios. There are four Launch Modes: *Trigger*, *Gate*, *Toggle*, and *Repeat*.

Figure 7.11
Launch Modes.

Trigger

Generally the default mode for launching clips, the one you have most likely been using thus far (Preferences Menu). *Down* position launches a clip. When the launch button is *released* (up), nothing else happens—up gesture is completely ignored.

Gate

Down and *up* positions both have an action. *Down*, the clip is launched, and *up*, the clip is stopped.

Toggle

A clip launch requires a single *click/press* to start and another single *click/press* to stop a clip.

Repeat

Launching is just like Trigger Mode, but with an added function. A single *click/press* launches a clip and *release* has no action. Additionally, holding down the launch button, *click/press + hold*, causes the clip to continuously re-launch back to its start position over and over until the Launch Button is *released* (up). The timing for the repeat action is determined by a clip's quantization setting. Settings vary from Global to 1/32. With shorter quantize settings, you can create perfectly tempo-synced stutter effects or beat repeats. Not only is this creative and fun, you'll find it useful when triggering clips from an external surface controller or trigger pads.

>  **Hot Tip** For a quick description by Ableton, hover your mouse over any Launch Box's Launch Mode chooser menu and their description will appear in Live's Info View. As mentioned before, this is the best way to refresh your memory when working in Live.

7.4.3 Legato Mode

Figure 7.12
Legato Mode
switch.

Just below the Launch Mode chooser is the *Legato Mode switch*. This affects how a clip handles the transition of playback from a previously playing clip in the same track when another clip is launched. By default, when you launch a new clip, it begins playing back from its start point, even when it follows or interrupts another clip. With Legato enabled, a new clip will take over playback from the current position of the already running clip located in the same track. This means that no matter where a clip's waveform or MIDI playback position is, when it is interrupted the new clip will pick up at that position of playback as set by the Quantization settings. Essentially, Legato Mode allows for smooth and accurate musical transitions between clips, especially if you want to merge elements from several clips in real-time.

Let's look at a quick example with Legato Mode on.

Suppose you have two four-bar loops (e.g., Clip A and Clip B) on the same track, one right after the other in the next Clip Slot. Let's say you launch clip A and let it cycle through two times. Halfway through the third cycle you launch Clip B in order to take over playback from A, but rather than B starting playback at its start point, it starts at its third bar because that's where Clip A was when it was interrupted by B. With Legato Mode off, B would have instead begun from its clip start point when launched.

7.4.4 Velocity Amount

An interesting clip launch setting is *Velocity Amount*. This slider determines a clip's sensitivity to MIDI input velocities when launching clips via a MIDI controller. *Drag* up or down on the slider to adjust the sensitivity. The higher the percentage, the more sensitive a clip is to MIDI input velocities. Sensitivity means that low-input velocities (soft playing) result in a lowering in volume of the clip's playback, and higher input velocities (hard playing) result in louder clip playback. MIDI velocity values range from 0 to 127, the higher values being the hardest or loudest. The percentage of effect is based on 0 to 100 percent, with the highest meaning maximum effect and 0 meaning no effect.

Figure 7.13
Velocity Amount.

7.4.5 Follow Action

One of Live's most unique features is *Follow Action*. This gives you the ability to assign an automated task (action) to a clip, such as telling it to automatically trigger other clips or even to auto repeat or stop, for example. The purpose of these actions is to automate clip launches, thus allowing parts of your song to sequence themselves in a predefined succession or a random order—per your instruction, of course. Follow Actions can be assigned to any number of clips. In order for a clip to affect other clips, they must be laid out as a group of clips— that is, two or more clips slotted one after the other on the same track (consecutive Clip Slots). To separate or break up clips into groups, use an empty Clip Slot. Create as many groups and as many Follow Actions as you like and let Live do all the work!

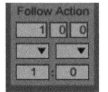

Figure 7.14
Follow Action.

For example, Clip A can be assigned to trigger (launch) Clip B after two bars. Clip B is then assigned to trigger Clip C, which is assigned to stop after four bars. In order for this all to take place, you must have a group of clips. Follow Action assignments are made in the "Follow Action" section of the Clip Box. The first row of input fields is used to define the *Follow Action Time* in: [*bars.beats.sixteenths*] and how much time (duration) will pass before an action takes place. In the second row is the *Follow Action A chooser* and *Follow Action B chooser* where you can assign up to two different Follow Actions for a clip. *Click* on the chooser and all possible Follow Actions will pop

Figure 7.15
Follow Action Group.

up from the menu. Just below the chooser menus in the last row is the *Follow Action Chance* A and B determinants. Assigning values here will determine the likelihood that either the A or B action will occur when the clip is launched. This means that each Chance control influences the probability that the other will occur. A setting of 0 means an action will never occur and the value of 1 will result in an action at every launch. The higher the value, the less often an action will occur when a clip is launched. When both Chance controls have a value greater than 0, the probability of actions for either one is affected. There are eight possible Follow Actions, all of which are clearly labeled in the Follow Action chooser menu, as you can see in the example. There are tons of possibilities with Follow Actions, limited only by your imagination and chance! For a creative explanation on utilizing Follow Action, refer to ▶ 7.10.1 for an example accompanied by downloaded content.

Figure 7.16
Follow Action A.

7.5 Sample Box

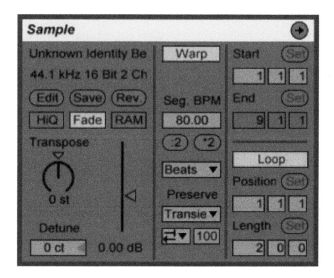

Figure 7.17 Sample Box.

The *Sample Box* is the heart and soul of an audio clip—*Notes Box* for MIDI clips. It is displayed within Clip View only when you are working with audio clips and contains the vital properties pertaining to the audio file viewed in the Sample Display/Editor. We'll discuss more about the Sample Display/Editor in the next section. For now, let's focus on the properties of the Sample Box.

7.5.1 Sample Properties

At the top of the Sample Box is the name of the audio file followed by its file type—sample rate, bit depth, and channels—referenced by the audio clip. Placing your mouse pointer over the file name will reveal a hyperlink and also display path and location of the sample in the *Status Bar* at the bottom of the main Live screen.

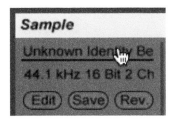

Figure 7.18
Click link to view sample in Places.

Figure 7.19 Status Bar showing sample location.

By *clicking* on the hyperlink, the sample will appear under Places within the Live Browser. From there you can audition the audio for playback. An additional benefit of seeing your clip in the Browser is that you can choose to display the date modified, location, size, file format, and rank. To choose what information to display, *right click or ctrl + click* at the top of the Browser content pane by the word "Name", then select the items you wish to display. Note that files not indexed in Places will not be viewable when you *click* their file link in the Sample Box.

7.5.2 Edit, Save, Reverse

Below the audio file's properties are three buttons: *Edit*, *Save*, and *Reverse*. All of these functions can be activated and used in real-time while your Set is playing or recording.

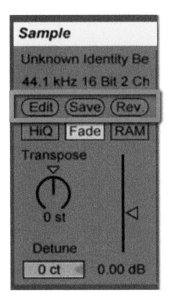

Figure 7.20 Edit, Save, Reverse.

Edit

Opens an external third-party sample editor, as defined in Live's *Preferences>File/Folder* tab, allowing you the ability to destructively alter the audio file and save it. *Click* this button to launch the editor application. It will then open up in its native user window.

Save

Should not be confused with Edit. Instead, it is used to save the clip's current settings so that every time it is used after that the same settings will be applied by default. This information is non-destructive and saved as a referenced analysis file (.asd) along with the audio file. These files streamline Live's use of external audio samples so that they can be used in real-time without hesitation.

Reverse

Does just what it says: reverses the audio file. When selected, Live creates a new audio file saved to the same location as the original, thus preserving the original audio file non-destructively.

7.5.3 High-quality Mode (HiQ)

The High-Quality Mode (HiQ) setting affects how well Live maintains the sonic quality of audio files when they are pitch-shifted (transposed). Unless your computer is struggling with processing power, you should be using this feature. Choosing anything less than the best quality for your audio productions should never be an option.

7.5.4 Fade

In order to avoid clicks or pops at the start and end points of a Session clip, use the *Fade* feature. This is especially useful for audio files that have not been cropped at a zero crossing (absolute silent point), or when looping a selected segment of an audio file. This could be a selection that you cropped out of a larger clip, for example. Fade is only available in Session clips and is automatically set by Live. The arrangement uses its own unique fades and crossfades as ▶ **6.4.5** .

7.5.5 Clip RAM Mode

This switch provides the option to load a clip's audio file into RAM rather than reading it directly from the hard disk. Use this when you need to reduce the load on your hard disk. Often when you have extremely large multi-track Sets, you may experience glitches, pops, or dropouts during playback if your hard disk is too slow, or simply becomes bogged down with processing audio, virtual instrument, and multiple plug-ins.

7.5.6 Transpose

The *Transpose* knob is used to pitch-shift a clip's playback by semi-tones (half steps), raising or lowering its sound. If *Warp* is active for the clip, transpose will not affect the clip's playback speed, just its *pitch*. This is ideal when you want to adjust the timbre of a clip while maintaining the integrity of the tempo. When deactivated, playback speed of the audio file contained in the clip will be altered based on the transposition amount selected—just like vinyl record players or the analog tape playback concept. In addition to simple transposition, it's possible to create some interesting and original effects for your looped clips using Transpose in conjunction with the various *Warp Modes* and *Preserve* settings. You should definitely try them out with your audio clips, especially using Beats Mode with "Transients" as your resolution! For more on creatively warping audio clips, launch to ▶ **Scene13**.

7.5.7 Detune

To fine-tune the pitch of a clip by cents, use *Detune*. This can help to lock up the intonation between two different clips or if you want to create a chorused effect—two or more clips deliberately detuned so that their pitch offsets, creating a beading effect or phasing effect. When either Detune or Transpose is activated, an orange *Reset Marker* will appear so you can quickly reset to the default setting by *clicking* it. Take note of this in the example. You can also *click* in the area of the parameter and then hit delete/backspace on your computer keyboard to return to the default setting.

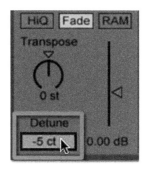

Figure 7.21
Detune.

7.5.8 Clip Gain

To change the output level (volume) of a clip, use the *Clip Gain* fader. This is independent of the track volume. Instead, it addresses the clip's audio file directly in a non-destructive manner. Clip Gain comes in handy if a clip's audio file needs to be balanced with others on the same track or it was simply a poorly recorded audio file. When Clip Gain has been adjusted, the displayed waveform will increase in size and the adjusted gain level will be indicated in decibels (dB) just below the volume slider. Be sure not to add too much or your clip will overload during playback, leading to distortion.

7.5.9 Warp, Master/Slave

Located in the center section of the Sample Box is the *Warp Switch*. When Warp is activated for a clip, Live applies time-stretching algorithms (*Warp Modes*) to the clip's audio file so that it will synchronize and follow your Set's tempo. In addition, the audio file can then be non-destructively manipulated to create, change, and quantize its timing and feel. Grid lines will appear behind the clip's waveform in the Sample Editor to represent its alignment to the rhythmically subdivided grid. You must activate *Warp* in order to loop audio clips in the Session View, even if

Figure 7.22
Warping
properties.

the original tempo is the same as your current Set, otherwise all clip looping features are disabled. When Warp is deactivated, the audio file will play back at its original beats per minute (BPM) and will not be looped. This is also known as a one-shot sample. You can, on the other hand, deactivate the *Loop* feature ▶ 7.5.13 so that a clip does not loop within a Clip Slot, yet Warp remains active. Most of the time your audio samples will be in Warp mode when they have beat/tempo-driven content.

When importing audio files, Live will automatically assign a yellow *Warp Marker* at the beginning and end of a warped audio file and identify each transient with *transient marks*—the prominent attacks, beats, or peaks in the audio. If you place your mouse pointer over a Transient Mark in the waveform timeline below the Scrub Area you will see a gray marker appear. This is a *Pseudo Warp Marker*, a handle designed to help identify Transient Marks when mousing over them. By *clicking + dragging* on a Pseudo Warp Marker, a Warp Marker will instantly engage.

Optimum Warping!

Automatic warping behaviors are defined by the Warp preferences in the Live Preferences menu. Feel free to change Warp preference settings, including the default mode that determines in which mode audio files are imported. When you change a Warp Mode from Beats to Tones, for example, you'll hear a noticeable difference in the audio file's sonic characteristics during playing. This is because each mode employs a different algorithm which dictates how Live analyzes and interpolates each transient. Tones is specifically designed for pitch-based audio material. Warp Modes are listed and explained in Clip ▶ 7.5.11 below. For a more advanced look at Warp Modes and how they apply to different styles of audio, launch to ▶ 13.2 .

By design, all warped clips play in sync with your Set's tempo—the master timekeeper if you will. When working with warped Arrangement clips, you can change this sync relationship by activating the *Tempo Master/Slave Switch*. This is set to Slave by default. Setting an Arrangement clip to Master overrides this sync relationship, causing the Set's tempo to sync to that specific Arrangement clip's original tempo—a little role reversal.

Surely you can imagine all of the uses for warping audio clips in the Session View. In general, Warp is ideal for tempo/beat matching, looping, and stretching audio material without a noticeable drop in fidelity. For everything you'll ever want to know about Warping, launch to ▶Scene13 .

7.5.10 Original Tempo

For every audio clip, Live automatically analyzes the contained audio file to determine its original tempo. It then warps it based on those results. The original tempo is then displayed in the Sample Box under the *Seg. BPM* (beats per minute) section. The displayed tempo may or may not always be exactly correct—often twice or half the intended tempo; therefore, it is possible to compensate for this by halving or doubling the original tempo. Just below the Original Tempo input field is the *Halve Original Tempo* and *Double Original Tempo*. *Click* the appropriate button to physically expand or compress the audio file by a factor of two, which will alter its playback

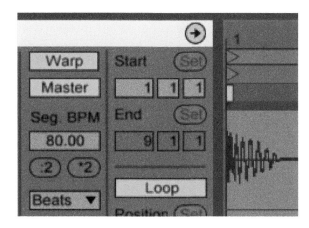

Figure 7.23 Tempo Master/Slave.

speed in relation to the master tempo of your Set. The resulting BPM will be displayed under Seg. BPM. Warp must be selected on the particular clip to execute this function! For finer adjustments, you may need to add and/or adjust the audio file's Warp Markers.

7.5.11 Warp Mode Chooser

Warp Modes are vital to how Live algorithmically divides and stretches an audio file for warping, quantization, and tempo synchronization. From the chooser menu there are six Warp Modes, each designed specifically for different types of audio content and for various types of transient/ rhythmic content detection. These modes are clearly labeled in the chooser menu, as shown in the example below. Choose a time-stretching algorithm that is best suited for the content of your clip(s). Try out the different modes. You'll be surprised at how well suited one may be over the other for your audio content.

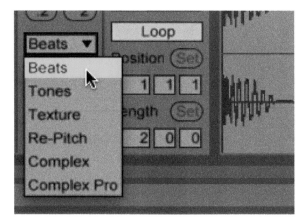

Figure 7.24 Warp Mode chooser.

Below the Warp Mode chooser you will see a variety of control parameters to fine-tune the selected time-stretching algorithm, whether it be resolutions, transients, grain, or formant-based manipulations.

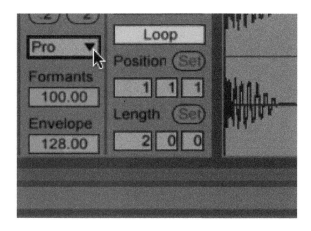

Figure 7.25 Below the Warp Mode chooser is its available parameters.

Each of these parameters is quite advanced and will be discussed in more detail in ▶*Scene13*. For now, we'll look at Beats Mode to get a feel for how Warp Modes and parameters relate to the Sample Box. Go ahead and select "Beats Mode" for now.

7.5.12 Sample Start/End

Within the Sample Box you can set an audio file's *Sample Start/End* point (Start/End Marker as seen in the Sample Display) and *Loop Position/Length* (Loop Start/End as seen located on the *Loop Brace* in the Sample Display).

The start point designates where the audio file begins playback when its clip is launched in the session or played back in the Arrangement View. The end point designates where the audio file will end. This is only relevant when the file is not set to loop (*Loop Switch* deactivated).

Figure 7.26
Sample Start/End.

To set the Sample Start and End, type or *click drag* inside the Start/End input field the desired *bars: beats: sixteenths* value. You can also move them directly in the Sample Editor using your mouse. Alternatively, they can be assigned in real-time using the *Set button*—that is, when a clip is running, *click* on the Set button to place the Start Marker wherever the play position is located at the current moment in time relative to the Quantization settings. Use this feature to establish a non-stop workflow. You can even map the Start/Stop points to a controller knob, slider, or key map ▶*Scene17*. Sample Start/End points also apply to *Clip Envelopes*, which will be discussed in ▶ 7.8 .

7.5.13 Loop Switch, Loop Position, and Loop Length

As you might guess, the *Loop Switch* engages looping for a selected clip. This can be for any portion of the clip's audio or the entire audio file, as determined by the loop selection in the Sample Editor. This means that it will repeat endlessly in the session until a related Stop Button is pressed. In the Arrangement View, it can be trimmed or *dragged* to loop endlessly. When deactivated, the clip will not loop at all; rather, the clip will only play once.

Loop Position and *Loop Length* (Loop Start/End) are seen in the Sample Editor as two triangles connected by the *Loop Brace.* This determines where in the audio file's waveform a loop selection starts and ends. The Loop Switch must be activated to loop a selection of the audio file. Loop Position and Loop Length adhere to the Marker Snap resolution settings.

To set the Loop Position and Loop Length, use one of the following three methods:

1. Use the Position and Length input fields.
2. Set them directly from within the Sample Editor by moving the Loop Start and Loop End markers with your mouse.
3. Use the Set button.

Notice that when manually moving the Loop Start/End, they will snap to the grid and also to Transient Marks as long as Snap to Grid is enabled. The more you are zoomed on the waveform, the more accurate the positions will snap. As mentioned, to make a quick loop selection, highlight a region of the waveform with your mouse pointer (*click + hold + drag*), then type ([⌘+L] Mac/[ctrl+L] PC).

Now, let's look at how Sample Start and Loop Positions work together …

For example, a Loop Position of "1" and Loop Length of 2 means that the loop starts at bar 1 and loops for two bars. As was shown earlier, the Loop Brace indicates where the sample's Loop Start/End points are located. It is very important to understand that the *Start/End Markers* are independent of *Loop Start/End*. Loop positions set the looping portion of a clip's contents, while the Start Marker establishes where in the file playback will begin regardless of where the Loop Brace is located. In other words the Start Marker is where playback starts and the Loop Brace outlines where the playback will cycle repeatedly (loop). As the example shows below, they can be separate starting points. Shown is the playback starting at bar 1, then cycling between bars 3 to 7, never playing bars 1 to 2 again unless the clip is re-launched.

Figure 7.27
Loop Switch,
Position, Length.

Figure 7.28
Two-bar loop
starting at bar 1.

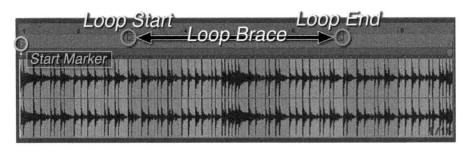

Figure 7.29 Sample Display/Editor: Start Marker, Loop Brace Start/End.

You can practice this concept by selecting a unique Start Marker position along your clip's waveform in the Sample Editor—this works for MIDI clips too. Then select a different section of the clip to loop using the Loop Brace. Now launch your clip and notice that once playback passes the Loop Start point, it will begin cycling as a loop between the Loop Brace—Start/End, regardless of where the Start Marker is located. This illustrates that the Start Marker only determines where the clip will start and has no bearing on the loop selection. Take note that Live references the Start Marker with two different names: Start Marker in the Sample/Note Editor and Start Position in the Sample/Note Box input fields.

7.6 Notes Box

The *Notes Box* is very similar to the Sample Box except that it addresses MIDI parameters. These properties are only available when a MIDI clip has been selected in either the Arrangement or Session View.

7.6.1 Transpose

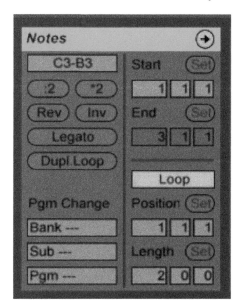

Figure 7.30
Notes Box Transpose.

The first section within the Notes Box is *Transpose*, which is a convenient way to alter the key (key track) of selected MIDI notes. When multiple notes are selected, the Transpose text field displays the range that the selected notes span. When no notes are selected this displays the range of all of the notes and subsequently controls all of the notes within the clip. To use Transpose, *type* a value to transpose by (+/-), a note name with octave number, or *click + drag* up/down in text field.

7.6.2 Play at Half or Double

Just below Transpose is *Play at Double Tempo* and *Play at Half Tempo. Click* on either button to physically expand or compress the MIDI notes by a factor of two in order to alter their playback speed in relation to the master tempo of your Set. This is a very useful way to increase the playback speed of a MIDI file without manually adjusting each MIDI note to play twice as fast or as slow. Often you will import a MIDI file (loop, riff, or song) and come to find that it appears to be programmed at half the speed it should be. Just hit the Double Tempo button to automatically conform it, in this way cutting a file's duration in half in order to fit four bars into two.

7.6.3 Reverse/Invert Notes

For a quick way to get creative, use the *Reverse* or *Invert* button to alter a clip's MIDI notes. Reverse physically sets the selected notes backward, so they will playback in reverse order. Invert physically flips the selected notes upside down, so that the notes on the bottom move to the top and vice versa. Using this with a drum kit MIDI instrument can yield some very interesting results since the notes will trigger different drum samples while maintaining the same rhythm.

7.6.4 Force Legato

In music the word legato is used to define a smooth playback style. Simply put, it means connected. The *Legato* (Force Legato) button forces selected MIDI notes to butt up against the beginning of the next note so they are nearly connected without overlapping. Some notes will be lengthened and others shortened depending upon their proximity to each other. This applies to both notes on the same key track and notes on different key tracks. Live will determine whether overlapping notes are intended to play simultaneously—like a chord—or if they should not overlap. Force Legato will really come in handy when you want to smooth out a recorded MIDI performance or correct overlaps that can cause an instrument to have too many notes bleeding into each other during playback due to long release times. In the same way, overlaps can lead to notes being muted during playback in some instruments.

7.6.5 Duplicate Loop

The *Duplicate* (Dupl. Loop) button is a very simple feature in that *clicking* it will double the length of the loop selection, therefore doubling the duration of the looped MIDI selection each time the button is pressed. Achieve the same result by *clicking* on the Loop Brace and typing ([⌘+D] Mac/[ctrl+D] PC).

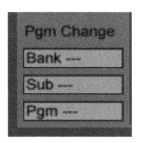

7.6.6 Program Change

Live handles all External MIDI Bank and Program Change Message information in this section of the Notes Box. Your External MIDI instrument Patch Lists (instruments/sound presets) and Banks (a catalog of instruments

Figure 7.31
MIDI Bank Select, Sub-Bank Select, Program Change choosers.

grouped as a set of presets) can be navigated and selected from the *Bank Select*, *Sub-Bank Select*, and *Program Change* choosers. They communicate directly with External MIDI devices such as synthesizers, samplers, sound modules, and third-party plug-ins when applicable. The actual bank and program details depend upon the specific device's protocol; but, in theory, you will be able to navigate patches via these menus. If you don't have any, don't feel left out. Current trends favor using virtual instrument plug-ins and Live's software instruments over external synths. To learn more, launch to External MIDI ▶ **15.9** and Virtual Instruments ▶ **15.8** .

7.6.7 Start/End, Loop Switch, and Position/Length

Within the Notes Box you can set a MIDI clip's (MIDI file) *Start/End* point and *Loop Position/Length* in the same way as described for the Sample Box. The Start point designates where in the clip playback begins when it is launched in the session or played in the arrangement. The End point designates where the clip will end when not looped. Loop Position and Loop Length determine where the actual loop selection begins and ends—the specific MIDI events selected for looping. To loop a selection within a clip, the Loop Switch must be activated.

Start, End, and Loop positions can also be assigned in real-time with their respective Set buttons. This will instantly drop a position marker at the current playback position, relative to the global Quantization setting—for example, at the next bar. To use the Set Start/End feature, the clip must be playing or paused (launch button green). Once active, the Set buttons can be *clicked*, thus setting the desired positions. Alternatively, positions can be set from within the MIDI Note Editor by manually *dragging* the Start/End, Loop position markers, or Loop Brace into the desired positions.

7.7 MIDI Note Editor

MIDI Editing has come a long way since the days of the event list—a long list of MIDI data used for programming MIDI events. Nowadays, MIDI is a graphical tactile experience, which we are all thankful for! However, there are DAWs that still offer an event list option for editing; but in Live you won't find an event list editor—instead, you will execute all of your MIDI editing in the *MIDI Note Editor*. This is often called piano roll in other DAWs since the piano keyboard map is the basis for the layout. The MIDI Note Editor is available in both the Session and Arrangement View for each MIDI clip. Select a clip—*double click* if not in Clip View—to reveal the Note Editor.

For the following description we will reference a MIDI drum loop assigned to playback with Impulse (Live's eight-sample Sample Player) located in the Instruments folder in the Live Browser. Launch to Clip ▶ **15.3.1** for more on Impulse. Let's take a closer look at what is accessible in the MIDI Note Editor.

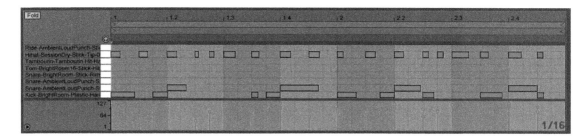

7.7.1 Basic Commands, Navigation, and Zooming

Figure 7.32
MIDI Note Editor.

Many of the commands in the MIDI Note Editor are the same or very similar to those you will find in most other music production software. Some of the common commands are as follows. Use your mouse to move a single note (*click + hold + drag*) or multiple notes (*shift + click*). You can also select multiple notes together (*click + hold + drag*). Create an *Insertion Marker* within the Note Editor by *clicking* on a position along the grid. An Insertion Marker is a vertical orange line that determines the point at which notes, time, selections, and pastes will begin. From here the Insertion Marker can be moved using the computer keyboard arrow keys. Other common commands such as cut, copy, paste, and duplicate can be executed in the Note Editor, each using the same key and menu commands that have been referenced throughout.

Navigate the MIDI Note Editor by using the *magnifier tool* to scroll or zoom. To scroll, place your mouse pointer over the *Beat Time Ruler* at the top of the Note Editor and then *click + hold + drag* left/right to navigate through the view. To zoom in or out, *click + drag* up/down with the magnifier from any position along the top of the Beat Time Ruler area. Feel free to use the *Clip Overview/Zooming Hot Spot* in the same way. *Double click* with the magnifier to zoom all that way back out so the entire clip is in view. Additionally, you will see a dark outline—the Zooming Hot Spot—in the Overview indicating what area of the MIDI clip is currently in view. From either edge of this outline, *click + drag* inward toward the center of the Clip Overview to zoom around the area in view. Depending upon which side you *click* on, a bracket handle will appear. This is called the *Left Zooming Edge* or *Right Zooming Edge*. For a shortcut to zooming, use the +/– on the computer keyboard number pad.

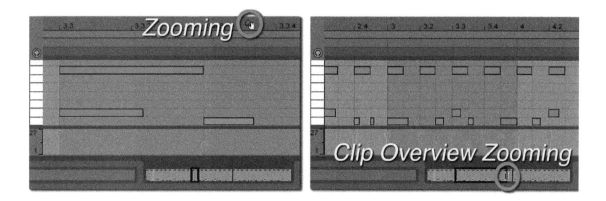

Zooming

Clip Overview Zooming

Figure 7.33
Zoom the MIDI
Note Editor from
the Beat Time
Ruler or Clip
Overview.

7.7.2 MIDI Note Ruler

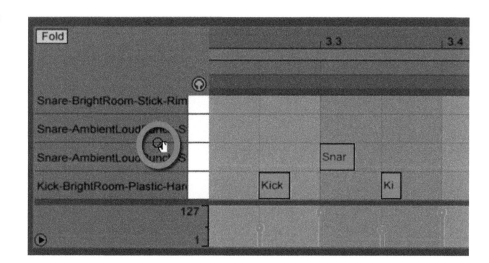

Figure 7.34
Zooming the
Note Ruler.

At the left of the MIDI Note Editor is the *Note Ruler* showing the MIDI key range and note names. In this particular case, the Note Ruler is listing Impulse's sample names as assigned to its particular keys (eight slots). *Click + drag* the Note Ruler to the right and left to zoom in and out on the MIDI notes—they will increase in size for better visibility—and up and down to scroll through the key range. *Click + drag* the Note Ruler up and down to scroll through the key ranges/octaves. The adjacent piano keys to the right of the key ranges make up what is called the *Piano Roll*. Slide your mouse over the individual notes to reveal specific note names (i.e., C3, D3, E3 ...). The names will also display in the Status Bar. If you need more real estate to work with and full visibility of the keyboard range, *drag* the divider (boarder) at the top of Note Editor upward to resize the window. This will allow for a larger view of the key range and MIDI notes. Hover your mouse pointer over the top black divider line between Clip View and Session or Arrangement View, then *drag* up or down to resize. Another option for increasing visibility is to *click* the *Fold* button at the upper left corner to hide all unused piano key tracks (rows). This will hide the Note Ruler names, leaving only the Piano Roll showing to the left.

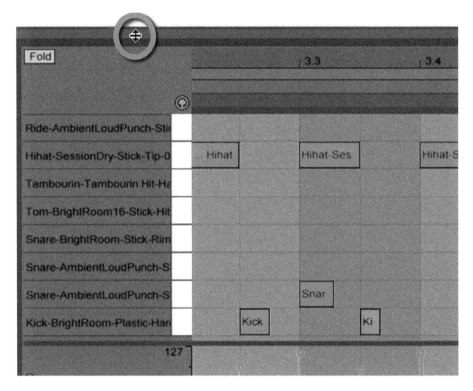

Figure 7.35
Drag the divider to resize the MIDI Note Editor or Sample Editor.

7.7.3 MIDI Velocity Editor

Located directly below the MIDI Note Editor is the *MIDI Velocity Editor*. From there you can edit the individual velocity of each MIDI note—nothing new if you've used a DAW before. In the lower left-hand corner you will see a triangle-shaped fold button that is used to hide the Velocity editor from view if you prefer not to see it. Each vertical *Velocity Marker* can be *dragged* up or down in order to alter a notes velocity. The taller the marker line, the stronger the velocity and greater the velocity value (1 to 127). As you increase velocity of a MIDI note, the actual MIDI note color will

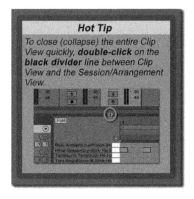

become darker and redder, or lighter and bluer as you decrease it (red = high; blue = low velocity). For quick access to velocity values, while mousing over a MIDI note, *hold* down the (⌘ Mac/[ctrl] PC) key to reveal a note's velocity value. With this key still held you can *click + drag* a MIDI note up or down to change its velocity. Placing your mouse over a Velocity Marker will cause the associated MIDI note to be highlighted. The opposite is true when mousing over a MIDI note. In this case the velocity marker will be revealed (highlighted in blue). Multiple Velocity markers can be selected for mass editing. The common commands apply. If you'd rather redraw velocity values with the mouse, enable Draw Mode 🖊 from the Control Bar, then use the pencil in the MIDI Velocity Editor window.

7.7.4 Insert/Edit MIDI Notes

Writing and editing MIDI notes in the MIDI Note Editor is a very popular way to work in Live—or any DAW for that matter—as opposed to using a MIDI keyboard controller. This is especially useful for those of you who do not have access to a MIDI controller or prefer not to enter notes on a keyboard at all. Not everyone is a keyboard player, but that is no reason not to create music!

Notes can be inserted with the traditional mouse pointer or the pencil tool. These are defined as *Edit Mode*—with the mouse—and *Draw Mode*—with the

Figure 7.36
MIDI Editor
Preview switch.

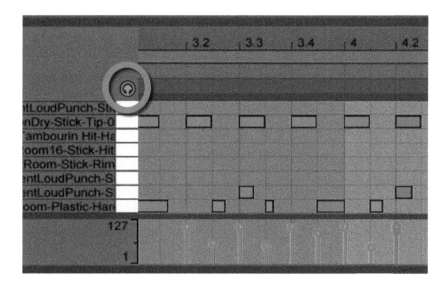

pencil. In Edit Mode all of the standard commands apply, such as copy, cut, paste, etc. Before inserting and editing MIDI notes, you should activate *MIDI Editor Preview* (head-phone icon), located at the upper left corner of the Note Editor above the Piano Roll.

This allows you to hear a note as you manipu-late it (*dragging* and *clicking*), provided you have an instrument assigned to that track to play it back. Notes can also be inserted or drawn during playback at any time in addition to recording them via a MIDI controller. Just keep in mind that notes can only be played if the Session Clip Slot is empty, otherwise *Session Record* must be enabled for MIDI overdubbing: there is no overdub feature for audio! A similar situation occurs when recording notes into existing Arrangement clips, thus the *MIDI Arrangement Overdub* feature. For more on Overdub Recording, launch to ▶ **8.2** .

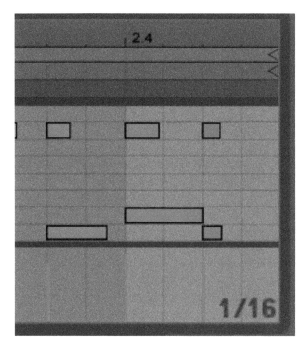

Figure 7.37 Grid/Marker Snap.

Hot Tip Be sure to activate Snap To Grid, if not already activated ([⌘+4] Mac/[ctrl+4] PC), while working in the MIDI Note Editor. You will find this helpful for snapping to different resolutions of time when moving notes, setting insertions, and making selections. You can also right + click or ctrl + click within the MIDI Note Editor window to access this feature. If the grid is de-activated, "Off" is displayed. When activated, a rhythmic value resolution is shown in the bottom right of the MIDI Note Editor (Marker Snap). Note that this is present in the Sample Editor as well.

Regardless of the input method you choose to use, all note insertions and edits are governed by the Marker Snap settings, *Adaptive* or *Fixed* grid. With Adaptive Grid, note starts and lengths are automatically snapped to the nearest grid lines upon insertion—wide, medium, narrow, etc.—which adapts to your zoom level. With Fixed Grid, you can choose specific rhythmic resolutions for snapping to the grid, which remains constant regardless of zoom level. Access all of these settings from the Options Menu or MIDI Note Editor contextual menu. To temporarily suspend note length snap restrictions set by the grid, *hold* down the (⌘ Mac/[alt] PC) key while inserting a note, and then *drag* to extend the note length as long as you desire. To permanently disable the Snap, unselect "Snap to Grid" from the Options Menu ([⌘+4] Mac/[ctrl+4] PC). To re-enable it, choose a new Marker Snap setting or enable it directly from the Options Menu. With Snap to Grid disabled you will be able to place notes anywhere you want. There are a few differences between Edit and Draw mode. Let's highlight a few specific functions of each mode.

Edit Mode

Adding and removing notes is as easy as *double clicking* on an empty spot in the MIDI Note Editor to add, and *double clicking* on a pre-existing note to delete. Use either the mouse or keyboard arrow keys to move notes around. Just select notes and *drag* or use the arrow keys. To lengthen or shorten notes, *drag* them from either of their edges in the direction you wish. If the Marker Snap settings begin to hinder your note placements, you can easily change the snap setting from the Note Editor contextual menu (*right click* or *ctrl + click*). For basic navigation without MIDI notes, the Left/Right navigates forward and backward. The Up/Down arrows navigate between existing MIDI notes, but only advances as far as the last existing MIDI note in the sequence. Up/Down also moves selected MIDI notes up or down in the Note Editor. If you want to move the Insertion Marker around in micro increments, *hold* down the (⌘ Mac/[ctrl] PC) key while using the arrow keys in any direction. This temporarily bypasses the Marker Snap, allowing you to insert notes at any finite position.

Transpose (move) notes either by octave or chromatically using a combination of key commands. To adjust notes by an octave up or down, select the desired notes and then *press* and *hold* the shift key then *press* the up/down arrow keys to move the notes up or down the key range within the MIDI Note Editor. To move the notes chromatically (one step at a time), select notes and use the up/down arrow keys to move them up or down. Move them forward and backward using the left/right arrow keys. In the same vein, shorten and lengthen MIDI notes in the Note Editor by *holding* the shift key and *pressing* the left/right arrow keys. These commands all adhere to the grid. Using the arrow keys will significantly expedite editing process. Remember to activate the MIDI Editor Preview (headphone icon in the upper left corner of above Piano Roll) to audition the notes while moving and editing them.

Draw Mode

Notes are inserted with the *pencil tool* when the *Draw Mode Switch* is activated. C*lick* to insert a single note or *click + hold + drag* to insert and draw out a succession of notes within the Note Editor. *Hold* down the shift key while inserting a note and you will be able to customize its length, otherwise the note length automatically snaps to the nearest grid line as set by the Marker Snap. To delete a note, *click* on it with the pencil and it will disappear. The one exception to this is if the start of the note precedes a grid line, then *clicking* on it will only delete the part of the note that falls within the grid line. Left behind will be a sliver to of the note that falls outside the lines. Turing off Snap to Grid will allow you to place notes wherever you want and with customized note lengths upon insertion. To do this, *click + drag* the note length left and right as you insert it. Using a Marker Snap setting while inserting notes can be really useful for quickly establishing note lengths and rhythmic patterns. As mentioned, note lengths automatically snap to the grid. To use this to your advantage, you should preset the Marker Snap to the resolution that you want your notes to adhere to. To really speed up your workflow, adjust the key command for the Marker Snap to toggle between grid resolutions on-the-fly ([⌘+1, 2, or 3] Mac/[ctrl+1, 2, or 3] PC). To lengthen notes after they have been inserted, simply place your mouse pointer at the either edge of the MIDI note and *drag*. If you don't want your notes to snap to the grid upon insertion, switch Snap to Grid off. In addition to note length manipulations, there is a convenient way to adjust velocity upon note insertion while in Draw Mode. Simply *drag* up and down to increase or decrease the velocity of the note you are adding. By default, its initial velocity will match that of the last note you drew in.

As you can se,e there are a number of ways to insert notes and make edits. Take some time to practice these features. Over time you will develop your own fast way of getting the job done, which will enhance your workflow. Definitely learn the keyboard commands. They will make toggling between each grid mode and inserting and editing quick, easy, and efficient. Beyond that, there are many more great time-saving keyboard shortcuts worth learning and using in Live for navigating and editing notes ▶ **Web** .

MIDI Stretch

When making a timespan selection in the MIDI Note Editor you will notice that a light blue highlight appears. If you look just below the Beat Time Ruler in the *Selection Area* there are two *MIDI Stretch Markers* at each end of this selection located in the *MIDI Editor Stretch Area*. These markers are used to stretch MIDI notes or envelopes similar to how Warp Markers affect audio file transients. This is essentially MIDI warping—unofficially. There are also *Pseudo MIDI Stretch Markers* that appear when mousing over MIDI notes between the Stretch Markers. They can be *dragged* to manipulate the MIDI notes within the selection, just like audio transients in warping.

7.7.5 MIDI Step Recording

For you MIDI aficionados out there, *MIDI Step Recording* is all about you. That being said, let's define it. MIDI Step Recording gives you to ability to use a MIDI keyboard and computer keyboard together to input MIDI notes into the MIDI Note Editor. This approach was made very popular by notation software as a speedy way to input notation without having to worry about timing on the keyboard. There is one catch with this: it can only be used with an existing clip. It doesn't matter if it has MIDI notes already; it just needs to be in a Session clip or Arrangement clip. With that said, here is a walkthrough in the Session View:

1. First, *arm* (record-enabled) the track you wish to input notes into in the Session View.

2. *Double click* in an empty Clip Slot to create an empty clip. You could also *right click* and choose "Insert MIDI clip". You must use *right click* or *ctrl + click* in the Arrangement View.

3. Enable MIDI Editor Preview (blue headphone switch)!

4. Using Edit mode (not Draw Mode), go to the MIDI Note Editor and place an *Insert Marker* at the beginning of the clip—or wherever you want to input notes—by *clicking* on a grid line inside the Note Editor. *The Insert Marker* appears as an orange flashing line.

5. Set the grid to "1/8" (just for this exercise). You can *right click* or *ctrl + click* anywhere in the Note Editor area to access and change the snap resolution.

6. Once this is set, *press* and *hold* the desired key(s)/notes on the MIDI controller that you wish to input, then *press* the right arrow key on the computer keyboard. Notice that when you pressed the MIDI keys nothing happened, but while *holding* them down and *pressing* the arrow key the notes were inserted.

7. Let go of the MIDI note(s). Repeat the last step again and you will see the same note(s) inserted on the next beat or grid line.

8. Do this one more time; but this time don't let go of the MIDI notes after you press the arrow key. Instead, *press* the arrow key a couple more times consecutively. This extends the note lengths (duration) beyond the visible grid to the next grid line and beyond.

To remove a note or shorten its length while amidst inserting it, keep holding that note's MIDI key down, then arrow backward over it and it will disappear. You can do this to any note to edit its length or delete it altogether as long as you don't let go of the notes. If you've let go, then use the mouse to remove notes, or set the Insert Marker at the point where you want to rewrite the notes or use the back arrow. When inputting notes, keep in mind that no MIDI notes can occupy the same space when they are the same note. Identical notes will always be overwritten when they overlap.

Inserting and editing notes and navigating the grid becomes quick and easy by using only the computer keyboard arrow keys. Once you find your desired input position, *press + hold* down the MIDI notes and use the Left/Right arrow keys to insert them. The Right/Left arrow key feature can be MIDI/Key mapped for mouse-less navigation and step input. Two arrows will appear in the MIDI Note Editor below the Fold button when in Map Mode. Assign these as you wish. Launch to ▶ **17.2** for more on MIDI/Key Mapping.

7.8 Envelope Box

What is an *Envelope*? In the world of DAWs, an *Envelope* is a linear or curved segment-based representation of various parameters (volume, pan, etc.) that are used to shape and manipulate the properties of, in the case, a clip's contents. These lines and curves are made up of adjustable breakpoints that enable them to take on shapes (ramps, curves, lines, etc.) by creating and adjusting multiple break points along a waveform or MIDI event. Such parameters and manipulations are linear and evolve over time. Therefore, envelopes can be automated and appended to a clip.

The *Envelope Box* provides a unique way of managing the automation of clips, device chains, and mixer parameters for both the Arrangement and Session View. It is used in conjunction with the *Envelope Editor* where all clip automation is carried out as *Clip Envelopes* ▶ **7.9** . Clip Envelopes can be set to affect the Mixer, the clip itself, and associated device parameters via the *Device chooser*. One point to keep in mind is that in Session View, Clip Envelopes can be assigned to affect clip parameters and Mixer parameters each inde-

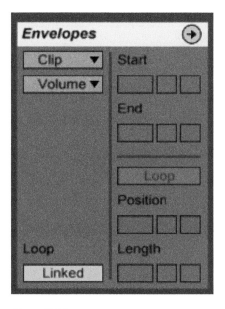

Figure 7.38
Envelope Box.

pendently. Therefore, Mixer *Envelopes* are created and manipulated via a Session clip Envelope Box and Envelope Editor. When *dragging* or recording Session clips into the Arrangement View, Mixer Envelope automation will be converted from Clip Envelopes into Track Envelopes/automation. In this way, Session clip automation is somewhat unique because no breakpoint envelopes are written to the Arrangement View tracks until globally recorded or moved directly from the session to the arrangement. At that point, the envelopes are written to Track Display and can no longer be viewed as a clip envelope; rather, they are attached to the track Mixer itself as envelopes. In either view, envelopes manipulate the same physical parameters, mixer or clip; but it should also be pointed out that certain clip parameters, such as Clip Gain or Transpose, are married to the clip and not to the track in the way that Mixer automation is in the Arrangement View. For more on automation, launch to ▶Scene10 ▶ 10.1 .

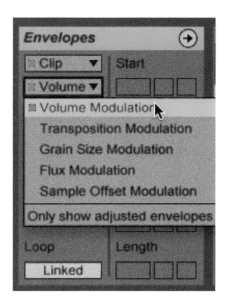

Figure 7.39
Clip Envelope control parameters.

7.8.1 Device and Control Chooser

To use envelope automation from Clip View, you will choose a device from the *Device chooser* and the specific parameters of that device from the *Control chooser*. Whichever device you choose, the associated control parameters will become available in the Control chooser. For audio clips, you will be able to select from Clip, Mixer, or a device in the track. For MIDI clips, you will be able to select between each device in the device chain, including each individual parameter or the Mixer.

7.9 Envelope Editor

The *Envelope Editor* is where all the magic happens. It's full of great and powerful automation features. The whole concept is very different than any other DAW you may have worked in and it will take some getting used to. To bring the Envelope Editor in view, click the Envelope's Title Bar or the Show/Hide Envelopes Box button. You should then see a red

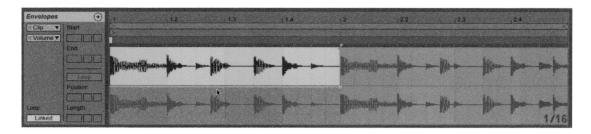

Figure 7.40 Envelope Editor.

dashed line through the audio waveform or MIDI notes, which is the envelope. In the Arrangement View, the Envelope Editor is also integrated as part of the Track Display when in Draw Mode.

7.9.1 Drawing and Editing Envelopes

There are two ways to create and edit envelopes—drag or drawing; but first let's identify what the envelope looks like. As you can see in our example, the envelope is represented by a red line and/or pinkish shaded area in the Envelope Editor.

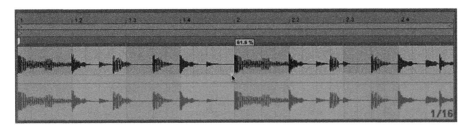

Figure 7.41
Clip Envelope.

The line itself represents the position at which the envelope is set and the actual shape of it. The line is the envelope referred to as a *Breakpoint Envelope* when there are little dots or anchors along the envelope. These are not viewable in Draw mode. To draw or edit clip parameters, just select a clip to show it in Clip View and then bring its Envelope Editor into view. From there, select the device or parameter from the Device and Control chooser that you wish to edit (automate); then, with your mouse, *click* the Breakpoint Envelope in various locations to add breakpoints, and then *drag* or just *click + drag* the envelope right off the bat. In Draw Mode, use the pencil to *click + draw* an envelope, shaping it to the desired position. For an in-depth walkthrough of clip automation and Envelopes, launch to ▶ **10.2** .

7.9.2 Link/Unlink Envelope, Start/End, and Loop Position/Length

When envelopes are set to *Linked* they loop with the clip, repeating the same automation over and over infinitely until the clip is stopped—synchronous with the clip loop, just as you would expect. When they are *Unlinked*, an envelope becomes independent of a clip's loop cycles and will affect the clip from the envelope *Start* position until the envelope *End* position regardless of what the clip loop is doing. This is indicated in the envelope *Start/End Point Box* in the upper

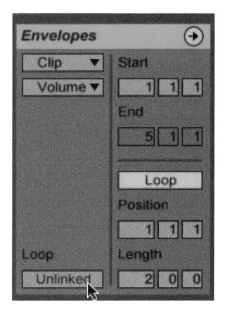

Figure 7.42
Unlink envelopes to create envelope shapes/effects that are independent of a clip's loop length.

right section of the Envelope Box. This allows an envelope to automate control over a clip for a specific duration or segment independently. An envelope can also be set to loop independently of the clip with its own loop length that differs from the length of the clip itself. This is an interesting way to vary or randomize the effect of envelope on a clip. For example, let's say you have a four-bar drum loop clip and you want the volume to fade down in bars 3 through 4 and then return to its original volume at bar 1 for each cycle (loop) over and over again. This would be a perfect time to use a linked Clip Envelope to automate the Clip Volume. Suppose you want the same effect, but your clip is only two bars. This is when you would use and unlinked Clip Envelope. Launch to ▶ **14.6** for a detailed walkthrough of looping with Unlinked Clip Envelopes. Once you grasp how this works, you'll understand the many great possibilities available with Clip Envelopes.

7.10 Musical Concepts

There are so many ways to create, produce, and perform with clips that there could be a whole other book just on that alone. Saving that for a rainy day, let's focus on some specific ways to enhance your music through Follow Actions.

7.10.1 Create: Rhythmic Loop with Follow Actions

A great way to create unusual and surprising results with Follow Actions is to load up a single track and fill it with seven to ten rhythmic-style audio beat loops in consecutive Clip Slots. Each clip should be no more than two bars in length for this example. Let's try this together:

1. After you load up your clips, set the tempo to 120 BPM.
2. Randomize each clip's transpose knob by two or three steps in alternating directions. Select one clip after the next and make the adjustment to each.
3. Adjust the Loop Brace for each clip to start and end differently from the other clips. Adjust the Loop Start/End or Loop Brace to a Fixed Grid resolution of 1/4 or 1/2.

4. Now adjust the Follow Actions for all clips at once using *shift +click* (multi-clip functionality) to these settings:

   • *Follow Action Time chooser:* **0–2-1**;

   • *Follow Actions A* (**Any**); *Follow Actions B* (**Other**);

   • *Chance A and B:* **0–7**.

   Again, it's important to use *shift + click* to select all the clips at once.

5. Launch one of your clips and sit back to see what happens.

6. Now increase the tempo by 15 to 20 BPMs to achieve a completely different sound and vibe with the rhythmic feel of the clips.

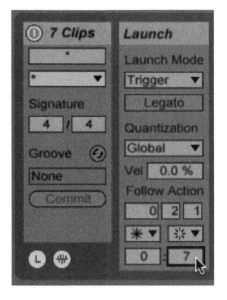

Figure 7.43
Follow Actions being applied to multi-selected clips.

This is just an experiment, but from here you use a delay on a Return track as a send effect for even more sonic possibilities. You can also experiment with reversing one or two of the clips and add even more random results. Now check your Set against the demo set from the companion website. Download "**CPP_7–10–1_FollowActions Project**".

When working with a Follow Action sequence that you generally like the sound of, you might consider resampling the audio from the Follow Action track into a new audio track (print to track ▶ 8.5). Using the track's Input chooser "Audio From", you can then create a new customized audio clip printed with the results of the randomized Follow Action parameters that you made in the first place. Save this to your User Library for later use. You'll be surprised at what you can create.

Keep in mind that Follow Actions can be fairly unpredictable depending upon how they are set. In some cases, no single Follow Actions will play or behave exactly like the time before. Therefore, when you do decide to resample the audio output of a unpredictable Follow Action sequence, remember that although it may have the same general rhythmic vibe and feel as it launches from clip to clip, you won't get the exact same pattern each time you launch the sequence. After all, this is one of the basic principles behind using Follow Actions in the first place: randomization and variation. For a more logical and predictable sequence of events, use the "Next" parameter within both the Follow Action's A and B chooser window in order for all your clips to play in order and cycle continuously. This is great for setting up predictably long and precise stage sets or for an "un-manned" visual performance that requires cues to be in sync and in an absolute order.

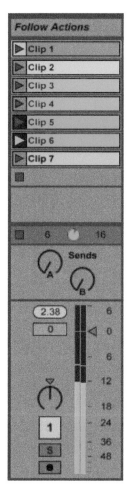

Figure 7.44
Follow Actions in action!

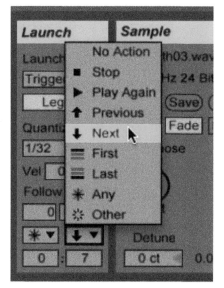

Figure 7.45
Try Follow Action "Other".

Figure 7.46
Try Follow Action "Next".

You will see in the two upcoming examples that both Follow Action A and B are in a randomized state "Any" or "Other" and in a more predictable state with the selection as "Next". Notice the other choices included in the chooser menu. By selecting "Any" or "Other", you'll find inspiring possibilities in your clip's behaviors.

Of course, all of this can be accomplished by simply aligning your clips in the Session View, launching them in order, and globally recording them as a sequence into your Arrangement View. For more on global recording, launch to ▶ **Scene 4**.

7.10.2 Produce: Dummy Clips

"Dummy Clips" not only have an unofficial and curious name, but also happen to be an intriguing and clever trick in Live. Think of Dummy Clips as a track of clips used to impose effects and automate the audio output of another track and its various clips. Dummy Clips make no sound—hence, dummy; rather, they impose an audible effect over other tracks using their envelopes to automate or imply specific audio effect parameters of devices or any Mixer parameter associated with the Dummy Clip track and Dummy Clips. To show this in action, we'll use the

Session View with two audio tracks to explain how these clips work. In this basic setup of Dummy Clips you will need one track to contain the audio clips you wish to enhance and another track for your Dummy Clips.

Create a New Live Set

1. Delete the two default MIDI tracks, then load up track "1 Audio" with an audio clip or multiple audio clips.

2. Launch the clip(s) to ensure playback, then stop all clips.

3. Change the Output Type chooser on track "1 Audio" from "Master" to "Sends Only".

4. Set "Monitor" of track "2 Audio" to "In" (input monitoring) and set its Input Chooser to receive audio from "1 Audio".

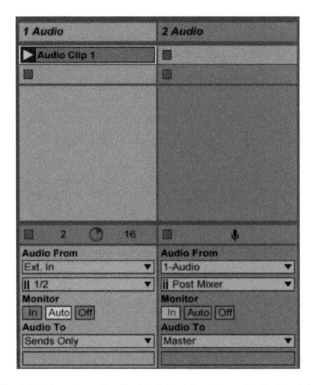

Figure 7.47 Track 1 with main audio clip, Track 1 with main audio clip. Track 2 set to Monitor Input sourcing from "1-Audio". Track 2 set to Input Monitor.

5. Test your signal routing by launching a clip on track 1. With the correct routing assignment you should be hearing and seeing your audio play through on

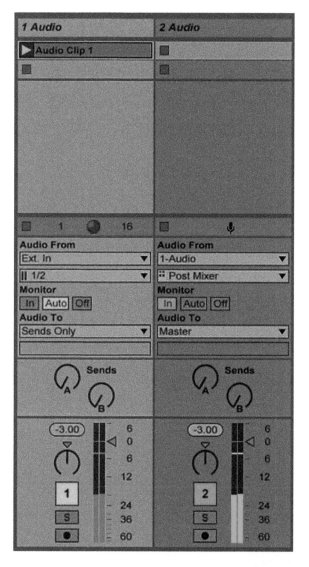

track 2. The audio levels meter on track 1 will still show a blue color signal, signifying that it is set to *Sends Only*, but no sound should actually be audible through track 1. Test this by temporarily deactivating track 2. If you hear nothing, then you are all set.

6. Now load up two basic audio clips from the Browser or your computer into track 2's first 2 Clip Slots. These are your Dummy Clips!

7. *Click* on the first Dummy Clip on track 2 in Clip Slot 1 to open its Clip View and select the Envelope button at the bottom of the Clip Box to open the Envelope Editor.

8. Select "Ping Pong Delay" from your Audio Effects folder in the Browser and *drag* it to track 2.

9. *Press shift + Tab* to go back to Clip View, then in the Envelope Box for Dummy Clip 1 select "Ping Pong Delay" in the Envelope Device chooser menu; then select "Dry/Wet" in the Control chooser menu below it. *Drag* the Dry/Wet Breakpoint Envelope downward to value "0.00%". The envelope will turn solid red ▶ 10.2 .

Figure 7.48 Blue signal meter indicates a Sends Only output signal.

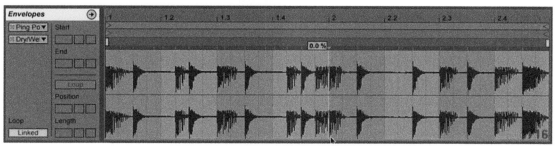

Figure 7.49 Drag the Dry/Wet horizontal envelope downward to 0.00%.

10. *Click* on Dummy Clip 2 on track 2 and repeat the same steps, but this time set the Control chooser Dry/Wet envelope automation to "90.0%". *Hold* down (⌘ Mac/[ctrl] PC) while *clicking* and *holding* on the envelope value line to work in small increments.

11. Launch your main audio clip 1 on track 1 if it's not already playing.

12. While clip 1 on track 1 is playing, launch Dummy Clip 1 on track 2.

Figure 7.50
Dry/Wet envelope automation to 90.0%.

Figure 7.51
Launch Dummy Clip 1 on track 2.

13. After a few bars' time, launch Dummy Clip 2 just below it on track 2. You'll now hear the Ping Pong Delay affecting the audio.

14. *Click* back on Dummy Clip 1 on track 2 and you'll hear just the main/dry signal.

To add even more effects and changes to your audio, add another Dummy audio clip to track 2 and proceed to edit and create new envelope automation parameters for each new clip. It's always a good idea to leave your Dummy Clip 1 in Clip Slot 1 designated and even renamed as a "Dry" Dummy Clip. Expand the Dummy Clip concept once more by adding Audio Effects Racks into the Dummy Clip column. By automating the Audio Effect Rack's various parameters in your Envelope chooser and Sub chooser menus, you can add infinite possibilities to your clip's audio output. Remember that you can always go back to your "dry" Dummy Clip in Clip Slot 1 to achieve a clean, unaffected signal. Also, always make sure that the Monitoring section is set to "In" on your Dummy Clip track.

Figure 7.52
Monitor section is set to "IN" on a Dummy Clip track.

By following the steps above you'll be creating innovative ways to add effects to your clips in no time. Don't forget that you can always save these new Dummy Clips as Live Clips to your User Library to recall and load them into your Set when needed. Now that you have completed this exercise, check your Set against the demo Set. Download "**CPP_7–11_DummyClips Project**" from the companion website and see how it works.

Recording

8.1 Recording MIDI Clips

After all of this talk about clips, you probably want to know how to record your own as you have with other Digital Audio Workstations (DAWs). Great news! Recording clips in Live is very easy and can be quite inspiring. To begin, let's take a look at the two ways to record MIDI clips in Live. You can either start out with an empty *Clip Slot* or an empty *clip*. We'll do both, but first let's go over the empty Clip Slot approach.

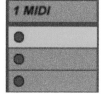

Figure 8.1
Empty Clip Slot.

8.1.1 Clip Slots and Clips

Open a new Live Set for this exercise. When you first open Live or create a new Set you will see a number empty Clip Slots. Remember, a Clip Slot is where clips are stored in the Session View. To begin, you need to first insert a software instrument in order to create sound.

Setup:

Figure 8.2
Empty clip.

1. Go to the Live Browser and navigate to *Categories>Instruments>Impulse*. With a MIDI track selected, *double click* on the "Backbeat Room.adg" preset or *drag* and *drop* it onto a MIDI track. Don't forget that the Device Drop area works too.

2. Now that you have a drum kit loaded, arm your new Impulse "Backbeat Room" MIDI track. *Click* on the Arm Session Recording button at the bottom of the track Mixer.

    You should notice that all of the Clip Stop buttons in every Clip Slot of the armed track changed to clip Record buttons.

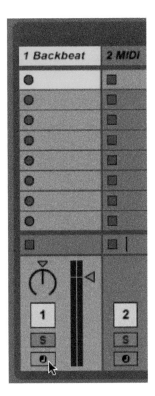

Figure 8.3
Arm your MIDI track.

Figure 8.4
The MIDI input channel meter lights up yellow, the track VU meter lights up green, and you should hear a percussive sound.

3. Now test out your MIDI drum kit to make sure it makes sound when you play a MIDI note. If you have a MIDI keyboard play the white keys C3 to C4 . If you don't, then use your computer keyboard, keys a to k (activate the *Computer MIDI Keyboard* on the Control Bar).

You should see the MIDI input channel meter light up yellow, the track level meter light up green, and should hear a percussive sound. If you are not hearing a sound when you play, then make sure the track is activated, armed, and that Monitor is set to "Auto" or "IN". The best way to troubleshoot is to determine where in the signal chain there is a problem (e.g., MIDI input or Audio output). Compare your setup to the example.

Figure 8.5
Computer MIDI
Keyboard On/Off.

4. Once your instrument and playback is working, you should setup *Record Quantization* so that you can record MIDI free of timing errors (Edit Menu). Record Quantization automatically corrects and alters your MIDI performance relative to the predetermined resolution and grid as you play—a real time saver and a necessity for performing on-the-fly! Let's set it to "Eighth-Note Quantization". This means that every note you play will lock to the closest eight-note grid line position. Always choose the smallest resolution (subdivision of the beat) that you intend to play.

Figure 8.6
Record
Quantization
settings under
the Edit Menu.

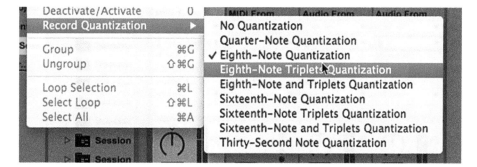

5. Activate the *Metronome* at the top left of the main Live screen and make sure that a count-in is enabled. *Click* the Metronome button and it will turn yellow. From the dropdown menu choose "1 Bar". If you need to adjust its volume, do so from the *Preview/Cue Volume* knob at the bottom of the Master track Mixer.

Figure 8.7
Metronome On/Off.

Figure 8.8
Preview/Cue Volume
knob controls the
Metronome volume.

Recording to an Empty Clip Slot

1. **Record:** let's record a four-bar drum loop. Any beat is fine, just keep it simple. *Click* a Clip Record button and begin playing after the count-in. Record only four bars!

2. **Stop Recording:** to stop recording, disarm your track by *clicking* on the Arm Session Record button while the clip is still running. You can also *click* the Control Bar's Stop button or *click* the Stop All Clips button. Each one has a slightly different effect. For this demo, just disarm the track.

Hot Tip You can select your count in value by right clicking on or clicking the Metronome dropdown button. Here you'll see a list of values to choose from, including 1 Bar.

This way you will hear it playback immediately—with perfect timing too! After it has played through it will start looping at whichever point you disarmed the track, most likely a few beats beyond a true four-bar loop, unless you were quick to disarm. If you are too late, Live may loop an extended length of incomplete or fraction of beats/bars. If you want to add more MIDI notes to your newly recorded clip (overdubbing), you will have to arm the track again and then use the *Session Record Button*. More on this in Clip ▶ 8.2 .

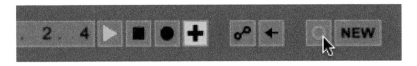

Figure 8.9 Session Record Button On/Off (MIDI only).

Figure 8.10
(below)
Set loop length to 4-bars in the Notes Box.

3. **Set Loop:** adjust the loop length in the MIDI Note Editor so that your new clip is a four-bar loop while playback is still running. You can do this by *dragging* the end of the Loop Brace to bar 5, or set the Loop Length to "4" bars in the Loop Length input field in real-time. For the adventurous user, *click* the *Set* button during playback when the *Playback Cursor* reaches bar 5. Enjoy and listen to your loop as it now plays back in perfect time. Stop whenever you are done listening.

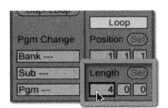

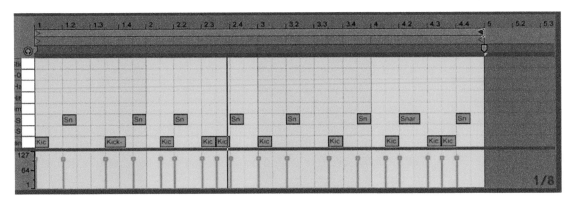

Figure 8.11 Set loop length to 4-bars by dragging the Loop End.

> **Hot Tip** Make recording even simpler by double clicking on a MIDI Clip Slot to automatically create an empty clip template. This allows for easy looping without punching out (disarming record) at the end of the bar. Duplicate these templates multiple times down the same track to fill out consecutive empty Clip Slots. This way you have plenty of ready-to-go MIDI Clip templates to record into. Easily increase the loop length in the clip's Notes Box Loop Length input field.

Now let's do the same thing, but this time with an empty clip (clip template).

Recording to an Empty Clip (Clip Template):

1. *Double click* in an empty Clip Slot below the one you just recorded in. This will create an empty clip template, which will appear in Clip View.

2. **Set loop:** this time we will preset the Loop Length. Either *drag* the Loop Brace to bar 5 or set the Loop Length field to "4" bars.

3. **Record:** arm your track, then *click* on the Session Record button on the Control Bar. After the 1 bar count-in, begin recording/playing your beat up until the end of the fourth bar.

4. **Stop recording:** although the clip is still recording, you can simply stop playing and listen to the clip playback without disarming anything. Disarm the track or disengage Session Record if you want, or simply enjoy the freedom of non-linear time in the session. The Metronome can become annoying, so turn that off. Stop playback when you are done listening (*press* a Clip Stop button then spacebar).

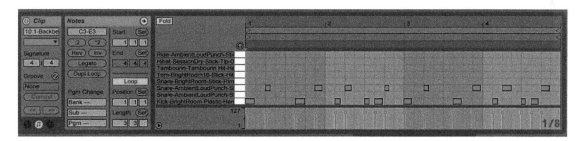

Figure 8.12 Your clip should look similar to the drum kit clip example.

8.2 MIDI Overdub Recording

There are two specific features for overdub recording in Live: the *Session Record button* in the Session View and the *MIDI Arrangement Overdub button* for the Arrangement View. The Session Record button ▣ located on the Control Bar has been designed for triggering record for all armed tracks in a Scene and for overdubbing MIDI and automation ▶ **8.2** . When activated, you can record multiple takes of MIDI notes into the same MIDI clip without overwriting the existing notes unless they are the same note. In that case, Live will formulate a consolidated version of the new and the old based on where they overlap. When deactivated, nothing can be recorded into an existing clip; rather, you can only record to empty Clip Slots. This is useful for going in and out of record mode in order to practice your takes while looping and monitoring input along with playback. In this way, consider the following process: with Session Record deactivated, Arm your MIDI track and then launch playback. Practice your take a few times; then, when ready, activate Session Record to record MIDI input into the clip. You can use this technique to add to that clip or to overwrite specific notes already in the clip.

> **Hot Tip** A great way to activate the Session Record button is to assign it to a key on your computer keyboard via the Key Map function. Launch to Clip ▶ **17.3** for more on this feature.

It's probably safe to say that you will spend most of your time MIDI overdub recording in the session, but from time to time you'll have a need to overdub a MIDI take directly in the Arrangement View. Maybe you need to overdub a section in your arrangement, or a little passage here or there to polish up your work. This is completely acceptable. The Arrangement View is intended to not only capture a performance from the Session View (global recording), but to be used as a traditional linear DAW. In fact, when it comes to recording, it is just like any other DAW. Now, in contrast to the way overdubbing occurs in the Session View, in the Arrangement View it's used in a more traditional sense ▶ **8.2.2** . In order to overdub MIDI in Arrangement clips, you must have MIDI Arrangement Overdub activated ⊞ . When activated, overdubbing functions the same as already described in the Session View. Notes are merged or mixed into an existing Arrangement clip, only overwriting identical notes that overlap on the same beat. When MIDI Arrangement Overdub is "Off", each recording is an entirely new take, with each record pass overwriting all of the existing MIDI notes instead of adding to them or simply doing nothing, as in the Session clips.

8.2.1 Session View Loop Recording

Loop recording is a really great way to take advantage of the Session Recording/ Overdubbing features in Live. Let's try it out:

1. Create a new empty clip template on your Impulse track. This will insert a new clip and bring up Clip View.

2. **Set Loop:** activate the clip Loop Switch, then set the clip's loop length to extend two bars, ending on bar 3. Use the Loop Length field in the Notes Box and set it to "2".

3. Turn the Metronome "On" and set Record Quantization to "Eighth-Note Quantization".

4. Arm Session Recording on your MIDI track.

    Now you are ready to record a simple "downbeat rhythm" for the kick drum:

5. **Record:** *click* the Session Record button to begin recording. After the count-in, record the kick drum on every downbeat (eight beats in all). After recording those beats, stop inputting notes and let the clip loop.

    Now add a snare backbeat rhythm (beats on 2 and 4). Let's play along without actually recording the notes:

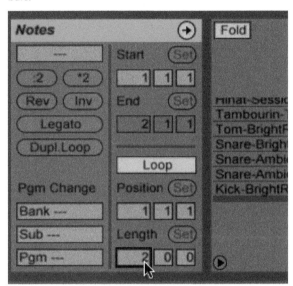

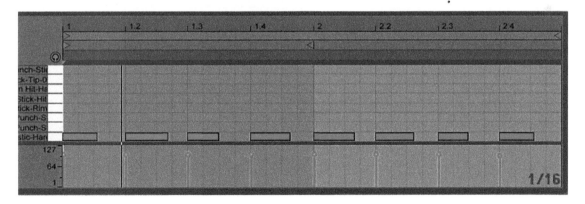

Figure 8.14 Record the kick drum on every downbeat.

6. While the clip is still running, deactivate Session Record. Begin performing the snare, along with the looping clip, playing on beats 2 and 4. After practicing a few loop cycles you should be ready to record the snare.

7. Reactivate Session Record and begin recording the snare (four notes in all)

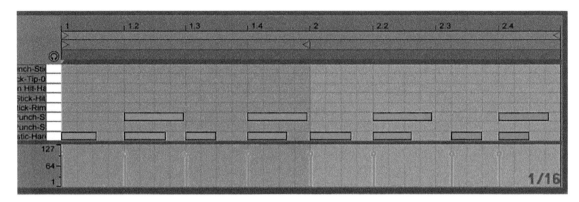

Figure 8.15 Record the snare drum on beats 2 and 4.

Let's make this super simple beat more interesting. While the clips are still looping, we'll add a ride cymbal. You can deactivate Session Record if you want to rehearse or just dive right in. It's up to you.

8. Record an eighth-note ride or hat pattern on every eighth note, (e.g., 1 + 2 + 3 + 4 + ... (16 in all)). Let this continue looping when you're done inputting the eighth-note ride pattern.

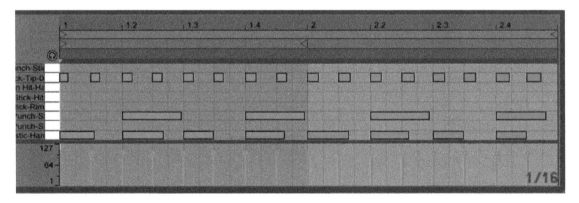

Figure 8.16 Record the eighth note ride pattern on every eighth note starting on beat 1.

9. Add a kick drum on the + of 3 in each bar. This is the eighth-note beat right after beat 3 (after the snare hit).

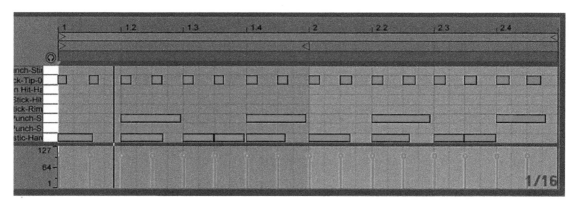

Figure 8.17 Add a kick drum on the + of 3 beat in each bar.

Turn off the metronome, if you haven't already, and listen to your perfectly quantized drum loop!

8.2.2 Arrangement View Loop Recording

For this demonstration we will use the same Impulse drum kit:

1. Toggle over to the Arrangement View and *click* the Back to Arrangement button to return playback priority to the Arrangement View.

2. Activate the Loop Switch on the Control Bar and set a Loop Brace at bars 5 to 7 along the Beat Time Ruler (two bars). You can *drag* the Loop Brace into position or highlight bars 5 to 7 in the Track Display then *type* ([⌘+L] Mac/[ctrl+L] PC).

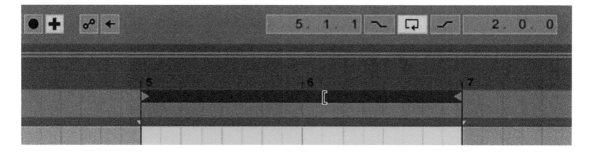

Figure 8.18 Activate the Loop Switch.

3. Turn on the Metronome and set Record Quantization to "Eighth-Note Quantization".

4. Arm your MIDI track and make sure MIDI Arrangement Overdub is activated.

Now, let's loop record a two-bar drum loop just like before. We'll do this in multiple takes or record passes (loops cycles), kick on the downbeats, snare on 2 and 4, and hat on every eighth (ride-like). To keep it stress free, let the loop cycle one time between each record pass:

5. Activate Arrangement Record on the Control Bar and off you go. After the count-in, begin recording the kick on the first pass, snare on the second, and hat on the third. Only record two bars for each, then listen back to the loop you just created.

6. *Click* the Control Bar Stop button to stop everything, or deactivate Arrangement Record to keep the Arrangement running.

There you have it. A simple two-bar drum loop created directly in the Arrangement View.

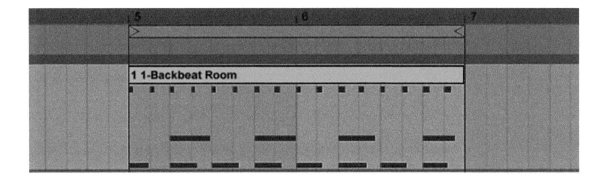

8.2.3 Takes

Now that we have walked through how to loop record with MIDI, let's narrow in on how Live actually handles all of the loop record cycles that occurred during recording. The overdub recording process in Live is very similar to that of other DAWs, but the way in which it handles takes or record passes is very different. With that in mind, *click* on the Arrangement clip (2-bar loop) you just recorded and you should see in the MIDI Note Editor a very interesting display of the clip. It should look something like the example.

Figure 8.19
Two-bar MIDI
drum loop clip.

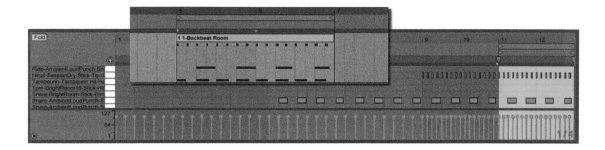

Figure 8.20
Each record pass (take) has been captured into the clip and stored off to the left in the grey area. The light grey area is the currently active playback area of the clip.

From the moment the arrangement begins to record, it captures every MIDI action/event. If you recall, on the first pass we recorded the kick, but then let the Arrangement clip cycle through one loop cycle without inputting new notes. Then we did the same with the snare, followed by the hat. As you can see, each take (record pass) was captured exactly in that order. What you see in the Track Display and in the MIDI Note Editor within the Loop Brace is the culmination of that overdub recording. The clip's Loop Brace is initially established relative to where looping was executed in the arrangement, but has no bearing on the actual looping in the Arrangement Playback. The clip simply generates its loop points based on the number of bars included in the loop record process as determined by the playback loop in the arrangement. Notice that the clip itself is not set to loop until its Loop Switch is activated. At that time, the Loop Brace determines the clip's loop selection—more on that in a moment.

Here is what is actually happening with takes …

A clip advances with each take (a loop record cycle), logging time as it loops through its loop length, whether MIDI note data is input or not, until you physically stop the recording process either with the Arm Arrangement Recording button or by *clicking* the Control Bar Stop button. Once stopped, your Arrangement clip in the Track Display remains focused on the final recorded pass, the current state of the last take. The last or final take is essentially a representation of all takes, showing all of the record passes as layers transparently placed (mixed or merged) on top of each other as shown on the Track Display and in the MIDI Note Editor (highlighted selection).

This is like a birds-eye view of the past and present. To choose a different selection for playback in the Track Display, you can move the Start or End Markers to encapsulate any segment of the clip from within the MIDI Note Editor. These markers indicate what segment will play in the arrangement. Think of a selection in the Note Editor as a moment frozen in time that you can choose from to be represented in the Arrangement View Track Display as an Arrangement Clip—

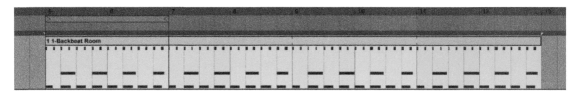

Figure 8.21 Two-bar clip with visible MIDI notes in the Track Display.

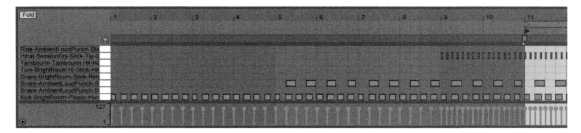

Figure 8.22 Two-bar clip with visible takes on the left and final take on the right in the light grey.

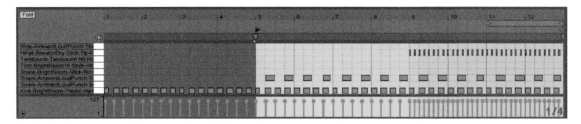

Figure 8.23 The clip's Start Maker position has been extended to the left six bars in length from within the MIDI Note Editor.

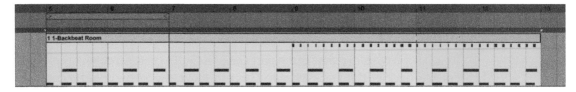

Figure 8.24 Notice that the clip's length in the Track Display reflects the newly selected Start Marker Position as made in the MIDI Note Editor.

hence, choosing a take. Move the Start Marker in the Note Editor to the left to add and make playable more of your takes in the Track Display, as seen in our example. Notice that the clip's length in the Track Display reflects the newly selected Start Marker Position as made in the Note Editor.

The last piece to this puzzle is activating the clip Loop Switch. When Loop is activated, the Loop Brace establishes the playback selection within the clip. The Start Marker is then only for assigning the actual start point of the clip's content. Nevertheless, if the Start Marker is moved outside of the Loop Brace, then it redefines the entire playback selection based on the Start Marker's new relative location to the Loop Brace. See in the example how the Start Marker's location affects the Arrangement clip's playback.

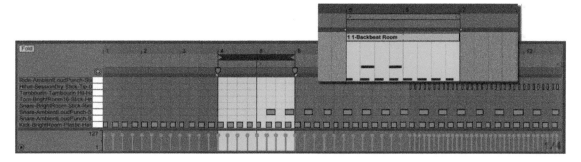

Figure 8.25 Move the End Marker in the MIDI Note Editor to adjust the end length of the Arrangement.

This entire take concept applies only to the Arrangement View. The difference is that in the Session View each take is added (mixed/merged) within the loop selection. This all occurs automatically during the overdub process. Any deviation from this must be set manually, post-overdub.

8.3 Converting Clips

MIDI clips are strictly instructions for a virtual instrument or external analog instrument to translate into sound. This makes MIDI extremely flexible in that its performance data can be manipulated. Notes can be transposed, rhythms changed, and the entire performance reassigned to a different instrument patch. That being said, there will be times when you wish your audio files had the same flexibility or even that they were MIDI files. At other times you will want your MIDI files to be audio files. Both of these can be done in Live with great success. First, let's look at converting MIDI clips into audio clips. Following that, we'll turn the tables and look at audio to MIDI.

8.3.1 MIDI to Audio

As you may recall from Clip ▶ 2.3.3 , we talked about using Freeze Track to conserve on central processing unit (CPU) and RAM, etc., but there is a whole other aspect to it. Using Track Freeze with MIDI clips allows you to turn them into audio clips in real-time; it also works for audio-to-audio clips, but let's focus on MIDI. We'll use our last take from the previous example to demonstrate this. This can be done in the Session View or Arrangement View.

In the Arrangement View, select your two-bar MIDI loop that you recorded, then *right click* or *ctrl + click* to bring up the contextual menu. From there, select "Freeze Track". Once the track is frozen it will have a white line through its title bar and its Clip View will be faded white.

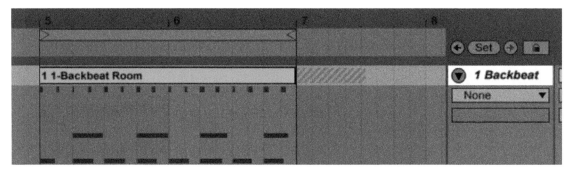

Figure 8.26 Frozen MIDI track.

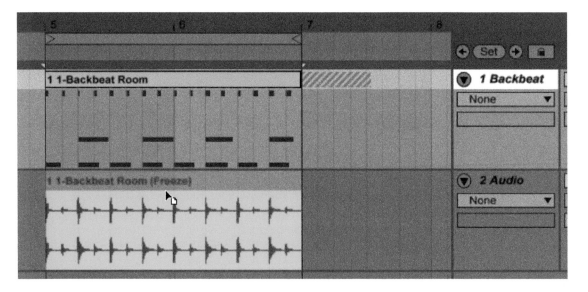

Figure 8.27 Copy drag the frozen clip and drop it on an empty audio track.

188

To convert it to audio is simple. Just *click + drag* the frozen clip and *drop* it on an empty audio track. It's a good idea to *copy drag* (opt/alt + drag) this in order to keep the original MIDI in place in case you want to make changes to it. See the example.

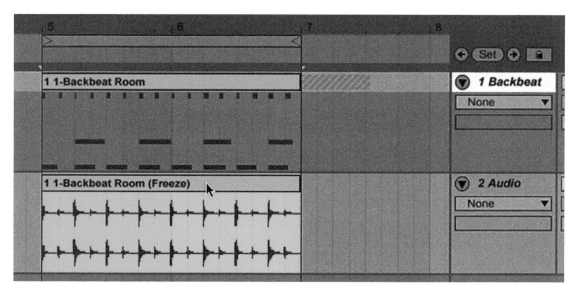

Figure 8.28 Frozen MIDI clip is still in place and a new audio clip of the MIDI in the track is created below.

Figure 8.29

Session View: Copy drag the frozen clip and drop it on an empty audio track Clip Slot.

The same process works with Session clips too. You can even *drag* frozen Session clips directly into the Arrangement View.

Your newly converted MIDI to audio clip is labeled as "Freeze". This feature makes life much easier! Imagine if you just completed a three-minute keyboard part as MIDI for your entire song. Why should you spend the time to print or export an audio stem of it? You shouldn't! Live will do it for you right in your Set. It's quick, easy, and a dummy-proof way to work and convert MIDI to audio, or render audio to audio. You have the option to "Flatten" your frozen track too. This is essentially the same process except it converts the MIDI clip, replacing it with the audio track and clip.

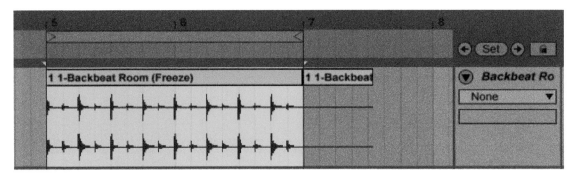

Figure 8.30 Flattened frozen MIDI clip.

> **Hot Tip** By dragging a frozen MIDI Clip to a new audio track as opposed to flattening, you will leave not only the original MIDI track's clip intact, but also the Instrument or Instrument Rack Device in place so that you can continue playing and recording on that track. Having your newly converted tracks ready to play at any time keeps the inspiration flowing.

At this point you're probably wondering what the extra clip tail is all about in the frozen clip and in the flattened clip in the arrangement?

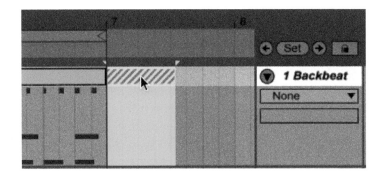

Figure 8.31 Freeze Tail clip.

This little region is called a *Freeze Tail Clip*. Its purpose is to include any audio carryover from a time-based audio effect such as delay or reverb so that the integrity of that effect is maintained. Otherwise any lingering effect would be truncated. This feature can be used in many ways, but is best used for playing back in its entirety along the timeline since the audio and effects are no longer being processed in real-time. In this way, the Arrangement View is used to play back frozen clips with Freeze Tail Clips. In our previous example there was not really any audio to carry over, so here is another example with a delay added to the Device. After you *drag* the frozen clip with tail to a new audio track, you will definitely see that it has an effect tail.

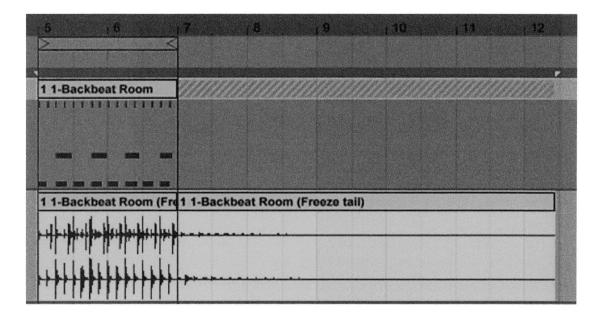

Figure 8.32
MIDI clip converted to audio along with its Freeze Tail.

Hot Tip Copy a Freeze Tail Clip from a frozen clip and place it anywhere in the Arrangement View or Session View to convert it to audio. You can also try reversing the converted freeze tail clip and place it just before the original clip you removed it from as a transition or build-up to the original clip. An audio effect such as reverb is perfect for this, creating that backwards "inhale" effect leading into the main clip.

When working with frozen tracks, keep in mind that editing clips while they are frozen may result in a surprisingly different playback effect if you unfreeze them. This happens when dealing with tracks that use delays, reverbs, or any other time-based processing. Effects such as these are dependent and influenced by note or audio events that obviously are now part of the freeze file and cannot be influenced by a new sequence of clips or edits, etc.

8.3.2 Audio to MIDI

There are some very exciting features in Live 9 to convert audio clips into MIDI clips. The idea is to take the audio information—pitch and rhythm—and generate the equivalent MIDI information (notes) so that a MIDI instrument can play and perform the data. The general concept is built around pitch and rhythm detection. You might be thinking: why not just warp the audio in order to manipulate it? That's a great option, but the goal here is to extract a performance and convert it into MIDI notes, playable and editable performance data that can be assigned to any patch or instrument. That being said, Live's convert to MIDI feature will not always perfectly translate your audio into MIDI, especially when working with more complex harmonic content or dense drum grooves. Try it for yourself and see what Live comes up with. There are four ways to convert audio to MIDI. Here we will look at only three: *Harmony*, *Melody*, and *Drums*. The fourth is not entirely a conversion of audio to MIDI, rather a hybrid approach called "Slice to New MIDI". Launch to ▶ 14.3 for more on that.

8.3.3 Convert Harmony, Melody, and Drums to New MIDI Track

Figure 8.33
Convert to MIDI.

The *Convert Harmony to New MIDI Track* feature is used to identify and parse out individual pitches amidst polyphonic audio content. This information is then translated into MIDI notes and placed into a MIDI clip on a new MIDI track.

The new track is automatically assigned an Instrument Rack called "Harmony to MIDI" and makes use of a piano sample-based Simpler instrument. Of course, you can swap out or make any changes to this instrument to meet your needs.

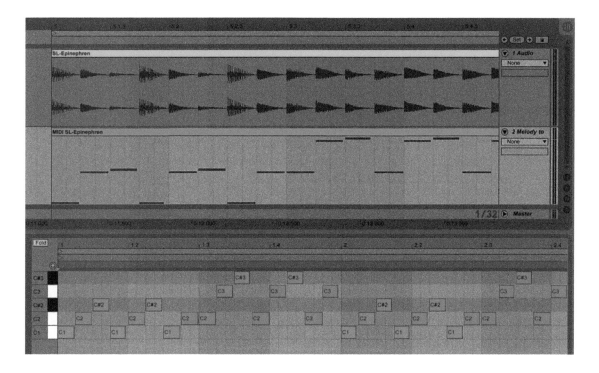

Figure 8.34
Audio as MIDI notes.

Convert Melody to New MIDI Track follows the same idea as harmony, except it is intended to convert monophonic (melodic) audio files into MIDI data on a new MIDI track. The other difference is that the Instrument Rack loaded called "Melody to MIDI" consists of two instruments stacked (layered) on top of one another: a synth and a piano. This is provided in such a way so that you can choose between a synthetic timbre, acoustic timbre, or a blend of the two for the sake of efficiency.

Convert Drums to New MIDI Track is designed to identify and convert drum sounds –non-pitched sounds such as snares, cymbals, kick, etc.—into MIDI notes on a new MIDI track. The major difference from the two other convert options is that the new track is assigned to a Drum Rack preset called "Drums to MIDI". This preset loads a preset drum kit for playback. This command does its best to determine the different drums and place them on the key range so that they trigger the appropriate and relative drum samples. For more on Drum Racks, launch to ▶ **16.3** .

There are a few options for executing any of the convert to new MIDI track commands. You can *right click* any audio clip located in the Live Browser and then choose the command that is most appropriate. This will create a new MIDI track with the default Device Rack. Another way is to *drag* an audio clip from the Browser or any location on your computer directly to a clip slot on a MIDI track that has a Device preset already instantiated. In this case, the MIDI clip will be created only. Another method, which comes with a bonus feature, is to *drag* an audio clip from an existing audio track to a MIDI track (preset must already be instantiated) or *right click* on an audio clip already on a track and choose the convert command from the contextual menu. The caveat here is that you can Warp the transients of the audio file prior to conversion, including deleting, moving, and adding Transient Markers. This can help to optimize and or customize the audio clip for conversion since Live uses the Transient Markers to discern the individual notes and events for conversion. For this reason, you might consider experimenting with the different convert commands on different audio content. You may, in fact, find some inspiration or really cool creative results. To learn about warping and Transient Markers, launch to ▶**Scene13**.

8.4 Recording Audio Clips

If you understand the concept of recording MIDI in Live, then recording audio is fairly simple. In fact, the Arrangement View is basically like any other DAW; it's the Session View that follows a unique way of recording audio in a non-linear environment.

Here are a few differences regarding audio in Live as a whole:

1. You cannot insert an empty audio clip in a Clip Slot. The clip template approach is used only for recording MIDI to a clip that has been preset with a Loop Length or Loop Brace selection in order to overdub record or for loop recording.

2. In the Session View you cannot record audio into or over an existing clip. The Clip Slot must be empty.

3. Overdub recording is not possible for audio recording, which is consistent with all DAWs.

On a side note, the Arrangement View does support punch-recording audio ▶ **8.4.4**.

That being said, Live's *Looper* Audio Effect does, in fact, incorporate a specific technique for overdub recording audio. Launch to ▶ **15.5.5**.

> **Hot Tip** Key Mapping is great for toggling MIDI Arrangement Overdub. Engage Key Map mode, then click "+"on the control bar and select a key on your keyboard to map it to. Now you can punch in/out by tapping a computer key.

Beyond these outlined differences, recording audio follows the same protocol as MIDI, and once you grasp the overall recording concept in Live, you'll be set to go. Of course, their input sources differ, so you must have some sort of audio source to record.

8.4.1 Session Clip Slots

When recording audio clips into the Session View, you must start with an empty Clip Slot. Try this example.

To begin, select an existing audio track or insert a new one and set it to the appropriate audio input channel based on your computer's audio setup. Since we are going to record a voice, set "Audio From" to "Ext. In" and the "Input Channel" to "1" (mono). If you have an audio interface and external microphone, that's great! Use it now. For simplicity, you can also use your computer's built-in microphone if it has one; just be sure to use headphones or turn your armed track's volume fader down about halfway or more so that there is no audio feedback!

Let's record two bars of vocal counting ("**1** … 2 … 3 … 4 … **2** … 2 … 3 … 4 …") since this is the simplest way to practice audio recording for the first time. We'll use the Session View for this first exercise.

Recording to an Empty Clip Slot

1. Turn the Metronome "On" and set the Tempo to 100 BPM.

Figure 8.35
Tempo 100 BPM.

2. **Arm Session Recording:** arm your audio track, then test to see that you are receiving an input signal by speaking or snapping your fingers into the mic. You should be seeing the mixer section VU meters light up green with signal. If not, double-check your audio configuration.

3. **Record:** *click* a Clip Slot Record button on your audio track and then begin speaking after the count-in. Record for two bars only!

4. **Stop recording:** *click* the Arm button to disarm your track (it will begin to loop), then *press* spacebar to stop arrangement playback. You could also stop recording by using the Control Bar Stop button ▶ **6.3.2**, or *click* the Stop All Clips button ▶ **5.3.2**, or, of course, any Stop Button on the record track. Each stop function has a slightly different result.

5. **Set loop:** the new clip recording should now be open in the Clip View; if not, open it. In the Sample Editor, adjust the clip's loop length so that it's set to a two-bar loop. You can do this by *dragging* the end of the Loop Brace to bar 3, or set the Loop Length field from the Sample Box to two bars. For the adventurous users, *click* the *Set button* during playback when the playback cursor reaches bar 3.

6. **Playback:** make sure the clip's Loop Switch is activated (Sample Box), and then listen back to your newly recorded audio loop. You'll probably notice that your timing was a bit off, but that's okay for now. Stop when you're done listening.

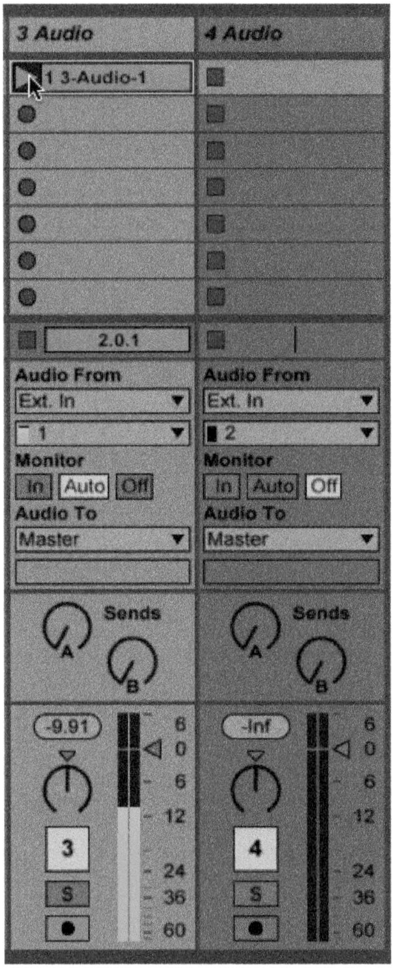

Figure 8.36
Activate a clip Record button.

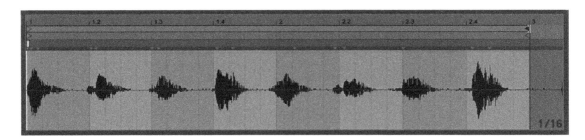

Figure 8.37 Set the Loop Length.

Now that you have successfully recorded audio into Live's Session View, you'll want to address any timing errors in your recording. Maybe you spoke a bit early on one beat and late on another? That's okay; but if you really want to fix these timing issues, you'll need to decide how to go about it. First, consider selecting the section of the loop that was performed the best—bar 1 or bar 2, in this case— and then set the clip's Loop Start/End positions to use that specific selection in the Sample Editor. Of course, this still won't solve timing problems that may still be obvious. In this case you have two options: first, move the errant words manually in the Sample Editor using Live's Transients and Warp Markers ▶*Scene13*; and, second, copy the clip into the Arrangement View track, edit the audio, and then bring the clip back into the Session View—probably not as efficient as the former, unless you have some complex edits to make. Then again, if that were the case, you would probably just record them in the Arrangement View in the first place. The idea is that you have options that will be dictated by your recording objective and the complexity of what you are trying to accomplish. To that end, recording audio in Live is not at all difficult. It, along with editing audio, can be done just as in other DAWs, including the added bonus of using warping to help get the job done! For all the information you need to adjust timing and quantize audio, launch to Clips ▶ **13.1** and then ▶ **13.3.5** .

There will come a time when you want to record multiple tracks at once that will make up all or a part of a Scene—for example, if you simultaneously launch multiple clips while recording on another track at the same time, or let's say a multi-mic drum set recording. The best way to accomplish this is by using Scene Launch buttons. This approach is called "Start Recording On Scene Launch". In this way you use Scenes to activate clip Record buttons for all of your tracks that are armed. You must turn on this feature in the Live Preferences *Record*, *Warp*, *Launch* tab. For more on this, launch to ▶ **9.2** . As an alternative, you can use Group Tracks to launch and record multiple tracks at once. Launch to ▶*Scene11*.

8.4.2 Arrangement Clips

Very few people can record a flawless performance in one take. Most of the time you'll need to overdub some element of the performance, or at least compile some elements together. For this reason, you will turn to Live's Arrangement View for "comping". Live does not have a formal playlist or take list features for compiling multiple performance takes into a single performance. Instead of this feature, you will need to record each take into different clips—session or arrangement—and then comp them in the Arrangement View using standard audio editing techniques (copy/cut + paste). Once you build your comp, you could then bring it into the

Session View or keep it in the Arrangement View depending upon which view you are working in. Remember, the arrangement is not only used to capture a performance from the session, but it's also a traditional linear DAW.

Let's continue with recording our vocal counting, this time in the Arrangement View directly as Arrangement clips.

Recording Audio Clips in the Arrangement

You have a few options at this point. You can set up a Loop Brace along the Beat Time Ruler and Loop Record your vocals, or you can just record them without a Loop Brace, stopping when you reach the end of your desired recording segment. Either way is really no different than any other DAW that you may have encountered before. For this demonstration, let's put Live in and Arrangement Loop playback (Loop Switch—On) so you can also choose between takes, and comp it all together when we're done recording:

1. Turn the Metronome "On" and arm your audio track.

2. **Set Loop:** in the Arrangement View, set the Loop Brace to two bars (bars 3 to 5) using your mouse to *drag* it into position. Once in position, *click* on the Loop Brace to select the loop region so you can record starting at bar 3—this is intentional!

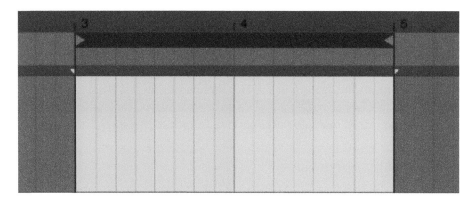

Figure 8.38
Set the Loop Brace in the Arrangement View.

Now, let's loop record a two-bar vocal count just like before, but this time we'll repeat it when the clip loops. All in all you will record your two-bar vocal counting two times through (two passes), starting your vocal counting over again on the second record pass (loop).

3. **Record:** make sure the loop brace is selected, then activate Arrangement Record on the Control Bar. Following the count-in, begin vocal counting out loud. Record two bars of vocal counting ("**1** ... 2 ... 3 ... 4 ... **2** ... 2 ... 3 ... 4 ..."). Remember to start over immediately following the second bar and repeat your vocal counting.

4. **Stop recording:** *click* the Control Bar Stop button to stop when you are done. You should now have a two-bar recording currently displaying the second record pass, as in the example.

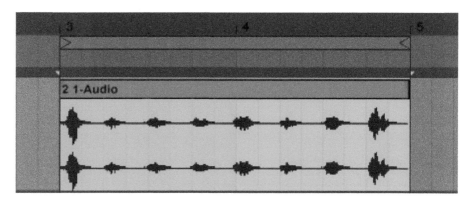

Figure 8.39
Two-bar recording currently displaying the second record pass.

When you stop recording there will be a break in the clip toward the front edge or wherever you stopped recording at. This happens when you stop a record pass mid clip, essentially creating an incomplete take. To remove this, *drag* and extend the edge of the complete take to the left so that the clip fills out all the way to the loop selection.

5. **Playback:** go ahead and listen back and see how you did. *Click* the Play button on the Control Bar or *press* the spacebar.

Now look at your clip in the Sample Editor. You will see the entire audio recording, including the first and second record pass. The first take will be darker gray since it is not currently being used in the Arrangement View. (Figure 8.40.)

As always, there will be some timing errors and/or speaking mistakes. For whatever reason, you may want to use one take over the other, choosing the one that you like best. In order to choose a different take, move the Start/End Markers in the Sample Editor to encompass the take you like. If it's the last take, then you can

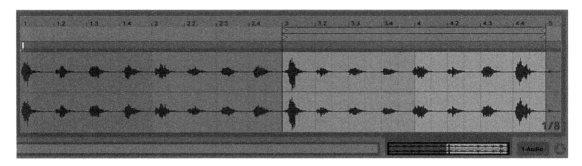

Figure 8.40 The entire audio recording including the first and second record passes with the first take grayed out, since it is not currently being used in the Arrangement View playback.

leave it alone. If it's the first take, then move the clip Start/End Markers to bookend the first two bars. In that case it should look like the example.

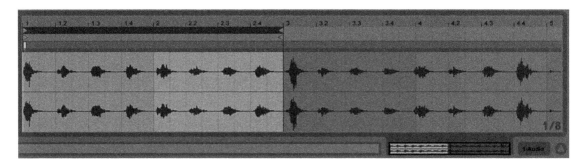

Figure 8.41 Choose the first take. Move the Start/End Markers in the Sample Editor to encompass take 1, bars 1–2.

Press opt/alt + *click* + *drag* on the clip's Start Marker to move the Start/End Markers together in unison motion. Wherever they are moved to, the visible clip region will update automatically in the Track Display. You can also do this using the Sample Box Start/End input fields.

If you want to loop your clip take to loop, then activate the Loop Switch and *drag* the Loop Brace to the desired take. As for timing errors, you can fix this with warp markers just as we explained for the Session View. (Figure 8.42.)

Last, as a little treat before moving forward, a pretty hip way to loop record audio on-the-fly is to use *Looper*. Looper provides some fantastic alternatives for real-time audio loop recording. Check it out!

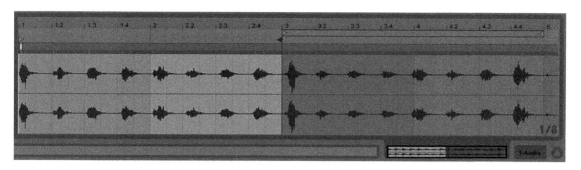

Figure 8.42 To loop your chosen clip take, activate Loop in the Sample Box and move the Loop Brace accordingly.

8.4.3 Comping

Comping has been a part of the recording and production process for ages. It's the art of compiling multiple takes—for example, vocal takes into one single and clean vocal recording. Here is one way to do this in Live's Arrangement View.

Back to our previous recording example: say you like the last take (second record pass), but you prefer the way you spoke "3" in the first take that you are not using as opposed to "3" in the second bar of the second take that you are using. That's okay; you just need to comp the two together in the Arrangement View. This is why you left room at the front of the recording instead of starting at bar 1:

1. In the Track Display, extend the edge of the clip out to the left until you see both takes. Now you should have four bars in view consisting of your entire recording in the Track Display.

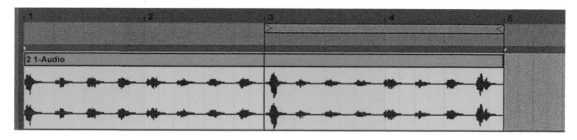

Figure 8.43 All four bars of your entire recording in view.

2. Since you like the way you said "3" in the first bar of the first take, make a selection around that recorded word along the grid. Set the Marker Snap to

a reasonable resolution, such as 1/4 or 1/8, or turn it "Off" so that you can select the audio segment perfectly. Highlight the specific segment with your mouse (*click + drag*), and then copy it ([⌘+C] Mac/[ctrl+C] PC).

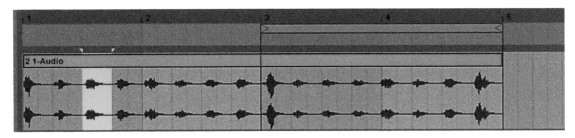

Figure 8.44 Highlight a specific segment and then copy the audio selection.

3. Now highlight "3" from the second bar of the second take and select paste [⌘+V] Mac/[ctrl+V] PC. When pasted, Live will automatically add crossfades ▶ 6.4.5 .

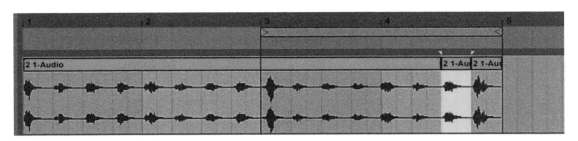

Figure 8.45 Highlight the selection that you wish to replace and paste the segment you previously copy selected from the other take.

4. *Drag* the left edge back into position so that only the second take is showing (the comp).

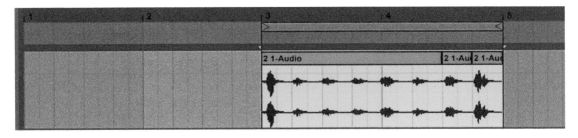

Figure 8.46 Resize your clip back to a two-bar take, the comp.

There you have it: your first comp consisting of three audio clips.

8.4.4 Punch Record

Now, let's say that you don't like any of your performances of the word "2" on beat 1 in the second bar. Well then, you'll just have to re-record it using Live's built-in auto Punch-in/Punch-out Switches or punch record it manually using the Arrangement Record button by itself. Let's look at both ways in the Arrangement View.

Auto Punch-in/Punch-out:

1. Set your punch-in/punch-out points using the Loop Brace. Set it to bar 4 to 4.2. *Drag* the Loop Brace with the mouse or use the *Loop Start/Punch-in* fields on the Control Bar.

Figure 8.47
Loop Start/Punch-In fields.

2. Activate the *Punch-in/Punch-out Switches* located on both sides of the Loop Switch. The Punch-in Switch prevents recording prior to the punch point and Punch-out prevents recording after the punch point. Make sure the Loop Switch is deactivated.

Figure 8.48
Activate the Punch-In/-Out Switches.

3. Place your Insert Marker two bars before the punch-in point to serve as your pre-roll (count it). When punch recording, Live does not give you a count-in.

4. **Record:** arm your track, then *click* Arrangement Record. Speak the number "2" at the punch-in to replace the original recording and Live will take care of the rest.

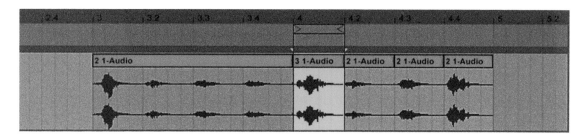

Figure 8.49 The resultant punch.

You may have noticed that if you're not precise with your "2" retake, the beginning of the transient may have been cut out of the recording. This is because you used the punch-in switch and set the punch-in point exactly at bar 4. To ensure this doesn't happen, or that the end doesn't get cut off on the punch-out, set your punch points a little wider, then trim the clip back down using the bracket tool with your mouse after finishing the punch recording. Extra room for recording is okay, even if it overlaps the part you were happy with. Live is non-destructive, but you can never get back what you don't record!

As an alternative to using the Live punch switches, you can simply do it manually using the Arrangement Record button.

Manual Punch-in/Punch-out:

1. Place the Insert Marker two or more bars before the point at which you want to record—in this case, around bar 2 for recording at bar 4.

2. Deactivate the Loop Switch.

3. **Record:** arm your track, then *press* Play. When you get to the punch point, *click* on Arrangement Record to punch-in and *click* again to punch-out when finished.

> **Hot Tip** Try setting up a custom Key Map to activate the Arrangement Record button, or use the default keyboard shortcut F9. Doing so will give you quick access to activate global recording on-the-fly for manually punching-in and punching-out, all without having to grab your mouse. Launch to ▶ 17.3 .

You could also reverse step 3 by using Arm to punch-in/punch-out while continually running Arrangement Record mode. Either way works.

8.5 Exporting and Printing

When you are finished working on your song or production, it's time to start preparing and exporting your full mix, stems, or clips to disk. This includes determining how you are going to handle MIDI tracks and effects. In general, the rendering/exporting process provides for many options, such as rendering an entire song, individual track, clip(s), or even video. So, what are your options?

Most of the time you will render your tracks offline, meaning that the process happens faster than the real-time it takes to play through the song or clip, and you won't hear playback during the rendering process. When rendering an External MIDI device, the process will always happen in real-time since it has to physically create sound to be recorded into Live. So, as far as MIDI is concerned, you have a few options regarding rendering (bouncing) or printing the source.

The ability to render your tracks offline as opposed to printing them to track in real-time brings up a great philosophical question. Which method should you choose to bounce your MIDI tracks down to audio, and when do you use it? Whether you render or print isn't really the question; that's really a matter of preference and convenience.

In ▶ **8.3.1** we talked about Freezing and converting MIDI to audio, so that is one option. A favorite! More importantly is the "when … ?" You could work through your entire project, then render your MIDI tracks at the end, or not at all depending upon the final delivery format of your production. If all you ever need is a stereo mixdown, then you may never bounce your MIDI to audio. If you're delivering audio stems to another producer or engineer, for example, then you could bounce them down to stems at the end.

Regardless, consider rendering or printing your MIDI—especially external MIDI—to track once you are done creating with them. In this way you have fewer devices to worry about and can work on your song away from your studio and for years to come. Otherwise, you may not have that instrument, plug-ins, or settings available elsewhere. Remember, MIDI is just instructions for your instruments to interpret and convert into audio. Everyone has a different approach and preference for rendering MIDI to audio.

The rendering process can be executed from both the Session and Arrangement View. The Arrangement View process is traditional in nature, designed more for exporting your entire song/Set or large segments of clips that extend over a period of time, such as multiple clips on the same track as well as across multiple tracks. The Session View is more discrete in that you can only render a single row of Session clips at a time: those that are to be played back over a set length of bars and beats.

8.5.1 Rendering/Exporting Audio

Whether rendering your MIDI to audio, audio to disk, a single clip, or exporting an entire arrangement, the process of exporting audio is virtually the same for all.

To Render/Export from the Arrangement View

1. Select a clip or highlight a time range in the Track Display that you want to render in the Arrangement View Track Display. This can be a whole clip or just a selection. For exporting from the Arrangement View, make sure that the Back to Arrangement button is not lit up!

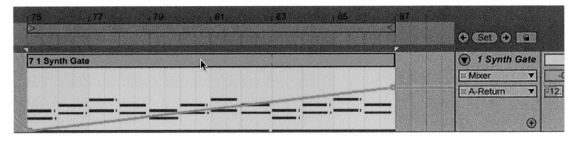

Figure 8.50 Select an Arrangement clip or time range you want to render.

In the Session View, launch your track clip or clips, then pause playback with the spacebar or Control Bar Stop button. Your paused clip's Launch button should remain green (in standby).

Figure 8.51
Green Launch button indicates a clip is paused.

2. Select "Export Audio/Video" from the File Menu ([⌘+shift+R] Mac/ [ctrl+shift+R] PC).

3. If you are rendering from the Session View, you will need to set the render "Length", but in the Arrangement View, length is not an option; your selection is your render length—"Length [*bars.beats.sixteenths*]".

4. Now set the track source you want to be rendered. Choose an output source to render: Master, a single track, or "All" tracks. The rendering of all tracks individually is also known as creating stems. If you are rendering your entire mix, then you will choose the Master Track as your rendered track and almost always use the Arrangement View; just be sure to select the entire time range or section of the song.

Use the Loop Brace as a quick selection tool for rendering in the Arrangement View. This is a good habit to get into. Set the Loop Brace to contain the beginning and end of your full arrangement or selection along the timeline. This ensures that all your information along the timeline and within your arrangement will be rendered and exported completely. (Figure 8.52.)

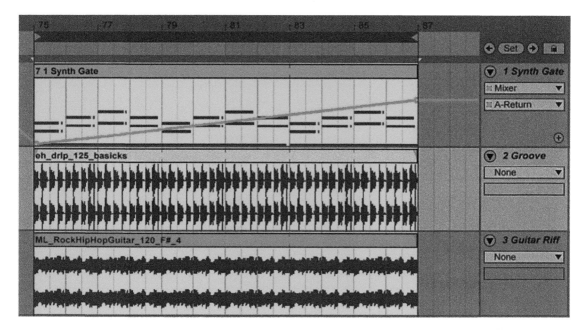

Figure 8.52 Use the Loop Brace when exporting from the Arrangement View.

5. There are a few destination formats we should review here as well: *Normalize, File Type, Sample Rate, Bit Depth*, etc.

6. Press "OK", then choose a save destination and select "Save".

Once the rendering process is complete, you can import the audio file(s) back into an audio track in Live to continue working with it or ship it off as a stem for the production phase. For additional information on exporting and rendering MIDI, launch to ▶ 3.7.1 .

8.5.2 Printing

Printing your MIDI tracks, Effects Returns, and other outputs to track is a valuable tool and often preferred when you don't want to export and then import those audio bounces (renders) back into your Set. It's also great if you want to print a submix (*internal mixdown of multiple tracks into one track*). This can also be done through Group Tracks ▶ 11.3 .

Let's print the MIDI Synth Track from the last exercise:

1. Create a new audio track and name it "Print Synth Gate".

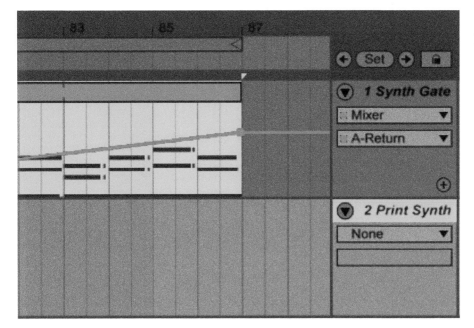

Figure 8.53
Arrangement
View.

2. Set the Monitor Section on your new audio track to "In". This will monitor audio input through the new track while in playback.

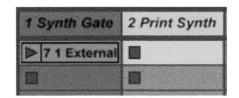

3. Route the Synth Gate MIDI Track's *Audio To* to your new "Print Track". (Figure 8.55.)

Figure 8.54 Session View.

4. Place the Arrangement Insertion Marker at the beginning where you want to print your clip. Make sure the Back to Arrangement button is not lit! In the Session View you would print to an empty Clip Slot.

5. Ensure that the Arrangement Loop Switch is "Off" or clip Loop Switch is "Off" for Session View Printing.

6. Arm the Print Track (Print Synth) and then activate Arrangement Record. In the Session, View, launch your clip.

Figure 8.55
Audio To chooser
routed to the
"Print Track".

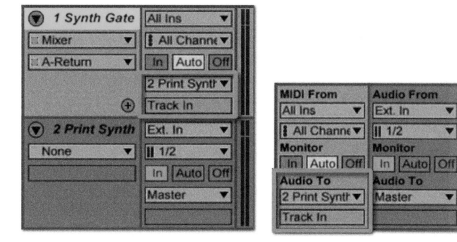

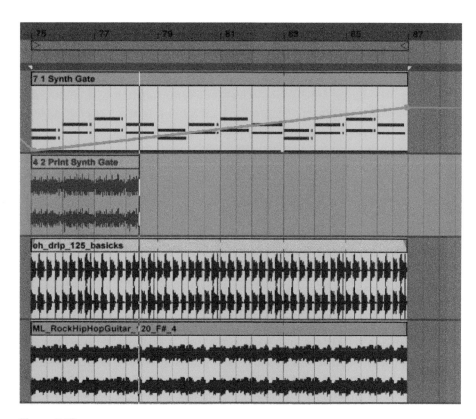

Figure 8.56
Synth Gate printing to the Print Synth track in the Arrangement View.

Of course you can also print MIDI from the Session View to audio in the Arrangement View. It follows the same exact process except that you'll be global recording the MIDI and printing the audio at the same time to their respective tracks.

Last, you might be wondering about the Send Effects. If you want to print them with the track, you will need to route the Return to the same print track, just make sure that all of your other Track Sends are not being sent to the Return at the time of printing. You could deactivate all of the other tracks while printing. You could also print an effects stem, but that's not as common. There are a number of creative ways to go about this. Experiment and see what works for you.

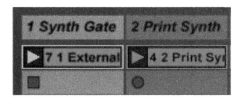

Figure 8.57
Synth Gate printing to the Print Synth track clip slot in the Session View.

Conveniently, Live has a feature called "Resampling" built into every audio track as an Input Type. When an audio track is set to Resampling, it will feed the output of the Master track for printing a stereo mix of your Set in real-time.

8.6 Musical Concepts

By now you should be familiar with Live's ability to function as a multi-track recording and production software package. Let's put it to the test and see how you can use it for traditional multi-track recording.

8.6.1 Produce: Real-time "Stack Tracks"

One very powerful technique when recording audio clips into the Session View is to use Live's real-time clip *dragging* capabilities while recording a performer (e.g., a guitar player or other session player) in order to create accompaniments or layering takes for cool double-tracked effects layers on-the-fly in real-time. Unfortunately, *dragging* clips while you yourself are recording makes this process a bit difficult since your hand would need to be free to control the mouse in order to *click + drag* a clip. This concept applies to working with a session player or additional musician while you are at the controls.

While recording clips to track, try *dragging* one of your previously recorded clip takes from the same track you are currently recording on to a new audio track while the current track is still recording. Now launch this clip to sync it up with your currently recording clip on-the-fly in real-time.

In this manner you can instantly create "Stack Tracks" or "double-tracking" on-the-fly in real-time! This is a powerful tool in the production process because it keeps

Figure 8.58
Dragging clip takes in real-time while recording.

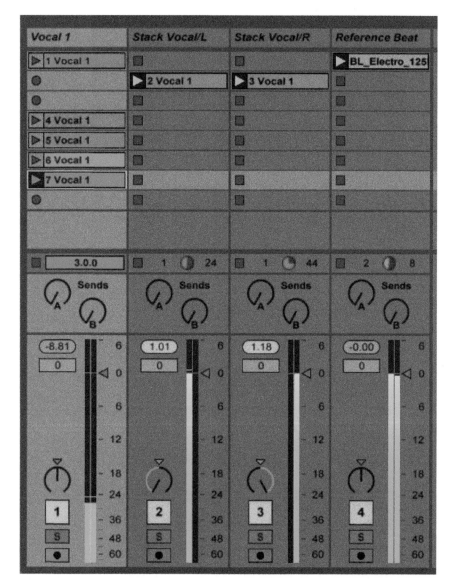

the workflow moving forward and the creativity flowing while never stopping to redirect a completed audio clip. As a matter of fact, you could designate two additional tracks in your Session View ahead of time and pan each track hard left and hard right. Once you record two new parts as clips, simply *drag* them one by one to the panned track for an instant "stack" or "duplicated" track, while keeping Live in play/record mode. There's nothing like the ability to launch and *drag* clips in real-time.

9.1 Musical Foundation and Structure

The communication of music and sound is most often dependent upon some sort of structure or form that makes up its musical foundation. Even in the most abstract genres there exists a unifying principle or nucleus of organization. Composer John Cage himself outlined a specific structure for his piece *4'33"—Four Minutes, Thirty-three Seconds*, which is perceived by the listener as silence. If we can agree that all music has some sort of musical foundation and structure, then we can agree that mainstream popular music is no exception. Most genres adhere to a specific song form. Such terms as *intro*, *verse*, *chorus*, *hook*, *bridge*, *vamp*, and *break* are household names. You could even go so far as to say that we have all even heard of sectional structures such as AABA, ABA, verse–chorus, or 12-bar blues. The point is that for centuries, musical foundation and structure have not only guided the composer, songwriter, producer, or performer, but have also served as musical vehicles for the listener. With this philosophy in mind, Live provides a clever method to creatively build musical ideas into foundations and structures in order to facilitate the organization of your music. This is accomplished with clips via *Scenes*.

Represented as rows in the Session View, Scenes are titled and launched from the column in the Master Track. All clips contained in the same row make up a Scene and can be launched together at once from their respective Scene Launch button. Only one Scene can play at the same time! Scenes are generally used to contain and launch song sections or launch a row of clips that form a part or section of a song (e.g., intro, verse, vamp, or break, etc.). They can also be used to launch an entire unit of clips that make up a multi-part instrument or vocal, such as a drum kit where the drum beat pattern (part) is divided amongst each drum of the drum kit across multiple tracks. Beyond organization, structure, and simultaneous clip launching, Scenes will allow you to apply a few very unique, on-the-fly tricks while working in the Session View.

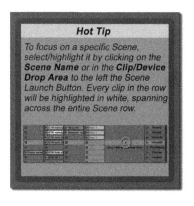

Hot Tip

To focus on a specific Scene, select/highlight it by clicking on the **Scene Name** *or in the* **Clip/Device Drop Area** *to the left the Scene Launch Button. Every clip in the row will be highlighted in white, spanning across the entire Scene row.*

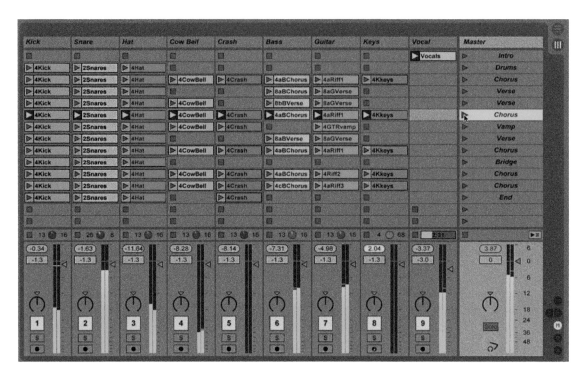

Figure 9.1 Launching a Scene in Live.

Using Scenes is very simple. They follow the same rules that govern clip launching. To launch, *click* the *Scene Launch button* just like you would any clip Launch button. To stop playback of a Scene, use the Stop All Clips button or any one of the other means of stopping playback discussed. If you need to delete a Scene, select it and *press* the delete/backspace key on your computer keyboard. To make launching Scenes even more fun and intuitive, color code them, then keep them organized by theme, musical idea, instrumentation, or whatever your imaginative mind desires!

9.2 Scene Launch Preferences

Making music on-the-fly is all about workflow, the ability to effortlessly make creative musical decisions without the interruption of a non-musical task. By now you know that Live makes this a priority. With that in mind, we'll take a quick look at a few Scene Launch preferences in order to streamline your creative workflow.

Master
▷
▷   Intro
▷   Drums
▷   Chorus
▷   Verse
▷   Verse
▷   Chorus
▷   Vamp
▷   Verse
▷   Chorus
▷   Bridge
▷   Chorus
▷   Chorus
▷   Outro
▷   End
▷

Figure 9.2
Color code your Scene List.

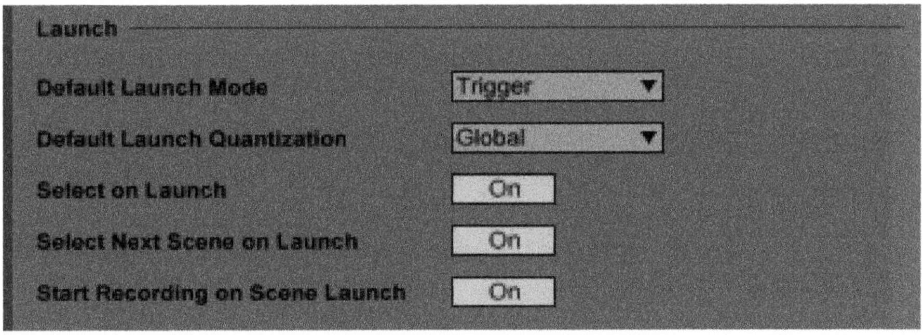

Figure 9.3
Launch
Preferences.

9.2.1 Select Next Scene on Launch

Under the *Live>Preferences>Launch* tab you can customize how Live behaves after a Scene is launched. This option is called "Select Next Scene on Launch". When this feature is set to "On", Live will automatically select the next Scene below the one just launched in order to prepare for the next Scene launch. This assumes that you are launching in your Scenes sequentially. This feature allows you to launch sequential Scenes without having to manually select them prior to launch, thus eliminating a step. The next Scene's launch button will flash until its downbeat launch timing has arrived, then the new Scene will launch and start playing accordingly. The Select Next Scene on Launch feature only works when using the return/enter key on your computer keyboard to launch Scenes. A common approach is to use the keyboard arrow keys to navigate between Scenes and use the return/enter key to launch. Alternatively, Scene launching can be assigned to a MIDI controller. For details on MIDI/Key Mapping, launch to ▶*Scene17*.

> **Hot Tip** Make your performances more intriguing by experimenting with various resolutions in the Quantization Menu. For example, change the quantization resolution between Scene launches amid playback using higher resolutions such as 1/16 or 1/8. With several tracks and multiple clips, this increases the complexity of your music. For ambient and non-rhythmic Sets, try selecting "None" in the Quantization Menu. This allows a freeform and legato approach to launching Scenes. Remember, the higher the resolution setting, the more precise you need to be when triggering your Scene launches. It's a good idea to practice with the Metronome to get a feel for this approach. Also, try using MIDI Map assignments for triggering clips and for selecting different quantization resolutions.

9.2.2 Record on Launch

Using Scenes for recording clips is essential to multi-track recording and for launching multiple track clips while recording on other tracks simultaneously —for example, when you want to record along with other clips that make up your backing tracks such as a keyboard part to accompany your rhythm section (drums and bass). This is common practice for recording since you'll usually want a count-in and then hear the music you are playing/recording along with. This means that you'll likely record with multiple clips lined up in a Scene. To make this process easy and efficient, you can set up Live so that when Scenes are launched they automatically trigger clip recording (auto launch clip Record

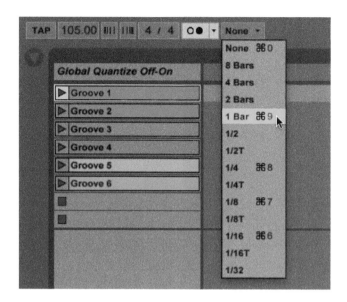

Figure 9.4
Quantization Menu and resolutions.

buttons) for any armed track. From the Launch preferences Tab, set "Start Recording on Scene Launch" to "On" to use this feature. When "Off", you will have to launch a Scene first, then *click* a clip Record button to record.

If you need to record on multiple tracks at the same time, *hold* (⌘ Mac/[ctrl] PC) *click* on the Arm button to arm each track. The Record on Launch automated feature streamlines the multitrack record/playback process into a one-click operation. Keep in mind that you can always step outside the Scene by launch-recording additional independent clip(s) on other tracks at any time after launching a Scene. The Scene will continue to loop while you manually select a new clip to record or overdub. This feature allows for great flexibility when recording parts on-the-fly. Not only does it give you the ability to multi-track record audio or MIDI in the Session View, but also provides you the creative freedom to record all of your clip ideas one Scene at a time and as your inspired ideas change from Scene to Scene in your Set.

As an alternative to Scene launch recording, you can use Group Tracks to launch and record on multiple tracks at once. Launch to ▶**Scene11**. Take this feature one step further by MIDI Mapping a controller to navigate and activate the Scene launch function. Launch to ▶**Scene17** in order to learn all about using mapping features.

9.2.3 Prepare New Scene for Recording

Now that you have a general understanding about recording with Scenes, there is one more feature to point out: *Prepare Scene For New Recording.* If you look to the right of the Control Bar transport, you will see a button labeled "NEW". Use this button to prepare a Scene to record new clips. When *pressed*, your selection will move to an empty and/or new Scene. Any clips playing in armed tracks will stop playing. Any others will continue playing from their current clip slot. The NEW button will be *clickable* as long as a track is armed and a clip is, or has just been, playing—in the Session or Arrangement View; otherwise it is grayed out. To that end, you don't have to be in playback to use this feature, as long as your Set meets the required conditions just stated. One practical use for this feature is to create a new take by advancing to a new empty Scene where a new clip can be recorded while maintaining the accompanying clips on other tracks that were not recording. You can also use this time to practice before recording a new clip to the Scene.

9.3 Capture and Insert Scenes

Scenes work well for organizing, building, and launching song sections, multi-track loops segments, vocal double-tracks, etc., not to mention the many ways of going about creating such elements. Since Live allows you to literally improvise your musical ideas, what better way to create Scenes then to just start firing off separate and randomly positioned clips to start building a nice hook or groove. Often times you will accidentally create a really cool song section,because you launched a clip inadvertently against another clip, but you'll have no idea what is making the clips groove so well together. If you stop all clips there is a good possibility that you will never know or have the ability to recreate that magic vibe or moment. Maybe you're in a live performance and you build up this incredible section on-the-fly and want to come back to it. Well, that's where Capture and Insert Scenes comes in handy. So, how does this work?

Just start firing off multiple clips until you have a groove you like. These can be spread across multiple tracks and over a number of different Scenes. Once you're satisfied with what you're hearing on playback, *right -click* or *ctrl +click* in any Scene while your clips are still running and select "Capture and Insert Scene" ([⌘ + shift + I] Mac/[ctrl + shift + I] PC). This will auto create a new Scene below the Scene row that you just *clicked.* Live will also number and/or copy the name of the Scene you selected to capture into. For example, if you select a Scene called "chorus", then Capture and Insert Scene, it will be auto named "Chorus1". If you selected an empty Scene, Live will name the next number based on the last number of the

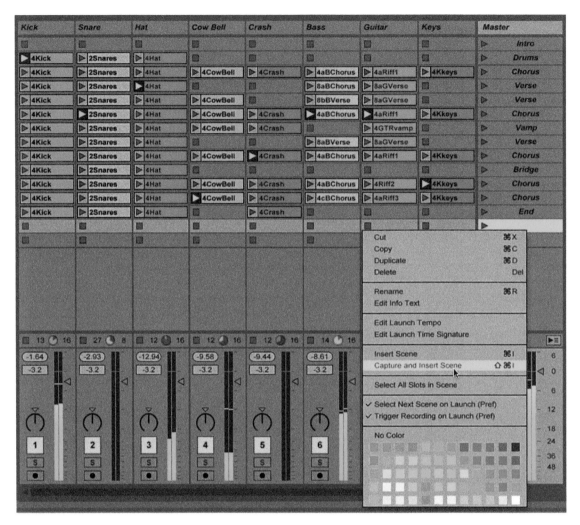

Figure 9.5 Launch any combinations of clips then select "Capture and Insert Scene" from the contextual menu.

Scenes in your Set. The new Scene will be made up of all the clips that were playing at that particular moment in time when you commanded them to be captured.

A really cool trick is that you can capture Scenes generated from the Arrangement View right into clips in the Session View in real-time. Whatever is playing in the Arrangement View can be captured into the Session View without ever having to launch a clip in the Session. Simply start playback from the Arrangement View—with Arrangement clips, of course—and then flip back to the Session View. While

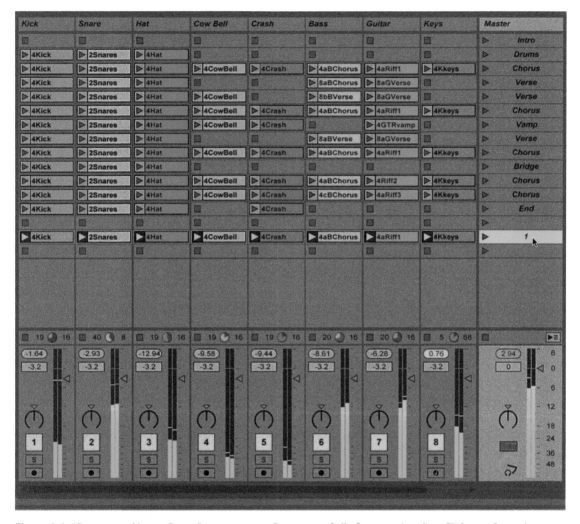

Figure 9.6 "Capture and Insert Scene" creates a new Scene out of all of your active clips. Click on a Scene (empty or filled) and then select the command.

the arrangement is playing, choose Capture and Insert Scene the same way as mentioned. A new Scene will then be created and playback will be taken over and generated from the Session View. The captured Arrangement clips are copied over just as if you *dragged* them, in regards to the clip length and their contents. Any clips, Session or Arrangement clips can be captured as a Scene!

In the studio, Scenes work great for auditioning groups of clips against one another, either while jamming alongside or rehearsing with a vocalist or session player.

With the Capture and Insert Scene function, you can keep moving forward in your musical flow without stopping to audition other ideas, thus allowing yourself or a producer to stay in the moment.

9.4 Tempo and Time

Working with tempo changes in your old linear Digital Audio Workstation (DAW) involves programming a tempo map of sorts, which is something you could never imagine doing in real-time. Sure, you could sequence out a song for a stage performance, but you are still locked to a linear map. First of all, Live's Session View is by no means linear and, on top of that, nothing you do has to be predetermined or remain permanent in a performance situation, or any situation for that matter. In Live, it's all about flying through tempo and time at the *click* of a Scene Launch button.

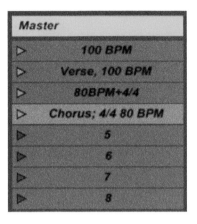

Scene Names are used to label or describe a Scene, but even more intriguing is the unique ability of a Scene name to store and impose a specific function upon launch. When a Scene's name includes a user-defined tempo followed by the acronym BPM (beats per minute), the indicated tempo will be imposed when

Figure 9.7
Scenes named programmed with tempo and signature changes.

the Scene is launched. The same holds true for time signatures. Simply enter the desired signature (e.g., 3/4, 4/4, or 6/8, etc.). Make sure to use the "/" (backslash symbol). To add a tempo or signature change into a Scene's Name, *right click* or *ctrl + click* and select "Edit Launch Tempo", or select "Rename" from the same contextual menu to rename a Scene with a tempo or time signature ([⌘+R] Mac/[ctrl+R] PC).

A Scene can be named in a number of different ways in regards to the actual wording of the name label. It can be just the launch tempo or can include both the name and tempo as long as "BPM" is included in the name (i.e., "Verse; 100 BPM" or just "100 BPM"). For time signature changes, Scenes should be labeled as a fraction (e.g., 2/4, 4/4, etc.). Both tempo and time signatures can be included in the same name in any order, but must be separated by at least

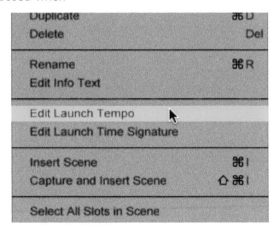

Figure 9.8 Select Edit Launch Tempo from the contextual menu.

one character or space (i.e., "80BPM+4/4" or "Chorus; 4/4 80 BPM"). Once a tempo or time has been added, the Scene's Launch button will turn orange to reflect the launch function. Once you've established a tempo or time into your Scenes, you will see how easy life becomes when you need to change up the beat or timing of your music. As always, you can rename a Scene on-the-fly to execute this feature in real-time while your Live Set is playing or recording. If for some reason you don't see the orange indicator, your nomenclature is incorrect.

Hot Tip A quick way to enter names and symbols in multiple Scenes is to use the Tab key to advance from one Scene to the next while the rename function is active. Enter a Scene name by ([⌘+R] Mac/[ctrl+R] PC). Don't press enter/return or navigate out of the Scene naming area just yet. After you type the new name, select copy ([⌘+C] Mac/[ctrl+C] PC) and then press the Tab key to advance to the next Scene. The next Scene name will be highlighted text input. Type in a new name or paste ([⌘+V] Mac/[ctrl+V] PC) to duplicate the name from the prior Scene. This is useful when you want to change a digit or letter to differentiate each Scene. Continue through the Scene names as needed using the Tab key. If you make a mistake or change your mind while typing, press the Esc key to revert and exit renaming.

9.5 Musical Concepts

Having the ability to call up a song section or group of clips intended to play together is brilliant. It all comes back to the non-linear approach provided by Live's Session View. Scenes often represent the essence of a musical structure, unifying the culmination of a musical idea that provides direction for a composition or song—a necessary stability for the creator, producer, or performer. Let's take a look at Scenes from these perspectives.

9.5.1 Create: Multi-track Instruments and Recording

Scenes allow you to link your clips of musical ideas so that they can be played simultaneously and recalled at any time. Scenes also serve as a neat and tidy way to launch your multi-channel/multi-track instruments and multi-track recordings. In this way a Scene contains a row of clips that make up one instrument or recording, such as a drum kit or a group of voices such as layered vocals. In this scenario, the instrument or recording is spread across multiple tracks because it requires the sum of two or more tracks to complete its overall produced sound—for example, the kick on track 1, snare on track 2 … or vocal layer 1 on track 1, vocal layer 2 with effects on track 2, and so on. You get the picture. The point to understand is that Scenes don't have to contain entire song sections; rather, they can be smaller parts of the whole, all the way down to a one-bar riff or a single instrument/vocal, if need be—literally anything that needs to play simultaneously.

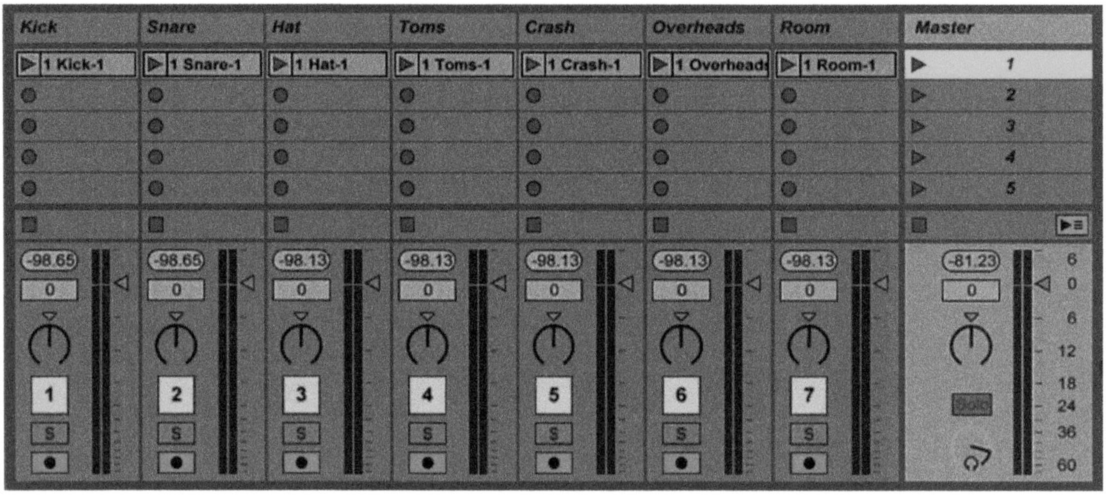

Figure 9.9 Here the tracks are all part of a single multitrack (multi-channel) instrument recording. This could be audio or MIDI. The Scene row represents each track part that will launch simultaneously to make up the whole instrument.

Here is the concept at the early stages of creating a beat from a composing/song writing standpoint.

When you first begin working with a MIDI drum kit or live studio kit, its parts will often be spread across multiple tracks (i.e., one track or channel per voice or mic input). The word "often" is used because you could sum the studio kit down to two-track stereo or route all of the MIDI channels to Omni—all or the same channel. This is not a new concept; but just to reinforce it: remember that each track is serving as one part of the whole instrument. In this way, while creating or recording your drumbeats, you will be launching and recording them as a row of clips across multiple tracks forming a Scene. The same concept applies to multi-voice or double-tracked vocals and instruments that you are trying to either develop or record to track. Once again their collective clips would be isolated as a Scene or as part of multiple Scenes that make up an entire song section.

As you advance through the production process, these examples will be incorporated as Group Tracks; but during the creative phase, it's quite possible that they are laid out as Scenes while being developed. There really is no right or wrong way to work with Scenes. What's more important is that you understand their potential. For more on Groups Track, launch ▶**Scene11**.

Let's take a closer look at how Scenes can be used to launch a multi-channel/track MIDI instrument as opposed to Scene launching entire song sections. Download Live Set "**CPP_Impulse-MultitrackMIDI Project**" from the companion website, or build your own similar instrument to follow along. Also provided are basic Impulse Kit MIDI files for importing for playback and for triggering your own original Impulse Drum Kit.

When working with a MIDI drum kit, you have a choice to work with all of the drums and MIDI notes triggering on one track/one clip or you can assign an individual track to each drum voice of the kit and then route each track to the same MIDI instrument—the playback sound source. The same goes for importing MIDI files— more specifically, MIDI drum loops. Many MIDI files you'll work with will be MIDI Type 1, meaning they consist of multiple vertically synchronous tracks, or simply multiple tracks intended to play in sync relative to the timeline (bars: beats). In the Session View, this will appear as multiple vertical tracks with clips laid out in a Scene. In the Arrangement View, this appears as multiple horizontal tracks and clips, which will play in sync vertically along the timeline—just as any other DAW does. Think of it as the kick on track 1, snare on track 2, hat on track 3, toms on track 4, etc. In many situations, using multiple tracks to trigger MIDI drum voices makes it easy to isolate and work with drum kits. As far as MIDI routing is concerned, this same concept applies to "Multi-Instruments" or "multis" (instrument devices that host multiple instruments within one interface that are triggered via multiple MIDI channels), often used with some of the popular third-party sample players, drum synths, and samplers. The concept and routing scheme is basically the same as the drum kit concept. Launch to ▶ **15.10.1** for the routing of multi-instruments.

You will see in the example two different ways that the tracks can be assigned to handle a drum kit played and triggered by Scenes. This is all about MIDI organization and dealing with multi-track MIDI file configurations. On the left, the kick track is located where Drum Rack has been inserted. The Snare, Hat, Tom tracks are simply routed to the Drum Rack "Backbeat Room" loaded on the Kick Track. In the example on the right, the Drum Rack has been removed from the Kick track and inserted on the "1 Backbeat Room" track. The Kick, Snare, Hat, and Toms are all assigned to the Drum Rack "Backbeat Room" track. Either way, it is up to you how you want to work. The important point is that all clips in the Scene are functioning together to make up the same beat as opposed to an entire song part, which also applies to audio tracks just as if the drums were a live drum kit with multiple mics.

If the MIDI instrument example was a keyboard instrument, you might program the left hand on one track and the right on another, although it is less common. The point is that each one of these tracks is routed to the same instrument, making use of a Scene in order to playback all of its track parts at once. A Scene can include as few or as many clips as you'd like. Just remember that the rows in the Session View are what constitute a Scene. In this case, the Scene handles a drum kit spread across multiple tracks.

Before you start importing a dozen clips for your multi-track drum kits or audio tracks, note that you can import multiple clips at once from the Browser or Finder/Explorer window and lay them in as multiple tracks/Scene rows as opposed to stacked into one track vertically (Session) or one after another horizontally (Arrangement). Simply *hold* (⌘ Mac/[ctrl] PC) while *dragging* them into the Session View or Arrangement View and they will automatically spread out to make a Scene, or place them across multiple tracks, as well as auto create new tracks if necessary.

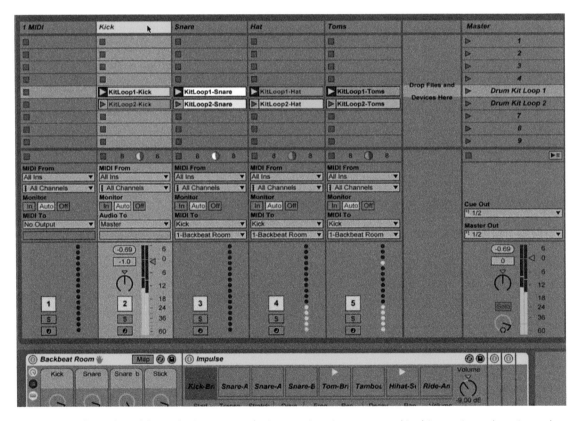

Figure 9.10 The Kick track hosts the instrument device, monitors its output, and in this case is used to trigger the kick sample.

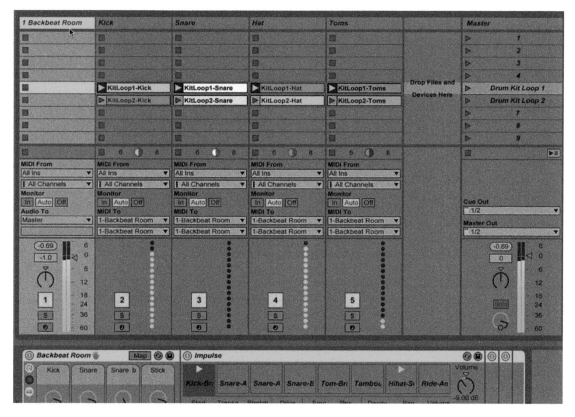

Figure 9.11 A separate MIDI track has been dedicated to host the instrument device.

9.5.2 Perform: Mapping Scenes

MIDI Mapping quickly comes to mind when taking your Scene launching functionality to the performance stage and beyond. Each Scene can be MIDI mapped to a controller key, control surface pad, knob, or slider, etc. This technique opens up a vast world of possibilities, especially when your Scene launches are tied in with the global Quantization Menu. By mapping a rotary controller knob to the Quantization Menu and a controller pad to your Scene launch function you can creatively launch your Scenes on-the-fly and in succession at every bar or down to the 1/32 value. Scenes keep your mind at rest and keep all your clips tidy and ready to go on a moment's notice. Some Live users use Scenes to launch cues on stage, which their band mates can use as a reference for timing and structure. Other performers use Scenes to color code and identify highly structured and involved song arrangements via clips. Launch to ▶Scene17 to learn more about mapping controls and launches for live performances.

Automation

10.1 Automation in the Arrangement View

Part of taking control of your Live Set includes automating various controls and parameters such as volume, pan, sends, device parameters, and song tempo. The purpose of automation is to provide hands-free control over the mixer, devices, effects, and global controls for programming musical effects and mixing. Automation control is different than Remote control in that automation is the process of moving and changing control parameters over time automatically without a physical human command. These movements are created manually or recorded into your song as a non-destructive element. The same consistent movements are made every time you playback the song. That being said, automation is often created via Remote Control. This gives you the power to not only Remote Control a device in a real-time performance, but also use Remote Control to record automation. Think of it like this: Automation is executed automatically by Live based on what you have programmed it to do. Remote Control is the process of physically controlling parameters with something other than the mouse.

Now, before we move on, keep in mind that automation in the Arrangement View is a linear concept and fixed against the timeline. This goes for Mixer automation and Arrangement Clip Envelopes. The exception to this is automation of Clip Envelopes in the Session View ▶ 7.8 .

10.1.1 Recording Arrangement Automation

In Live, automating a control parameter is a quick and painless process. No special assignments need to be made. Just activate *Automation Arm* on the Control Bar to prepare tracks for automation recording when you're ready. *Click* the Arrangement Record button and start moving faders and knobs with the mouse or a MIDI controller. Live records all of your actions while in Arrangement Record mode. This makes automating volumes, sends, track activators, device parameters, and other controls quick and easy. You can automate any parameters by changing or moving them with the mouse, MIDI controller, or even computer keystrokes.

Once a control has been automated, a red LED indicator will appear next to it informing you that the control or parameter has been automated. You will see these indicators in the Arrangement View, Session View, and in various other views. Although you can record Mixer automation into the arrangement while viewing the session, and frequently that's the preferred way, the actual Breakpoint Envelopes will write in the Arrangement Track Display fixed against timeline at the time and place in which they were created. That is, unless you are using Clip Envelopes to manipulate specific clip parameters. The session will display automation indicators, but aside from that you will rely solely on the arrangement for viewing and editing track automation and envelopes—unless you make another record pass to write new automation.

Now, before we get too deep into automation, it should be pointed out that there are two types of automation: Mixer and Clip. When working with Arrangement Clips, clips use strictly Clip Envelopes to automate clip-specific parameters. All other automation is Mixer or Track automation, such as Mixer volume or effects parameters, etc. With Session clips, you can choose between Mixer and clip automation via Clip Envelopes. More on that later.

At any point, if an automated parameter is manually adjusted (i.e., moving the track volume fader while mixing), that particular automation will be overridden and no longer active. To reinstate that automation, you must *click* the *Re-Enable Automation* button located next to the Automation arm button on the Control Bar. It will illuminate orange when any automation has been overridden. This applies to both Session and Arrangement Views.

Figure 10.1
A red LED indicator means that a control or parameter has been automated.

10.1.2 Automation Lanes

In the Arrangement View, envelopes represent the automation of a control or parameter. They indicate the state or value at which the control or device parameter is set at any given time. By default, envelopes will appear as a red horizontal line that extends the length of its track. This is called *Automation Lane*. When a control has been automated, the envelope will reflect the value changes. Use the *Device Chooser* in conjunction with the *Automation Control Chooser* to view track envelopes. This menu is located below the Track Name. Set the Device Chooser first to see the automated device and then choose the parameter you wish to view. If a parameter is already automated, you will see the red LED (indicator). If not, then *click* on any automated control parameter to bring its envelope into view. If you wish to view multiple envelopes at a time, or an envelope in a separate lane, *click* the *Add Automation Lane* button below the control chooser to the right ⊕. This will add an additional lane containing the current parameter's envelope. Add as many lanes as you need, with each lane set to a specific envelope. Fold lanes to hide them from view using the fold button below the Control Chooser. To remove a lane from view, *click* the *Remove Automation Lane* button, the same button you used to add a Lane ⊖.

10.2 Automation in the Session View

Automating devices and parameters in the Session View is not all that different than in the Arrangement View, except that envelopes are created and managed only via clips. Session Clip Envelopes can address clip specific parameters, track devices, and Mixer parameters. The big difference here is that Arrangement Clip Envelopes only deal with clip specific parameters. We've already discussed the basics of Clip Envelopes in ▶ 7.8 , so now let's narrow in on how to put Session Automation into action.

10.2.1 Recording Session Automation

Recording automation in the Session View is not really much different than recording audio or MIDI.

Let's step through the process:

1. *Click* the Automation Arm button to enable automation recording in the Session.

2. Import a clip and arm its track so that you can record automation into its clips. You could arm multiple tracks if you wish to record the same automation to more than one clip simultaneously. In Live's preferences under the *Record/Warp/Launch* tab, you can set how Live handles recording Session Automation. Your choices are *Record Session automation in* ... "armed clip" or "all playing clips". For this tutorial, we will stick to armed clips.

3. Once everything is enable and armed, *click* the Session Record button. This will immediately launch the Record button for the selected clip—multiple clips in the same Scene if their track is armed. There will be no count-in, so be prepared or wait for the clip to loop around, then start moving the parameters you want to automate. Alternatively, you can change the record behavior by *right clicking* or *ctrl + clicking* on the Session Record button. From there, uncheck "Start

Transport with Record"—you can also *shift-click* to temporarily override this preference. When deactivated, you will have to *click* clip Record buttons to begin recording automation. There still won't be a count-in!

4. As you manipulate parameters, recording automation to the Session clip, the envelopes will display in the Envelope Editor within Clip View. When you're done recording automation, feel free to stop recording however you like. If you want playback to continue, simply disarm or deactivate any one of the three features used to prepare recording: Automation Arm, Session Record, or track Arm.

Once automation has been recorded to a clip there will be a red LED next to the automation parameter. At any point, if an automated parameter is manually adjusted (i.e., moving the track volume fader while mixing), that particular automation will be overridden and no longer active. To reinstate that automation, you must *click* the *Re-Enable Automation* button located next to the Automation arm button on the Control Bar. It will illuminate orange when any automation has been overridden. This applies to both Session and Arrangement Views. Now, just because you're done recording automation doesn't mean it's permanent. At any point you can rearm and record additional automaton or overwrite previous automation on-the-fly during playback or by starting the same process over again. If you really dislike the automation, delete it altogether. Simply *right click* or *ctrl +click* on the clip in question and choose "Delete Envelopes".

If you do, in fact, wish to overdub changes or new automation over the existing automation, activate Automation Arm (if not already) and Session Record, then move the desired parameter you wish to overdub. Something worth noting is that Live does not have a dedicated automation modes, such as Touch, Latch and Write. Instead, when your mouse moves to automate a parameter, automation occurs like Touch mode would: automation stops recording once you let go of the mouse command. When using a MIDI controller or control surface, automation records as if it were in Latch mode. This is not the case in the Arrangement View.

10.3 Working with Envelopes

Recording automation in real-time can bring a more humanistic feel to your music; but the process is not always as precise as you might hope. There are many music genres where precision automation is the only acceptable kind. In that case, envelopes must be as accurate as possible—perfectly aligned to the grid and timeline. In any case, at some point you will find yourself drawing custom

Clip Envelopes or envelopes in the Track Display to either correct recorded automation or to create new automation. This will occur as clip- or track-based automation, depending upon which view and parameters you wish to automate. As you can see in the example, an envelope is represented by a red line and/or pinkish shaded area in the Sample Display/Note Editor for Clip Envelopes and in the Track Display. Simply select the device and/or parameter you wish to automate in the Clip Envelopes Box or from the Device Chooser in the Arrangement Mixer, then with your mouse *click + drag* or *click + draw* the envelope, shaping it to the desired position. The envelope line represents the position at which the associated parameters are set.

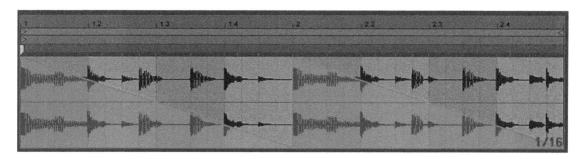

Figure 10.2 Custom Clip Envelope.

10.3.1 Draw Mode

Draw Mode allows you to literally draw in envelopes in either linear or curvilinear shapes without making a selection or adding breakpoints. To draw envelopes, turn on Draw Mode from the Control Bar. Select a track or clip and choose the device you want to automate from the Device Chooser and then select a control parameter. With your mouse, draw (*click* or *drag*) across the Envelope Editor area in the automation lane to create a custom envelope as whatever shape you like. Note that when in Draw Mode, the Track Display or Sample Display/Note Editor becomes the Envelope Editor. The new envelope will be drawn as stair steps based on the visible grid and snap settings. Choose a different Marker Snap resolution to alter this behavior.

Figure 10.3 Draw custom envelope automation as a Clip Envelope or in the Track Display.

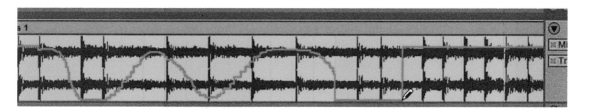

As an alternative, you can *hold* down the opt/alt key when drawing to override the Marker Snap. To draw curvilinear shapes, *hold* down the opt/alt key while *dragging*, or turn off the grid Marker Snap. *Right click* or *ctrl + click* in the Envelope Editor to bring up a contextual box, then select "Off" below the Fixed Grid settings to turn off Snap to Grid. Each grid setting dictates the increments at which a breakpoint segment is created (drawn/inserted)—good for precise drawing. Once you have created your envelope, you can redraw it as you wish, or switch out of draw mode and use your mouse pointer to *drag* breakpoints or larger line segments around. For fine-tuning at micro values, *hold* (⌘ Mac/[ctrl] PC) while drawing. To remove a Clip Envelope, *right click* or *ctrl + click* in the editor and select "Clear Envelope". In Automation Lanes, simply highlight automation and *press* the delete/backspace key on your keyboard.

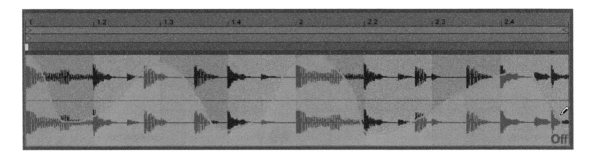

10.3.2 Editing

Although draw mode can provide high-resolution automation (curves made from very small line segments), it is not always the most efficient or fastest way to work with automation. There are times when you want gradual and smooth diagonal envelopes or fewer line segments, etc. For this reason, you may choose to create and edit envelopes by working directly with *Breakpoints* rather than drawing and redrawing them. Breakpoints are the little dots or anchor points along an envelope, fixing it to a value. This is probably the most common way to edit envelopes once they have been recorded or drawn, but will ultimately depend upon how you like to work. Take note that breakpoint editing does not snap to the grid.

If you place your mouse pointer directly over the envelope line, you can *click + drag* it up and down. This affects the entire clip. Notice that from the point you *clicked* a breakpoint was inserted. Use multiple breakpoints to create automation in segments along a track. They can be *dragged* in any direction you desire, up or down, forward or backward. To add a breakpoint, *click* anywhere on the envelope. To delete a breakpoint, just *click* on it. You can also *click + drag* it upward and

Figure 10.4
Use Draw Mode to create curves.

downward to change values. You can also automate envelopes segments by making highlighted selection inside the Arrangement Track Display or along a Clip Envelope (Session or Arrangement View) with your mouse then *click + hold* on the line and drag upward/downward in the editor.

Figure 10.5
Make a selection and drag the envelope up or down.

Taking this a step further, you can create ramps in the envelope between breakpoints. Simply *click + drag* a breakpoint to wherever you want. Do this to create interesting types of automation shapes such as long volume fades, swells, ramps, shaping, and effectual volume-based stutters, among other things.

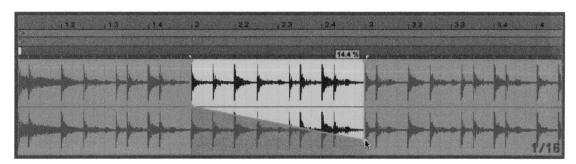

Figure 10.6
Create envelope ramps and other linear line segments.

Breakpoint Envelopes don't have to be linear. If you want a curved envelope, hover your mouse over the segment you wish to curve until it turns blue, then *hold* down the opt/alt key—you will see a curve icon appear at your mouse pointer—while *dragging* a line segment up or down. To remove a curve, *hold* opt/alt and *double click* on the curved segment. To create and automate a segment of an envelope, make a highlighted selection inside the Track Display or along a Clip Envelope (Session or Arrangement View), then hover your mouse over the envelope line until the line segment turns blue. Then you can *click + drag* the envelope up or down in the display. If you want to move all or a portion of your automation while maintaining the relative breakpoint relationships, highlight the desired area and

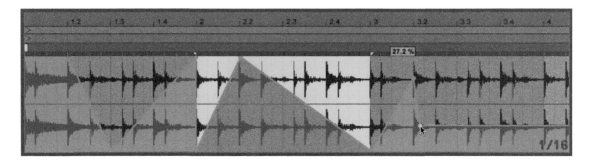

then *click + drag* on a breakpoint, moving up or down as you wish. The entire selection must turn blue before you move it.

To quickly delete automation, in the Arrangement View, highlight the envelope by making a selection in the Track Display, then select *Delete Envelope* from the clip's contextual menu, or in the Sample Display/Note Editor, *right click* or *ctrl +click* and choose "Clear Envelope". Alternatively, you can *click + drag* on any breakpoint while *holding* the shift key to erase over breakpoints *en mass*. As in Draw Mode, use the (⌘ Mac/[ctrl]) PC to fine-tune.

10.3.3 Commands

Envelopes can be manipulated in many ways—locked, cut, copied, pasted, duplicated, and deleted. To access these commands, highlight a selection of the envelope where the specific automation has been created, then *right click* or *ctrl*

Figure 10.7
Use breakpoints to create precise automation in segments.

Figure 10.8
(left) Unlocked envelopes.

Figure 10.9
(below left) Locked envelopes.

+ *click*. This will bring up a contextual menu where all of these commands are located. Each command has an associated keyboard shortcut. Take a moment to familiarize yourself with them to speed up your workflow. Of course, the traditional copy, cut, and paste commands are accessible as a keyboard shortcut or from the Edit Menu; but it should be pointed out that they function differently than the contextual envelope commands. Envelopes cannot be individually copied or duplicated with the traditional copy, cut, and paste commands. Instead, you should use the envelope commands to cut, copy, paste, etc. for isolating envelopes. To lock envelopes, choose the Lock button located to the right of the scrub just above the track names, or from the envelope contextual menu. When lock is "On", envelopes are locked to the timeline and remain unaffected by clip editing. When unlocked, envelopes are bound to their clips, following them when they are moved.

10.4 Song Tempo Automation

Clips and tracks aren't the only automatable features in Live. You can also automate global parameters, such as the *Master Track volume*, *Song Tempo*, *Crossfade*, or *Global Groove Amount*. To learn more about Crossfade, launch to ▶ 5.6.3 . More on Groove: ▶Scene12 . Besides automating the Master Track volume, Song Tempo is probably the most common automated global parameter, so let's spend a moment looking at it. The process is not much different than automating anything else, but enough to focus on. First off, the Song Tempo envelope is located in the Arrangement View under the Master Track's Device Chooser. To create automation for Song Tempo, you will either draw or create breakpoints, or you can globally record while moving the Tempo value up or down. You do not need to use the Automation Arm feature. Since Song Tempo automation lane lives in the Arrangement, Song Tempo automation is strictly linear. This creates a tricky relationship with the Session View because to use it in this way you are undermining the non-linear concept of the session. In effect, you have locked it to a timeline. If you don't care about that and still want automation tempo accelerations and decelerations to go along with your session, you can, but you will have to be creative with the Arrangement Loop Switch. With a strategic loop length you could achieve a successful result, but a better way to handle tempo automation when working in Session View is to launch tempos via Scenes. You won't get the exact same result, but you will enjoy the freedom of non-linearity ▶ 9.4 .

Putting the Session View out of mind, working with Song Tempo automation in the Arrangement View is straightforward and makes logical sense for producing and performing arrangements. Get creative and customize your Set with automation however you like. Try automating any of the global parameters and Master Track features in Live.

Grouping Tracks

11.1 Group Tracks

By now you have spent a lot of time launching and recording various Session clips and rows of clips with Scenes. This has, no doubt, been a fun and inspiring time, but not without limitations. Although there are an unlimited number of ways to use Scenes, the one thing you can't do with Scenes is launch only a selection of clips (multiple tracks) within the same Scene all at once when they are part of a larger row of clips. The only way to launch clips in a Scene is either one at a time or the whole Scene. Now, you could come up with a few workarounds to launch some clips and not others, but that's not an efficient way to spend your creative energy. That's what *Group Tracks* can do. With Group Tracks, it becomes possible to launch multiple clips within a Scene separately from within the Scene. Therefore, you can treat a multi-track instrument—the drum kit analogy or stacked vocal recordings—as a single track with a single Launch button. Groups may not be revolutionary themselves considering the concept of groups is a common everyday feature in other music production software, but Live's Group Tracks are a quite unique and really make non-linear music production viable and successful because they are tracks, not just groups. To that end, Live's Group Tracks have some convenient features. Whereas groups usually serve as an organizational tool and a method for global mixing and editing, in Live they also function as pseudo audio tracks or track containers, meaning that they come complete with mixer controls and the ability to host audio effects. This makes them great for submixing and for control over groups of tracks. Does this sound a bit like an Aux track to you? Just a bit! There are definitely some similarities, but more like a mutant super hero aux track! In a more practical sense, Group Tracks are great for organizing your workflow environment. They will come in handy for hiding all of those tracks that can get in your way when working with a large Live Set.

11.1.1 Groups Tracks in Session View

Groups Tracks are especially interesting in the context of the Session View. Here they have more to offer than the plane old groups found in other DAWs (Digital Audio Workstations). Sure, they are great for the common and practical features you'd expect from grouping tracks, but in the session they are a launching vehicle, actual tracks in and of themselves for firing off musical ideas as multi-tracks. This is how it works: you can combine as many tracks, audio and MIDI, together into the same group. Tracks can be added to an existing group from within a group, as well as deleted or moved out. To create a Group, select the tracks that you wish to combine, then *right click* or *ctrl + click* and select "Group Tracks" from the contextual menu ([⌘+G] Mac/[ctrl+G] PC). To ungroup, select "Ungroup Tracks" from the same contextual menu.

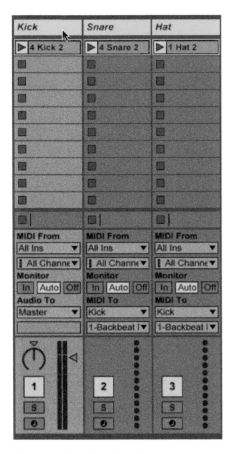

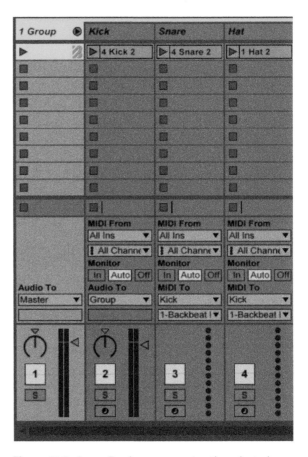

Figure 11.1 Multi-select the tracks you want to group and choose "Group Tracks" from the contextual menu.

Figure 11.2 Group Track encompassing the selected tracks.

Once you group a set of tracks, a new track column will appear. This is the Group Track Unfolded in view to the right of the Group Track, which will be all of your grouped tracks. Group Tracks are made up of *Group Slots*. These may look like Clip Slots, but they're not. In fact, they cannot hold clips. Instead, what you will notice is that Group Slots display a small color-striped box to the right of the slot. This box indicates the presence of clips in clip slots within the grouped tracks. Group Slot colors are derived from the first grouped track going vertically left to right. Use the *Group Unfold button* to the right of the Group Track's title, which extends as a bracket over the group, to show or hide the group. When hidden, you are left with only the Group Track.

Notice in the Session View example that the entire Group Slot is filled up with the color of the grouped track clips. When multiple colors are involved, the Group Slot color will be that of the left-most clip inside the group. Be aware that there are no fancy Group Track colors in the Arrangement View.

11.1.2 Audio Routing

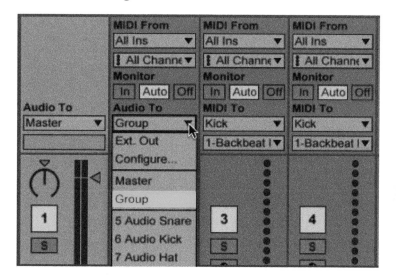

Figure 11.3
Colored Group Slot represents the presence of clips within its grouped tracks.

Figure 11.4 When tracks are grouped their "Audio To" is automatically routed to the Group Track.

When you group audio tracks, their "Audio To" is rerouted to the Group Tracks' output unless they have been assigned specific custom routing prior to being grouped. This is why Group Tracks are ideal for submixing and adding group-based audio effects. Nevertheless, this signal routing configuration is not mandatory.

You are free to route the audio outputs of each grouped track to any available audio track or the Master track. One reason for using custom routings is to use a Group track simply as a way to organize your tracks, or a way to launch certain clips simultaneously within Scenes on your own terms. From the Group Track, "Audio To" is routed automatically to the Master track, although it too can be rerouted. Since a Group Track is, in fact, a track with a Mixer Section, it is afforded all of the luxuries given to an audio track, except an Arm button.

11.1.3 Groups Tracks in Arrangement View

Based on the parallel mixing principles shared by the Session and Arrangement View, Group Tracks are carried over between both views, as mentioned. In the Arrangement View you will see the same groups just in a horizontal layout, organized top to bottom. As expected, there are no Group Slots and therefore no color coding.

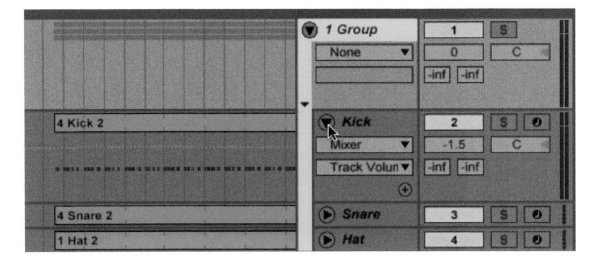

Group Tracks and grouped tracks function in a traditional sense—like a traditional DAW, if you will—and are ideal for submixes and organizing your Arrangement View workspace. You will really appreciate how much space is saved when using Group Tracks to fold up your clutter.

Figure 11.5
Group Track in the
Arrangement
View.

A Group Track will display a representation in its contained track clips. When folded, the clips are represented as an overview in horizontal lanes. When unfolded, clips appear as gray transparent lines in the Group Track Display. When a Session Group Track is unfolded it will be unfolded in both the Arrangement View and Session View, and vice versa.

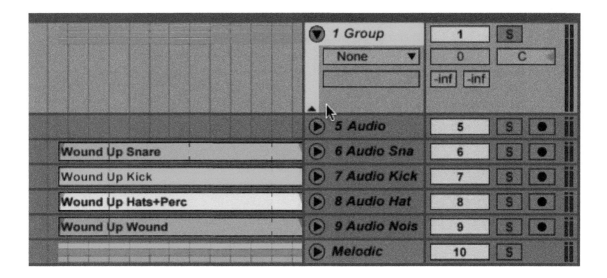

Figure 11.6
Fold Group Tracks
to save space.

11.2 Launching Group Clips

As mentioned earlier, one advantage of Group Tracks is the ability to launch a group of clips, which when played together make up a single instrument, such as a multi-track drum kit in the Session View. Launch to multi-track instruments and recording: ▶ 9.5.1 . At times, a row of clips will collectively make up an instrument or multi-track audio recording. Group Tracks make it easy to launch all track clips with the *click* of a single Launch button. To launch all of the contained tracks within a Group Track, activate the *Group Launch* button. These look and behave just like Clip Launch buttons. Stop your entire group by *pressing* the *Group Stop button*—an easier way than using individual Clip Stop buttons or the Stop All Clips button!

Now for the fun part! Once you have launched a Group Track, you can then launch other Session clips in various tracks and Scenes, as well as stop or launch the

Figure 11.7
Group Launch
Button.

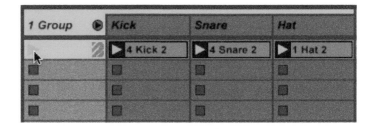

individual clips from within the Group Track. This can all happen while other clips play on or advance from clip to clip within or outside the Group Track. Taking this a step further, you could even launch additional Group Tracks on the same Scene: launch if you've got 'em!

11.2.1 Groups versus Scenes

"Nesting rows of clips in a Scene ... "

One of the most exciting aspects of grouping tracks is the ability to simultaneously launch multiple clips in the same Scene without having to launch the Scene itself, which is helpful when, for example, a row of clips doesn't make up an entire Scene. This is great for both performance and recording. With this knowledge of Group Tracks you can rethink how you might produce and perform your Session clips. Now, you may launch an entire Scene, then launch a Group Track from another Scene without having to launch all of the other clips in that Scene. Here is a comparison: without using Group Tracks, Scenes would be used to launch multiple track clips at once. In that way, each Scene contained specific clips that were generally designed to play at the same time. Group Tracks, on the other hand, allow you to use a single Scene row to launch groups of clips rather than requiring each group of clips to have a unique row (their own Scene) to avoid inadvertently launching other clips. You also have the ability to stop clips more efficiently, with the grouped track clips all stopping when the Stop Group button is *pressed*. This concept is important for your multi-track instruments and layered song parts. As a bonus, your workflow environment will become organized and clutter free with all of the folding and submixing you'll start doing.

> **Hot Tip** If you find that you really like how a few of your individual clips/tracks are grooving together, why not group them and use them as a modular idea that you can launch whenever you want?

Here's a look at how Group Tracks work. There are three different Group Tracks in the Session View, all of which have clips in the same row. To start out, the first group "Kit" that consists of an Impulse-based Instrument Rack has been launched. Without groups, these three tracks within the Kit would have been a Scene of their own.

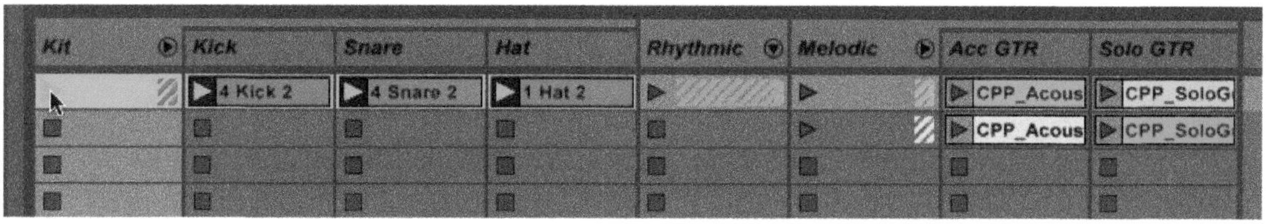

Kit		Kick	Snare	Hat	Rhythmic	Melodic	Acc GTR	Solo GTR
		▶ 4 Kick 2	▶ 4 Snare 2	▶ 1 Hat 2	▶	▶	▶ CPP_Acous	▶ CPP_SoloG
▢		▢	▢	▢	▢	▷	▶ CPP_Acous	▶ CPP_SoloG
▢		▢	▢	▢	▢	▢	▢	▢
▢		▢	▢	▢	▢	▢	▢	▢

Figure 11.8
"Kit" Group Track launched.

After activating the Group Launch button, the next group that resides in the same row will be launched: the "Rhythmic" Group Track.

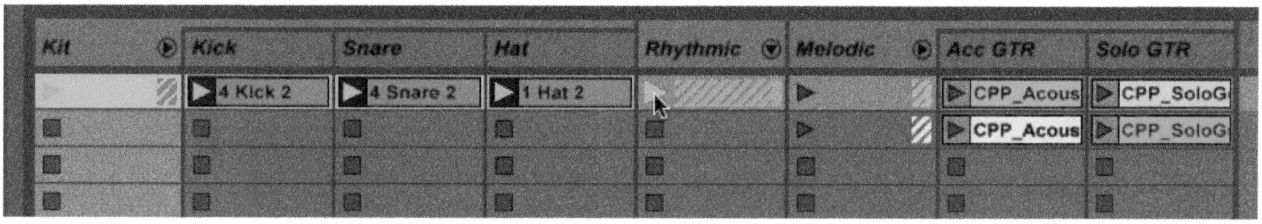

Kit		Kick	Snare	Hat	Rhythmic	Melodic	Acc GTR	Solo GTR
		▶ 4 Kick 2	▶ 4 Snare 2	▶ 1 Hat 2	▶	▶	▶ CPP_Acous	▶ CPP_SoloG
▢		▢	▢	▢	▢	▷	▶ CPP_Acous	▶ CPP_SoloG
▢		▢	▢	▢	▢	▢	▢	▢
▢		▢	▢	▢	▢	▢	▢	▢

Figure 11.9
"Rhythmic" Group Track launched.

Still working from the same row, the third and final Group Track "Melodic" is launched. Once again, all of this has been done without leaving the row or launching a Scene. While all three groups are playing back, they can be folded up. See in the example that Group 2 is folded up and the next Scene for the Melodic Group can be launched, while leaving the Kit and Rhythmic Groups running.

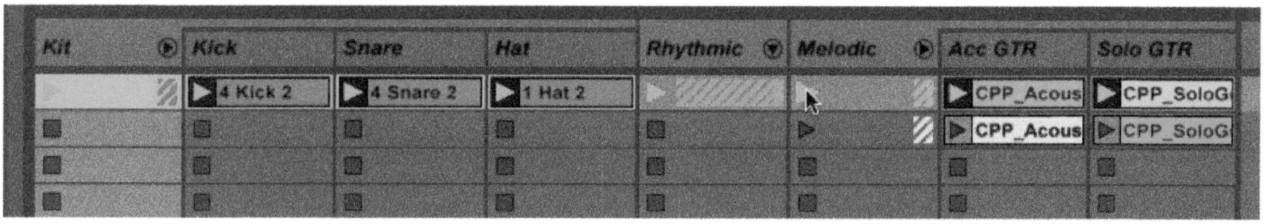

Kit		Kick	Snare	Hat	Rhythmic	Melodic	Acc GTR	Solo GTR
		▶ 4 Kick 2	▶ 4 Snare 2	▶ 1 Hat 2	▶		▶ CPP_Acous	▶ CPP_SoloG
▢		▢	▢	▢	▢	▷	▶ CPP_Acous	▶ CPP_SoloG
▢		▢	▢	▢	▢	▢	▢	▢
▢		▢	▢	▢	▢	▢	▢	▢

Figure 11.10
"Melodic" Group Track launched.

To do this, the second Group Launch button is *clicked*. This activates all of the clips on that row, leaving the other groups alone. If the Scene Launch button had been used, all clips playing from other rows and tracks would have been stopped. The only way to prevent this would be to remove the Stop buttons on those tracks below the clips so that the Scene Launch wouldn't trigger their Stop buttons. Of course, you will still want to remove Stop buttons for clips that you wish to continue playing within a Group Track so that Scene launches or Group Slot launches don't stop those clips you wish to remain playing ▶ 5.4.1 . Note that you cannot remove Group Stop buttons.

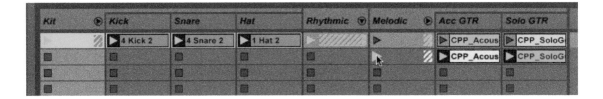

At this point, don't take for granted the notion that Group Slots are similar in function to traditional clips. They are to some degree, but there is an anomaly—that is, at times you will see more than one Group Launch button lit up on the same track at once. This would appear to break Live's golden rule for clip launches on a track, but it doesn't. You still can't launch more than one Group Slot at once in the same track; but Live uses the Group Track/Group Slots to also indicate what rows have active clips, as you will see in the example.

Figure 11.11 Launch a new row of clips without launching a Scene.

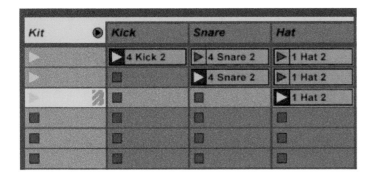

Figure 11.12 Multiple Group Slots activated indicating active clips on different rows within the Group.

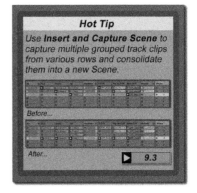

11.3 Mixing Concepts

Mixing with Group Tracks makes Mixing, Submixing, creating group-based effects, and Automating tracks very easy and super-efficient. As mentioned, you can add audio effects to the Group Track, which processes all of its contained tracks (grouped tracks). You can also use Group Track sends as you would non-grouped track sends. In this case it would be a group-based Send. Mixing and automating is a piece of cake with Group Tracks. All you have to worry about is the Group Track instead of all of the contained tracks, unless you want to create individual automation within the group and tweak your mix. If so, dial in your internal mix, then use the Group Track to mix your overall tracks in your Set.

11.3.1 Submixes

Figure 11.13
Use Group Track to submix grouped tracks. The Group Track/Submix itself can then be sent to Return tracks for Send effects if desired.

We talked a bit about submixing with the Return Tracks in Clip ▶ 5.8 . You'll find that with Group Tracks you have little need to submix with Returns or any other audio track for that matter, but feel free to do so if you wish. The general concept is to create a mix of all tracks to be grouped—the internal tracks. For example, try balancing the drum kit. Mix the kick, snare, cymbals, overhead mics, etc., then group them and control their overall mix for your Set with the Group Track fader (group them before creating the internal mix if you wish). This is a very common production technique and is useful in lieu of mixing down drum kit multi-tracks to two-track (stereo audio file). This way you can submix and print them when needed either as a performance or for a traditional multi-track mixing session within the arrangement.

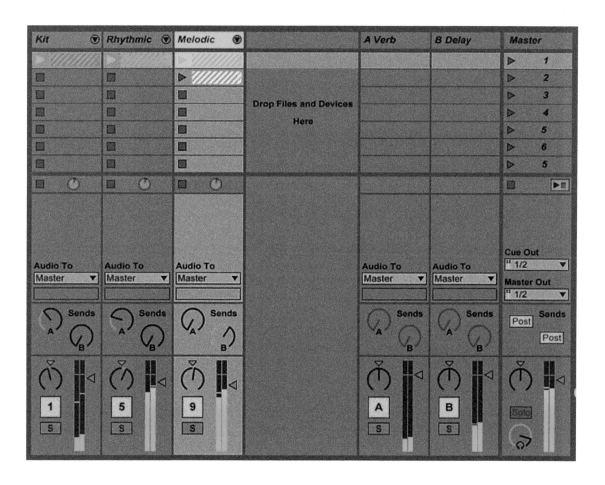

Submix your tracks in the Session View or Arrangement View and don't forget how simple it is to automate them in real-time or to draw a few custom envelopes to submix your tracks ▶Scene10.

Let's try this submixing example using the Set "**CPP_11–3-1_Submixing**" from the companion website:

1. **Pre-mix:** adjust the volume and other various track parameters of each drum track to a mix you like. Balance volumes, EQ (equalization), etc.

2. Select all the drum tracks and Group them ([⌘+G] Mac/[ctrl+G] PC). Rename the Group Track "Submix Drums".

3. Place a Multiband Dynamics or EQ8 and Limiter audio effect on the new Group Track, "Submix Drums".

4. **Mix:** tweak the mix a bit more using the Multiband Dynamics and the EQ to compress the overall drum mix or use the EQ8 and Limiter to achieve a similar result.

5. You then could adjust the volume of the "Submix Drums" Group Track accordingly based on the overall mix and you will instantly notice how your drums begin to blend with the mix in a much tighter way.

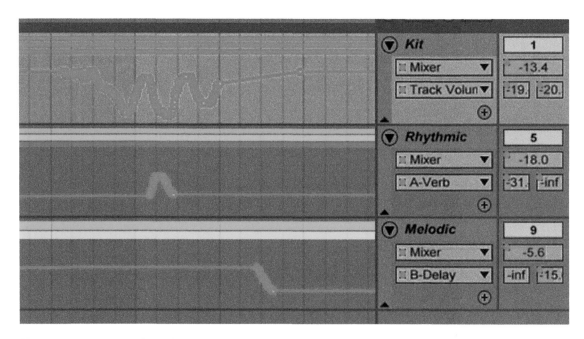

Figure 11.14 Group Track envelope automation.

6. **Bonus:** send the snare track from within the Group Track to a Reverb Return and conservatively adjust the send amount to taste. Don't overdo it, unless you're going for that big room snare of the 1980s!

> **Hot Tip** *Routing multiple Group Tracks' outputs into an audio track (or Return) with applied audio effects can really tighten up your mix in Live. Try the power of multiple Group Tracks submixed into a single audio track—nesting them together.*

11.4 Musical Concepts

Group Tracks will play an important role in your future experience with Live and beyond. Ableton has truly found a way to merge the concept of an Aux track, audio track and a traditional group into one track type. With Group Tracks, the Session View becomes that much more unique since Group Tracks have both a use for mixing and performance.

11.4.1 Create: Printing Group Tracks

Printing Group Tracks is a necessary part of the production process: since they function as a submix of their contained tracks, they are perfect for building and creating audio stems to be delivered to a mixer or additional producer for adding new layers. This is also important for sending your mix stems for use in another DAW software application. One of the most common uses of stems is for a multi-track studio drum kit that must eventually be mixed down to two-track stereo. When that time comes it's great to be able to print your entire group for use in your Set or someone else's. The general process is very quick and easy. The hard part is deciding which method works best for you. To print a Group Track's output to track—Session or Arrangement View—route the Group Track's audio output to a new audio track and then record the Group's output into the new track or use the Export Audio/Video function to export the submix and then re-import the render back into Live if needed to continue working with it. For step-by-step printing and exporting, refer to Clip ▶ 8.5 .

Groove

12.1 Introduction to Groove

In a musical sense, a groove is the rhythmic feel of a song or beat, but more than that, it is a real-world phenomenon that occurs when a human performs a piece of music live (onstage) or in a recording session. It is the breath of life that makes music come alive, breathing with the beat. Each performer or band generates their own groove when they play and perform, in turn stamping their own music or style with a signature. Their respected fans and admirers often seek after this "signature sound". In an effort to recreate this experience in the Digital Audio Workstation (DAW), Live has its own groove technology called *Groove*. The goal of groove technology is to breathe life into clips when they lack the authentic feel of a real performance or to conform two different clips so they have the same feel—in other words, to either liven up a vanilla beat that doesn't groove and is lacking soul, or to reshape the beat to emphasize a different feel than originally performed. For this reason, Live includes a number of grooves as part of the Core Library. Applying one of these grooves to your audio or MIDI clips changes its timing and feel, conforming it to the new imposed groove. But how, you ask? This is achieved by altering where the beats fall in time and applying emphasis (accents) to the beat by altering the intensity or velocity of specific beats. A prime example of this is the recreation of a swing feel (shuffle). A real swing feel is not something you can mathematically quantize in the digital realm. It is truly a human experience that doesn't have an exact formula. With Live's groove technology, you can take that real inexplicable human feel and apply it to a straight beat. Even better, extract it from a real-world recorded performance or studio recording. Who needs the mathematical equations when you can get the groove direct from the source! This is all managed at the Clip Groove chooser menu available from the Clip View Clip Box ▶ 7.3.2 . That's already been discussed, so let's take a close look at how Grooves and the *Groove Pool* work. Then we'll discuss how to *Extract* and *Commit Grooves*.

12.2 Grooves

Grooves are located in the Live 9 Core Library and are accessed from the Live Browser under Places in the Packs section. You'll notice right off the bat that some of the more popular swing and groove styles, such as the MPC (Music Production Center) and Logic grooves, are included as presets located in the *Core Library> Swing and Groove folder*. There are a few different ways to load and work with these grooves. One way to apply a groove to a clip is to *double click* one or *drag* it directly from the Browser onto a clip. It will instantly alter the clip's feel in real-time. Once you've placed the groove on the clip, you can then preview other grooves for that clip by using the *Hot-Swap Groove* button in the Clip Box. Hot-Swap Groove is used to quickly select, change, or audition grooves on the fly. *Click* the Hot-Swap Groove button in the Clip Box, then try out other groove presets located in any groove folder. This is one way to quickly load and audition the various groove presets.

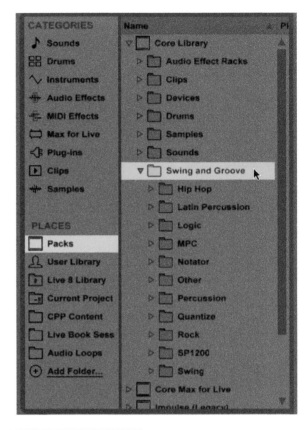

Figure 12.1
Core Library
Groove presets.

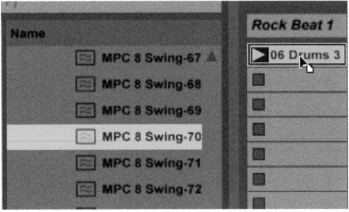

Figure 12.2 Drag a groove preset onto a clip.

12.3 Groove Pool

After a groove has been *dropped* onto a clip, it will be stored in the Groove Pool located below the Live Browser. You can also *drag* grooves directly into the Groove Pool from the Library. To show the Groove Pool, *click* the *Groove Pool Selector* ≈ just above Info View or choose "Open Groove Pool … " from the *Clip Groove Chooser* in the Clip Box. This is where all grooves being used or previewed in your Set will appear. It is also where you will view and edit those grooves. Keep in mind that these grooves are not audio files. They are files containing performance information (timing and emphasis) that has been extracted from an audio or MIDI clip. More on extracting grooves in Clip ▶ 12.5 .

Figure 12.3
Groove Pool.

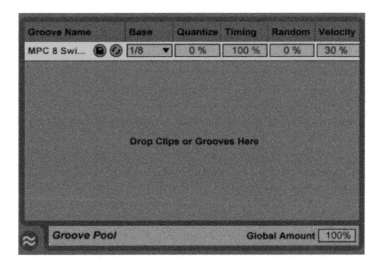

Figure 12.4
Grooves in the Groove Pool are accessible in the Clip Groove chooser.

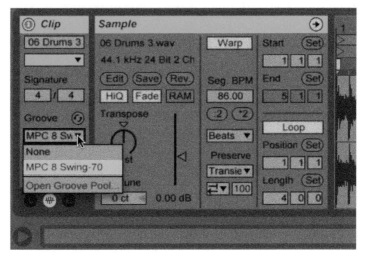

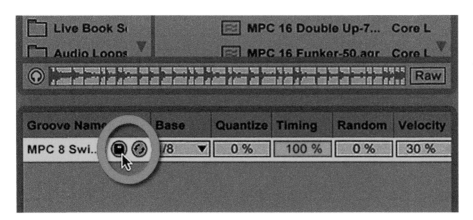

Figure 12.5
Save and Hot-
Swap grooves to
the User Library.

Once grooves have been added to the Groove Pool, they can then be accessed via the Clip Groove Chooser. From the Chooser menu you can select and apply any of the grooves that are stored in the Groove Pool to a selected clip. This integrated list of grooves is always available and even accessible while working and performing in real-time. This gives you an incredible amount of flexibility by having the unique feel and timing of various grooves available at your fingertips. Grooves in the Groove Pool can be Hot-Swapped with other groove presets in the Core Library or saved as presets to the *User Library>Grooves* folder. There are many options; you have the ability to Hot-Swap clip grooves with groove presets stored in the Groove Pool, Core Library, or User Library. Each groove in the pool has its own Hot-Swap button and Save button located next to its name.

12.3.1 Groove Pool Parameters

Within the Groove Pool Menu are built-in groove parameters for tailoring a groove's effect on its assigned clips. There are five parameters for each available groove in the pool: *Base* (Resolution Base), *Quantize*, *Timing*, *Random*, and *Velocity*. Additionally, the Groove Pool itself has a *Global Groove Amount* slider that controls the overall effect amount that grooves will have over all clips with a groove assigned. Each clip groove parameter in the Groove Pool determines how a groove affects its assigned clips. When clips are assigned to a groove, the groove's parameters will become active. Here is a description of each parameter:

Base: sets the subdivision of the beat that the groove and assigned clips are measured against. This is related to the smallest beat value expressed in the groove and clip, such as the lowest common denominator.

Quantize: the amount of quantization Live applies to assigned clips before they are re-quantized by the groove's feel. This is standard quantization to the nearest division of the beat as determined by the base.

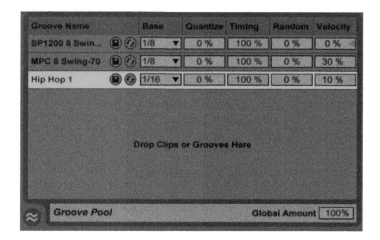

Figure 12.6 Groove Pool parameters.

Timing: how strong of an effect the groove will have on all its assigned clips.

Random: the level of fluctuations (randomness or humanization) applied to the timing of assigned clips.

Velocity: how much affect the groove's beat accents will have over the velocities or beats in the clip. This can be applied with an inverse effect based on positive or negative values.

Global Groove Amount: controls the overall percentage of strength of Timing, Random, and Velocity that will be applied for all grooves. This can be automated in the Arrangement. Interesting … huh?

Let's take Groove for a test drive!

1. Load up some drum loops from the companion website or your own into Live and then *drop* some groove on them to see what will happen. Now that you are working with various grooves, take a few minutes to experiment with the Quantize and Timing parameters in the Groove Pool.

2. Once you have a groove loaded in and applied to a clip, adjust its Timing percentage to 0, and then adjust the Quantize percentage to influence—clean up or straighten out—the clip's feel as it plays. If you *drag* the Quantize percentage all the way up to "100%" you will surely start to hear the effect, especially when applying straight 1/16 or 1/8 values to a swing-style beat.

3. Assuming you have placed the Quantization value in the middle say, "50% to 60%", gradually start adjusting the Timing value. Timing will utilize the groove's feel and flow to start pushing and pulling the original beat placements off the grid while maintaining a tighter quantized feel around the beats not affected by the groove. This is controlled by the Quantize percentage. Cool stuff, right?

This is a great way to begin learning how to use groove. The key is customization. Once you find or create a new groove you like, save it in the User Library on-the-fly and keep on working. From the Groove Pool, *click* the save button to the right of the groove name. It will then be saved in your User Library.

12.4 Commit Groove

Auditioning and experimenting with grooves on your percussive clips can be pretty fun. You can do it all in real-time, trying as many as you like. This is possible because it's a non-destructive process—that is, until you *click* the Commit button. Commit is used to permanently apply a groove assignment to a clip, which writes the groove to the clip by adjusting a MIDI clip's notes or an audio clip's Warp Markers. This is a somewhat destructive setting, meaning that when you commit a groove to a MIDI clip, its MIDI notes are moved (quantized) to match the groove—passed the point of no return! When applied to audio clips, Warp Markers are added and adjusted in order to quantize the clip to fit the groove, which is non-destructive. Until you commit a groove, it is only temporary. This means that Live processes a groove in real-time until you commit it. After you select Commit, "None" will appear in the Clip Groove Chooser, signifying that the groove has been permanently written. Of course, you can always use undo if you haven't made a million other changes to your Set.

> **Hot Tip** If you commit a groove to an audio clip, then realize that you don't like the feel, what are you going to do? You can always delete all of the Warp Makers added by the groove engine, then re-warp the sample back to its original state, or just call up the original clip from the browser.

For MIDI, you can use the quantize feature or manually move the MIDI notes back into position or reload it if you have it saved elsewhere. This is a little bit more cumbersome than the fix for audio clips. Here is an example of the results of the Commit Groove feature. In the first figure you will see the original straight clip, then a resultant grooved clip to "8th-note swing". Remember a swing feel drags back the eighth note upbeat, delaying its timing and compressing its proximity by moving it closer to the next downbeat.

You can see in the before and after example that the eighth notes were assigned a Warp Marker and were slid back toward the next downbeat. The similar process happens for MIDI clips.

You can see that the MIDI notes have been changed, both position and some lengths. Once again the eighth note upbeats have been pushed back. As you use the feature with your own clips, you'll also start to notice the velocity affect that is imposed on your clips.

Figure 12.7 Swing groove chosen from the Groove Pool.

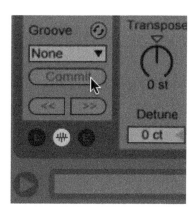

Figure 12.8 Commit Groove.

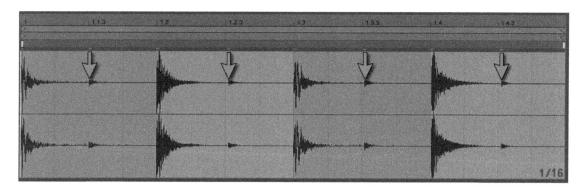

Figure 12.9 Original straight feel audio clip.

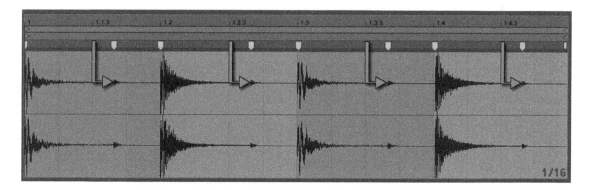

Figure 12.10 Swing groove committed to the audio clip moving the eight-note back.

255

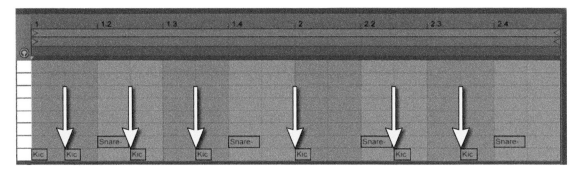

Figure 12.11 Original straight feel MIDI clip.

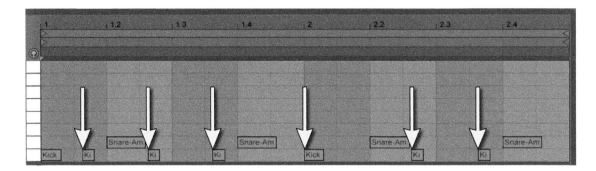

12.5 Extract Groove

Enough with using groove presets, let's make our own! If you have a killer drum beat or loop, extracting and adding it to the Groove Pool is very easy. Simply *drag* any clip or audio/MIDI file directly into the Groove Pool and Live will analyze and extract its groove. It then becomes available in the Groove Pool for applying to clips. This is very useful for conforming your clips to another song or beat. Alternatively, you can extract grooves by selecting the "Extract Groove(s)" command from the clip contextual menu. *Right click* or *ctrl + click* on clip and select it.

This can become quite addictive as you may find yourself extracting everything you can get your hands on—old vinyl records, vintage sample libraries. You name it. All is extractable and that's what makes the groove function so much fun and enjoyable to use. Applying a customized extracted groove to a clip that may have been out of the pocket or felt incompatible in a particular song can jumpstart your work tenfold.

Figure 12.12
Swing groove committed to the MIDI clip moving the eight note back.

Hot Tip

*Quickly open and browse the Groove Library. **Right-click** inside the Groove Pool area or on a Groove to bring up a contextual menu and select "**Browse Groove Library**".*

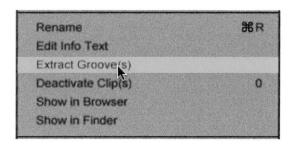

Figure 12.13 Extract Groove from the clip contextual menu.

12.6 Musical Concepts

Since a large majority of studio albums are at some point produced in the box, it's important that they sound as lifelike or humanized as possible. Even if your next record will be recorded with all live session performers, don't forget that there are many musical genres where pre-production begins as a mock-up of samples and loops. Even the big artist albums or film scores spend time in the land of MIDI using virtual instruments and sampled loops. To be honest, a large majority of TV, film, video games, and Internet audio content are completely conceived in the DAW without a live performance, some of which never see a live musician. The point is, groove attempts to remedy this by providing the ability to use samples, live performance, and MIDI altogether so that they feel right and organic, not only for sample-to-sample integration, but also for grooving along with real studio musicians.

12.6.1 Create: Grooving Your Backing Tracks

So far, you've seen many ways to utilize groove functionality in Live. Now that you've had some time with it in action, you'll find that it's easy to get caught up grooving with your drum tracks and extracting your favorite drummers. Don't forget, this feature also works great with bass lines and guitar phrases too, not just drum- or percussion-based rhythmic loops and recordings! One way to work with guitar and bass phrases is to dial in your drum track's groove so that it feels exactly how you want it to. Once your drum track clips are grooving—Clip Groove in place—select the same groove for your bass and/or guitar track clips. Since your drums are already using the groove file, it's very easy to select it through the Chooser Menu in the Clip Box. Go ahead and do this for each additional instrument you want to conform. Now, before you launch all of your clips or Scenes, you'll want to spend some time working with the timing for your bass and/or guitar clips by dialing in the perfect Quantize and Timing values in the Groove Pool. Your ears should tell you fairly quickly what is and is not working as you massage your bass and guitar clip's groove and feel against your drum clip. Since this affects all drum clips as well, you may have to make some compromises to get the feel to work across the board. If you really want to push things to the limit, try mixing and matching a straight quantized groove for a bass clip against a swing groove on the drum clips.

Warping Your Mind!

13.1 Elastic Time

There was a time when the only way to change the tempo of an audio recording was to alter the playback speed, or if you were an audio ninja, you could conform a performance with hours of editing magic. Altering the sample rate unfortunately changes the pitch of the audio and can degrade the quality of the playback. As an alternative to this, software companies have developed and implemented time-stretching/pitch-shifting algorithms into their software. This allows you to lengthen or shorten an audio file to fit the tempo of a song by altering the sample rate and pitch shifting it back to the correct pitch.

In addition to all these advanced algorithms, developments in this area have led us to the acclaimed REX/ReCycle technology by Propellerhead that allows you to slice up an audio file into little slivers (chunks or pieces of sound) at each transient point so that the tempo can be changed without affecting the quality of playback or the pitch. There is no shifting or stretching of the sample slices, thus preserving the quality and pitch of the audio. When the file is sped up, the slices trigger faster, truncating and fading the slices at their end points. When it is slowed, the transients are spread out, leaving little tiny gaps, which can be covered up with fade outs, or the tails can be stretched—okay, a little stretching, but only the tails! With this technology, you can alter the tempo of your loops without degrading the audio. Audio files can then be locked to a grid and effortlessly looped in a DAW (Digital Audio Workstation). Although this is a powerful and useful technology, unfortunately ReCycle is a third-party application and requires you to slice up your own loops, then import them into whichever DAW you use. REX files can be imported into other DAWs, but they are focused around short loops, not whole performances, songs, complex harmonic elements, or non-rhythmic material. That's a specialty of Live's Warp technology. Not only does Live support the use of REX files, it successfully liberates audio from the historical confines of time and tempo. Where once the pitch of an audio recording or loop was married to its original tempo, Live has the ability to change the tempo and pitch independently and even allows you to reference the feel from one performance and apply it to another. In this way, audio in Live is truly elastic. In all fairness, slicing is still a fantastic way to effectively create and use tempo-synced loops; but time-stretching algorithms have come a long way since their inception, not to mention their ability to quantize audio!

13.1.1 Warping

Working with audio in Live is extremely flexible in that you can quickly sync sample content to your song tempo or let it play back at its original tempo. This is very important considering that not all samples are intended to be time-stretched, as is the case with ambient textures, effects, or one-shot percussive impacts, etc. When rhythmic synchronization is important, activate the *Warp Switch* located in the center section of the Sample Box to engage Warp for the desired audio clips and let Live time-stretch your audio to match the tempo of your Set.

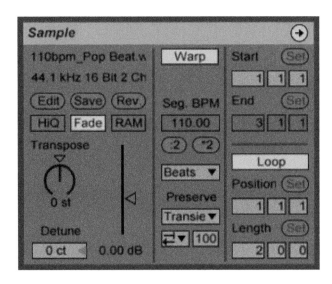

Figure 13.1 Activate the Warp Switch in the Sample Box.

Once activated, audio clips will follow the Set's tempo by employing one of Live's time-stretching algorithms to the clip's sample (audio file). Warping allows the sample to be manipulated and/or quantized to alter its playback, timing, feel, and physically loop clips. Of course, each sample may require some tweaks in order to get the most out of the various time-stretching algorithms (modes). When deactivated, the audio clip will play back at its original tempo. Even when working with content intended or created to play at the current tempo of your Set, it's a good idea to switch Warp On. Warp enables the ability to loop your clips—super important! Beyond that, manipulating your audio loops to various tempos is a common task; therefore, all of the necessary Warp settings are configured from the Sample Box. To streamline the assignment of Warp mode to various clips, tell Live when to automatically assign Warp Markers to samples and to what mode they should be set by default. You will find these settings in Live's Preferences under the *Warp/Record/Launch* tab.

> **Hot Tip** With "Auto-Warp Long Samples" turned "On" in the preferences, Live will automatically warp your audio in files—with a total timeframe of approximately 45 seconds and longer—to the best of its ability. By turning this preference "Off", you must initiate this warping process yourself, which can be a time saver when you already know exactly where you want to place your Start Marker, where to warp from, or when you want to work with a specific selection within your long sample. Try both methods. The logic behind this is that long samples represent whole songs, which a DJ might want to warp, for example. Therefore, you want it "On" automatically. Conversely, a long sample might be an ambient sound effect or just something that you don't want to warp or loop. You will run into the same issues with short samples too. Live has a customizable setting for how they are handled.

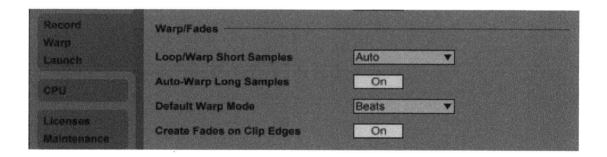

Figure 13.2
Warping
Preferences.

13.1.2 Transients

An audio file's waveform is made up of a number of transients. These transients can be thought of as the peaks amongst the valleys in the waveform. A transient is defined as the identifiable peak or surge in volume caused by a sudden impulse in the acoustic energy, such as a drum hit or intensive harmonic inflection.

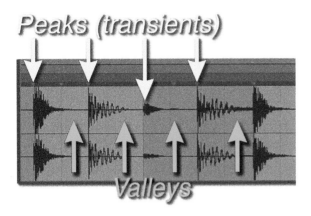

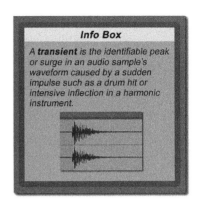

Info Box

A **transient** is the identifiable peak or surge in an audio sample's waveform caused by a sudden impulse such as a drum hit or intensive inflection in a harmonic instrument.

Figure 13.3 Transients as seen in a waveform.

The attack transient is the onset of these peaks, the start phase of each peak within an audio file most easily identified as the first sound created from a sonic event. In its simplest form, a transient could be where a note or percussive hit begins (e.g., a drumbeat or a keyboard chord). On a grander scale, it could be the periodic dynamic accents, events, or rhythmic pulses of an entire song.

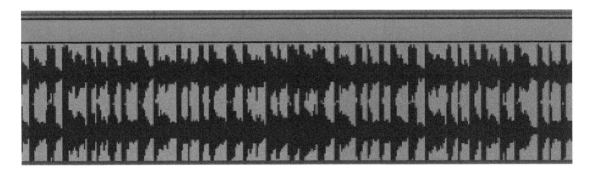

Figure 13.4 Transients of an entire song waveform.

Live's warping technology is centered on transient analysis. Transients (*Transient Marks*) are what it uses to warp audio files. When you first import an audio file, Live analyzes the waveform to determine where all of the attack transients are located. Based on this analysis, Live determines the original tempo of the audio and references it against the timeline based on the calculated attack transients.

Figure 13.5
Transient Mark in
the Sample Editor.

These transients are then marked out as little gray vertical lines along the time ruler at the top of the display. Most of the time, Live will place the Transient Marks accurately. If not, they can be moved, deleted, or added. All of these functions can be accessed under the Sample Editor's contextual menu or as keyboard shortcuts, except the move function. To move a Transient Mark, *click* it, then *hold* shift + *drag* the Transient Mark to the desired location. This will move the Transient Mark without adding a new one or moving an existing one (described next).

13.1.3 Warp Markers

When you mouse over a Transient Mark a *Pseudo Warp Marker* will appear. This looks like a gray handle or anchor. Keeping in mind that the Transient Markers represent how Live has derived the tempo of an audio file, Pseudo Warp Markers act as virtual handles for moving transients.

Figure 13.6
A Pseudo Warp
Maker appears
when the mouse
pointer is placed
over a Transient
Mark.

You can move them forward or backward from grid line to grid line along the time ruler, thus converting them into *Warp Markers*. *Double clicking* on a Pseudo Warp Marker will also create a Warp Marker. Once moved from left to right, the Pseudo Warp Marker will automatically turn yellow, so no need for *double clicking* to assign a Warp Marker. *Double-click* a Warp Marker to delete it. You should also notice the square-shaped Warp Markers—usually at the front and end of a warped sample. This indicates that they were placed automatically by Live's warp engine.

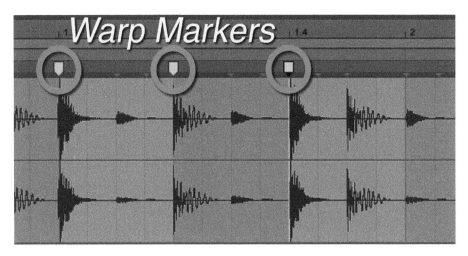

Figure 13.7
Drag or double-click a Pseudo Warp Marker to convert it into a Warp Marker.

Think of a Warp Marker as a physical anchor holding a transient or any part of an audio file in place against the timeline/grid. They are most commonly used to bind a transient to a beat or rhythmic value within a bar. Moving a Warp Marker will move its anchored point and transient. Moving one earlier in time (to the left) compresses the audio preceding it, while stretching/expanding the audio that follows it. This is because the audio will remain anchored to the closest adjacent Warp Markers. In the same way, if a Warp Marker is moved later in time (to the right), the audio preceding it will be stretched and the audio behind it will be compressed. This is the case for all Warp Markers that are between two adjacent Warp Markers: bookends, if you will. To anchor surrounding Transient Marks, hover over a Transient Mark to bring up a Pseudo Warp Marker, then *press + hold* (⌘ Mac/[ctrl] PC) and you will see the adjacent Pseudo Warp Markers appear.

Figure 13.8
Press + hold ([⌘] Mac/[ctrl] PC) to bookend a Transient Mark with Pseudo Warp Markers.

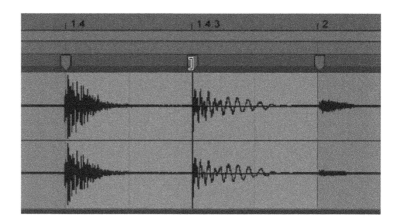

Pressing (⌘ Mac/[ctrl] PC) + double click will auto create the appropriate adjacent Warp Markers in place of the adjacent Transient Marks. You will now have three Warp Markers.

Figure 13.9
Press + hold ([⌘] Mac/[ctrl] PC) + double-click to convert the bookend Pseudo Warp Markers into Warp Markers.

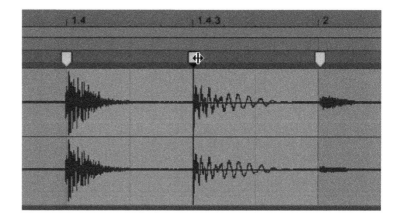

Conveniently, Warp Markers can be automatically carried over with an audio file into another Set. All you must do is save the Warp Markers to the clip. This is done in the Sample Box. *Click* the Save button and Live will save the Warp Markers in the analysis file (.asd) of that sample so they are automatically recalled when you import it into another Set.

Figure 13.10 Save Warp Markers to the sample's analysis file for recalling later.

265

13.1.4 Master versus Slave

Warping is ideal for tempo/beat matching, looping, and pitch/time-stretching audio material without a noticeable loss in fidelity. By default, all warped audio clips play in sync with your Set's tempo. When working with Arrangement clips, activating the *Tempo Master/Slave Switch* located in the selected Arrangement clip's Sample Box can change this. Initially, all Arrangement clips are set to "slave", meaning that they are warped so as to conform to the Set's tempo. When set to "Master", the Set's tempo adjusts to the clip's original tempo. For example, if the Set is at 100 BPM (beats per minute) and the clip's original tempo is 120 BPM, the set will automatically switch to play back at 120 BPM—the clip's original tempo. You can set as many warped clips to master as you wish, but whichever clip is at the bottom of the Arrangement View has priority. If for some reason you decide to delete the tempo-master clip, but want the set's tempo to remain unchanged, you must *Unslave Tempo Automation*. This is located from the Control Bar's Tempo contextual box (*right click* or *ctrl + click* on the tempo field). Once unslaved, the clip will return to slave mode and the Set's tempo will remain at the adjusted tempo.

Figure 13.11
*Tempo Master/ Slave switch for Arrangement clips only.

To get a better understanding of how flexible the Tempo Master/ Slave feature is within the Sample Box, try working with an audio clip containing sample material of live drums that are a bit loose and out of time. Switch the Slave button to Master, then unfold the Master track at the bottom of the Arrangement View. Select Song Tempo in the Control chooser and examine the fluctuations in the Song's Tempo automation lines (envelopes) and breakpoints, noting the radical changes in tempo along the timeline. Launch to ▶**Scene10** for more on automation and envelopes.

Hot Tip

Use the **Tempo Master/Slave** feature located in the Clip View Sample Box to force a Set's tempo to adjust to the tempo of a clip. This will allow for tempo fluctuations inherent of a live recording to be maintained when a clip is warped.

Fluctuations can be viewed in the Master Track as tempo automation.

13.2 Warp Modes

Warp Modes are vital to how Live algorithmically divides and stretches an audio file for warping and tempo synchronization. There are six Warp Modes, each designed specifically for different categories of audio content such as rhythmic, monophonic, polyphonic, harmonic, or melodic. These modes have been provided in order to achieve the highest quality of time-stretching possible without noticeable stretching artifacts or degradation. Choosing the best mode and setting for your audio file will help Live with transient analysis and detection.

To achieve warping, Live uses a form of what Ableton calls "granular resynthesis techniques" for some of their Warp Modes. To get into the specifics of granular

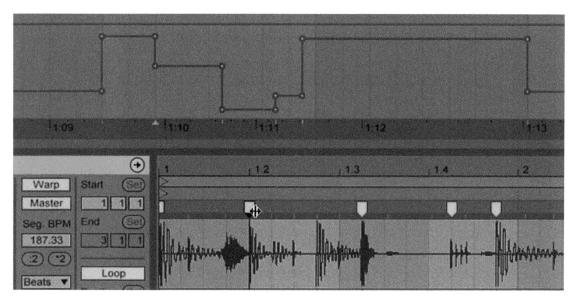

Figure 13.12 Song Tempo derived from a clip's Warp Markers that was set as Tempo Master.

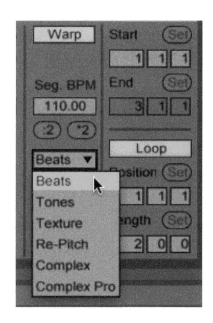

Figure 13.13
Warp Mode chooser.

synthesis is a bit much for this book, but granular means grains, like pieces of sand. Sand is made of refined earth granules. In audio, a grain is a very small duration, or segment of an audio file, often the smallest sliver/slice possible. When an audio file is broken down into these snippets of grains (slivers), it can be expanded or compressed at a micro-level. This means that segments of grains can be omitted for compressing time, or expanded like an accordion for the expansion of time, and can then be looped in rapid succession for sustainability. Granular-based Warp Modes use a unique algorithm to determine the method and quantity of grains selected, how they overlap, and how grains fade or crossfade.

Warp Modes and their associated parameters are selected from the Sample Box Warp Mode chooser. There you can assign the algorithm (method or process) that is best suited for the content of your audio clip. Each is clearly labeled in the chooser menu based on the goal of the algorithm's analysis process. There are six modes to choose from: *Beats*, *Tones*, *Texture*, *Re-Pitch*, *Complex*, and *Complex Pro*. Each mode enlists its own set of parameters and settings located below the chooser.

13.2.1 Beats Mode

Beats Mode has been designed for predominantly rhythmic and percussive content, as you might have guessed. It works especially well for beat loops and other beat-driven content. Beats Mode comes with multiple parameters: *Preserve: Granulation Resolution*, *Transient Loop Mode*, and *Transient Envelope*.

Figure 13.14
Beats Mode.

The *Preserve: Granulation Resolution* chooser offers seven different options that can determine how a sample is virtually divided into metronomic chunks that align to the tempo and grid. Generally speaking, the algorithm divides up each transient or beat division of the audio waveform into segments. This is the concept behind warping in general. The *Preserve* control determines how a sample is virtually divided and bound to the grid, the goal of which is to preserve the transients or beat divisions. When using "Transients" as the Granulation Resolution, the sample will be divided by its transients to determine how it should be warped. Resolution can also be set to a fixed note value, which forces a division upon the sample regardless of its transient content, thus preserving the beat divisions. The default setting is Transients, meaning that the time-stretching algorithm will divide the sample by its transients to warp it. This can also be set to fixed note values, as seen in the example.

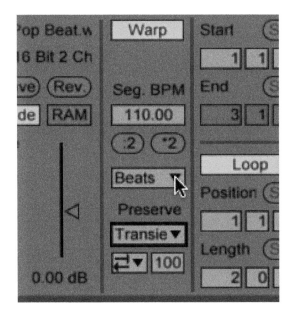

Figure 13.15
Set to Beats Mode.

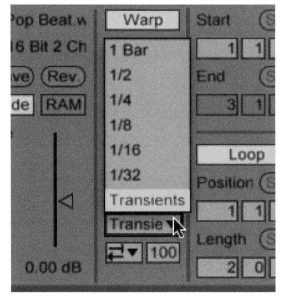

Figure 13.16
Preserve setting: Granulation Resolution - Transients.

The *Transient Loop Mode chooser*, located to the lower left below the Granulation chooser, helps to manage how the stretching algorithm handles the endpoint of each transient or division segment in the sample as determined by the Preserve setting. This sets the way in which the audio content between segments (transients/divisions) is handled when each segment reaches its endpoint. Since a warped sample has been divided and sectionalized, the warp algorithm must fill in any missing information or gaps between them, which were created when the sample was expanded (stretched). The Transient Loop Mode determines how these micro-gaps are handled by the warp engine, whether they are looped forward, back and forth, or left as is.

There are three modes that can be assigned to handle this process: *Loop Off*, *Loop Forward*, and *Loop Back-and-Forth*. When set to "Off", the audio between each divided segment will play to the end of its content, then stop. Any remaining time before the next segment will result in silence. This can sound like stutters or a gating effect during playback when the playback tempo is set to be slower than the tempo of the clip. With Loop Forward chosen, the audio will begin to loop, rather than just stop, once it reaches the end of the decay until it is time to play the next transient. With Loop Back-and-Forth the audio will start looping in the reverse direction when it reaches the endpoint of the transient decay. It will exclude the attack transient, rather than play back to the very beginning of the loop segment each time it gets to its endpoint.

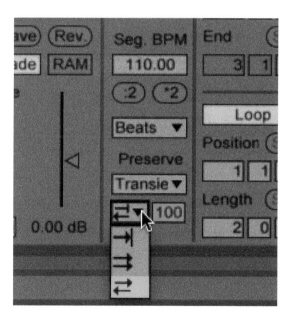

Based on the stretching percentage you apply, loop results are usually undetectable—for the most part; the Transient Loop Modes chooser provides discrete and exacting control over these gaps, giving you additional control and flexibility. To the right of this chooser is the *Transient Envelope Box*, which allows you to control the amount of fadeout at the end of each divided segment. This is theoretically designed to smooth out and minimize any potential pops, clicks, or other audible artifacts in the audio playback while applying warping to an audio sample.

Figure 13.17
Transient Loop Mode.

13.2.2 Tones Mode

Tones Mode has been designed for melodic or monophonic audio content. This includes instruments such as leads, vocals, guitars, and basses, etc. that are characterized by distinct notes/pitches. Tones mode has one parameter called *Grain Size*. This give you control over the average size of the grains used for time-stretching. The size should be based on the content of the waveform. A larger grain size can help the time-stretching algorithm smooth out stretching when pitches are somewhat blurred, but the tradeoff is an audible granulated effect—this can be undesirable.

Figure 13.18
Tones Mode.

13.2.3 Texture Mode

Texture Mode has been designed for warping audio content that is less focused around single pitches and more around polyphonic/harmonic content. This includes dense harmonic content, ambient soundscapes, sound design, and noise-based non-harmonic frequency content. There are two parameters associated with this mode: *Grain Size* to determine the size of each grain used and *Flux* (fluctuations) to generate randomness in the grain selection process.

Figure 13.19
Texture Mode.

13.2.4 Re-Pitch Mode

Re-Pitch Mode has been designed to alter the playback speed of a sample in order to sync directly with a song's tempo like a tape machine or record player. Therefore, this is a frequency-based pitch shifting effect rather than a true time-stretching effect. The slower the tempo, the lower the sample's pitch becomes as it remains relative to the tempo and vice versa. Examine this in real-time by adjusting your Set's Tempo while Re-Pitch is selected for a clip and you'll really hear a difference.

13.2.5 Complex Mode

Complex Mode has been designed for warping entire songs and tracks, and inherently consists of a multitude of musical elements and frequencies. It uses its own unique algorithm for identifying a broad spectrum of content such as melody, bass, polyphony, harmony, rhythm, ambience, or all of the above simultaneously. When using Complex Mode, keep in mind that all of its warp power comes at the expense of central processing unit (CPU) power. While this tax on horsepower may not be an incredible amount, it must be noted as such. If artifacts or other audible discrepancies are noticed during

Figure 13.20
Adjust your Set's Tempo with Re-Pitch Mode.

playback, adjust your Buffer Size in the Preferences to a larger numerical amount for the specific project at hand. Review these settings in Clip ▶ 2.3 . Unfortunately, you won't have this mode if you are using Live 9 Intro. Same goes for Complex Pro in the next section.

13.2.6 Complex Pro

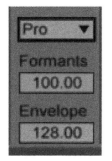

Figure 13.21
Complex Pro.

Complex Pro is a more advanced CPU intensive version of Complex Mode. It has been designed for warping the same types of audio content as Complex, but is intended to be superior to it by using "élastique Pro V2", a time-stretching engine by zplane (zplane.development). This provides sample-accurate pitch shifting designed to preserve formants. In theory, this is better for complex polyphonic material due to its ability to minimize artifacts. With Pro you have additional parameters not available in the regular mode. The *Formants Slider* helps to maintain the audio file's original tonal quality, the formants being the characteristics of a sound that are not naturally transposed. For example, you'll prevent the effects that cause "munchkinization" or that chipmunk effect in the human voice. When set to "100%", the formants should remain unchanged from their original state. This should allow a sample to undergo drastic transposition (pitch) while maintaining the tonal quality and characteristics of the sound. The *Envelope Slider* should also help to shape the spectral content of the warped sample, theoretically improving the effect warping has on the sample. You'll also find that Complex Pro can be a friendly tool for sound design situations and other audio-altering projects you may find beneficial. Complex Pro is a great warp mode enhancement for complete experimentation in addition to just warping your heavily concentrated stereo tracks.

13.3 Warping Samples

Audio files come in all shapes and sizes and are usually broken down into various samples or full-length songs and tracks. Some are short and some are long, while others are perfectly cut loops and others are not. No matter what type of audio files you are importing, they can all be warped in Live. With warping, non-looping audio files can be turned into loops and songs can be matched to sync with other beat loops or other songs. On top of that, any audio file can be quantized, adjusted, and re-grooved using Live's warp engine. When it comes to tempos and warping, Live has been designed to do a lot of the guesswork for you. It makes educated guesses and presumptions that help to guide and dictate how a sample's tempo is determined and handled within a Set. This is all a part of the analysis process, which is stored in the audio's .asd file. These files can also store clip settings for future use. If you prefer not to have Live create .asd files, then turn off this function from within the *Preferences>File/Folder tab*. The first time you import an audio file into Live, it will undergo analysis. This process will temporarily disable the file from playback. If you want to *drop* in audio files on-the-fly, you should analyze your audio content before you perform, etc. To do this, add the folder containing your audio files to Places in the Live Browser. From there, simply *click* on each audio file and they will be analyzed and then ready for immediate import, playback, and editing. If "Create Analysis Files" is active, then .asd files will appear, not in the Browser, but in the same folder as the samples.

13.3.1 Loops

Dropping perfectly cut audio loops into Live is the fastest way to get your Set grooving right away. This is because Live does a good job of analyzing short samples and loops to determine their original tempo automatically. Once a loop is imported to your Set, Live will place Warp Markers in an audio file—often just at the start and end—based on the Auto Warp settings and file length, anchoring it to the grid. The loop should then play back in sync with the Set's tempo. If the loop has been imported into a Clip Slot, it will automatically loop when launched. Occasionally Live will miscalculate an audio loop's original tempo by a factor of two, being either twice or half the correct tempo. For this reason, you can quickly double or halve the audio file's original tempo by *pressing* the *Double* or *Halve Original Tempo buttons* in the Sample Box below the "Seg. BPM" section, as seen in our example. If you know what the tempo should be, then you can type it into the input field. *Click* on input filed in the Sample Editor and then type the correct tempo into the BPM field.

When an audio loop is not cropped to a perfect loop length, then you'll have to do some adjustments in the Sample Editor, moving Sample Start and Warp Markers around to make it work perfectly as a loop. The most common edits are removing silence from the front or end of the audio file, but often you will need to isolate a loop region from within an audio file waveform. This will be the case when the sample is longer than necessary or if you want to choose a more desirable section to loop. A common case is a lead-in to a drum groove or when a sample file is not cropped. Now, before you actually start warping and fixing these minor issues, let's talk about the Warp Commands.

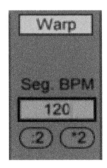

13.3.2 Contextual Warp Commands

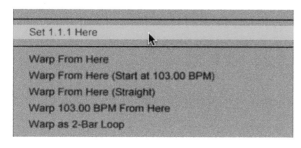

Figure 13.22
Type in the correct Tempo.

What can Warp Commands do for you? Let's say that you import a song and it plays in sync with the tempo of your set, but the downbeats are off. A quick remedy is to place a Warp Marker at the first down beat of your audio file. Then open the contextual menu from the sample editor and select *"Set 1.1.1 Here"*. Now your song should align correctly on downbeat with the Metronome. If the song has a pickup or lead-in to the downbeat—incomplete measure/bar at the beginning of the song leading into the first downbeat—move the clip's Start Marker backward to include that portion of audio.

Warp from Here

Probably your most frequently used Contextual Warp Command will be "Warp from Here". This warps all the audio from the selected Warp Marker to the end of the audio file just as you did in the previous example. To reiterate its function, let's say that you imported a song or loop and it's just not in sync. So, you will have to take matters into your own hands. It's best to listen closely and look for the downbeats of the song and identify the Transient Marks for them.

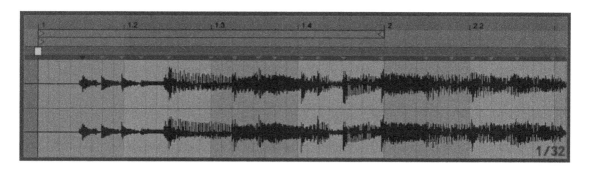

Figure13.23 Select "Set 1.1.1 Here" to align your warped sample with the grid-based downbeats.

In this example, before using the "Warp from Here" command, the first step is to align the start of the loop with "Set 1.1.1 Here" since there is silence at the front of the loop.

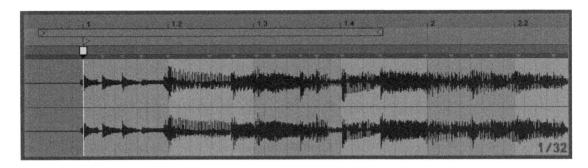

Figure13.24 Downbeat out of sync.

Now we can use the "Warp from Here" command.

Figure13.25 Looped synced.

After you have executed that command, listen to the song and see if it's in sync. If not, you may have to add and/or adjust Warp Markers. Starting from the beginning, you will want to manually add or move Warp Markers to the nearest downbeat on the grid. Once you have set the first few you can try the auto warp command "Warp from Here" from the last downbeat Warp Marker you just added or tweaked. Don't be surprised if the song drifts out of sync over time. To fix this, just find the area where it starts to drift and add/adjust the Warp Markers to the correct grid lines, then "Warp from Here" again. Keep repeating this as necessary. The good thing is that when you are all done with this, you can save the Warp Markers so you will not have to do this again.

Warp from Here (Start at ... BPM)

This command tells Live to begin the auto warp process on the file using the current tempo of your Set.

Warp from Here (Straight)

This is very similar to the previous command, except that "Straight" tells Live that the song has one consistent tempo. You might find this works better for those genres or songs that adhere to these types of tempo maps—for example, Dance, Techno, Electro, and House.

Warp ... BPM from Here

If you already know the actual tempo of your audio file, then you may choose "*Warp ... BPM from Here*". This will make Live warp the file precisely to the tempo that you have entered, in this case the exact tempo of the original recording.

Warp as ... -bar Loop

The last and probably most straightforward command is "Warp as ... -bar Loop". Live will automatically recommend a loop length based on the analysis of the file's tempo. Based on this information you can tell Live to warp the clip to the loop length it has suggested. This works well for warping perfectly cropped audio loops, especially those that extend beyond the traditional loop lengths (e.g., eight bars).

13.3.3 How to Fix "Out of Sync" Audio Files

Ok, now to fix an out of sync audio loop. To make the necessary changes you will need to identify the first downbeat and the beat where you want the sample to start.

> **Hot Tip** Keep the key command short cuts in mind while warping. They will make manipulating and warping samples more expeditious. Here are two to keep in mind: first, press + hold shift then click + drag to move Transient Marks without moving or creating Warp Markers; and, second, press + hold opt/alt then click + drag to move Warp Markers free from the grid snap.

1. *Double click* to delete the first Warp Marker that was assigned by Live automatically when you imported the clip. (Figure 13.26.)

2. Insert a new Warp Marker at the Transient Mark located on the first beat where you wish to start sample playback from. (Figure 13.27.)

3. *Right click* or *ctrl* click on the Warp Marker you just created and select "Set 1.1.1 Here".

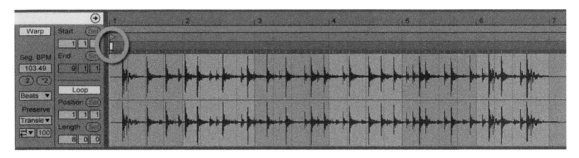

Figure 13.26 Locate and delete the first Warp Marker assigned by Live.

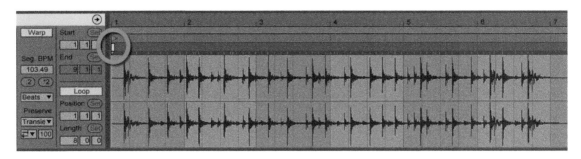

Figure 13.27 Insert new Warp Marker.

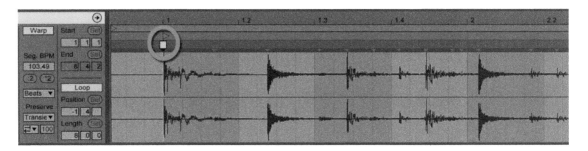

Figure 13.28 Set 1.1.1 Here.

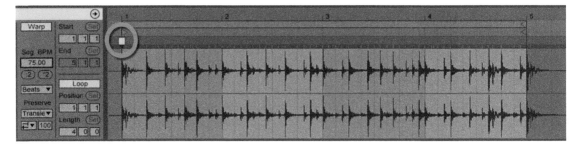

Figure 13.29 Warp from Here.

4. *Right click* or *ctrl click* again on the same Warp Marker and select "Warp from Here". Live will then re-warp the sample based on the new Sample Start point exactly where you inserted the Warp Marker. This will adjust the Sample Start and Bar 1 to be in alignment at your new Warp Marker position.

5. Adjust the Loop Brace so that it is set to cycle playback of the sample waveform as a perfect loop, choosing a specific selection that you desire. Bars 1 to 4 would be a full loop, as seen in our example. When you move the Loop End point it will move the Sample End Marker too.

6. Make sure the Loop is activated and then launch your clip. It should now play back in sync with your Set's tempo. Double-check this by turning on the Metronome. If playback sounds too slow or too fast, try doubling or halving the loop's original tempo as this is often the cause. If Live has been precise with your sample it will be a perfect loop and will show a rounded number in the Original Tempo Box. If not, then you may need to fine-tune the last Transient Mark at the End Marker because it's not exactly on the beat.

13.3.4 Songs/Tracks

Before you start warping every audio file you have on your computer, understand that there is no perfect trick for warping full songs or tracks. A long audio file/sample or song can often breathe and fluctuate rhythmically over time. That's what makes music human. In any case, some files will be quick and easy, while others will require some clever work to warp the file from start to finish. To that end, do not be misled by an algorithm and what it is supposedly designed for because you never know which one will work the best—regardless of what Live might suggest. At this point, Beats Mode is always a great option to start with. Don't hesitate to try it, just use your ear and make sure you don't start to hear artifacts or granular effects.

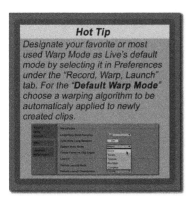

Hot Tip

Designate your favorite or most used Warp Mode as Live's default mode by selecting it in Preferences under the "Record, Warp, Launch" tab. For the "**Default Warp Mode**" choose a warping algorithm to be automaticaly applied to newly created clips.

So far we have primarily dealt with importing short audio samples or loops. So let's start talking about long audio files such as full songs and tracks. When a long audio file is imported into a Set, Live will auto-warp it under the presumption that it is a song/track or long non-loop based sample. Of course, this setting can be changed in Live's preference, but this is the default setting and will work fine for now. Now go ahead and import one of your own songs/tracks—mp3 is fine too. Play it back with the Metronome turned on. If it is warped correctly, then you're ready to keep on grooving away with your music. Unfortunately, full songs are complex and Live cannot be right all of the time. Sometimes a mismatch is as simple as the downbeat being off, while the tempo is correct. Other times mismatches are due to silence at the beginning of the file or any number of other things. In this case you'll have to make some tweaks to get the timing right. For the first example using a full song track, Live created only one Warp Marker, but the second example track was not so lucky.

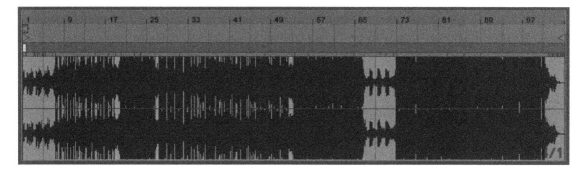

Figure 13.30 Full audio track (song) in Live's Sample Editor/Display with 1 Warp Marker.

The second track has tons of Warp Markers spattered all over the place like abstract art, which doesn't mean anything. The real issue is that Live didn't detect the Start Marker right and started late in the song, skipping a few bars at the beginning of it.

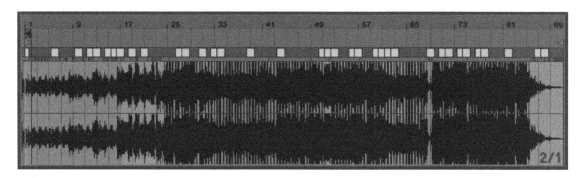

Figure 13.31 Full audio track (song) with multiple Warp Markers and errant Start Marker position.

No need to worry, this can be fixed. Remember, all of the necessary commands are available from the Sample Editor's contextual menu. Bring this up by *right clicking* or *ctrl clicking* in the Sample Editor to command all of your warp manipulations.

Now the track is in tempo, but it doesn't start at the beginning of the song … ? This is an easy fix!

1. Zoom in to the very beginning of the track and see what's going on with the Start and Warp Markers.

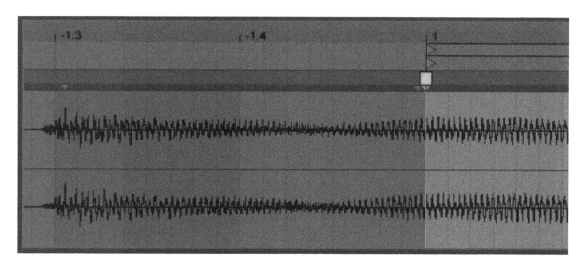

Figure 13.32
Zoom in on the Start Marker to see the correct starting position.

2. Since the song is in tempo, we'll just add a new Warp Marker where the start of the song should be.

3. Make an insertion with your mouse at the exact point, then *double click* to insert a new Warp Marker. If the grid is preventing you from selecting an exact position, then turn it off from the contextual menu Fixed Grid "Off" ([⌘+4] Mac/[ctrl+4] PC). Now you will be able to *click* and move selected Warp Marker freely left or right along the timeline. You can also override grid snapping using the *opt/alt* key.

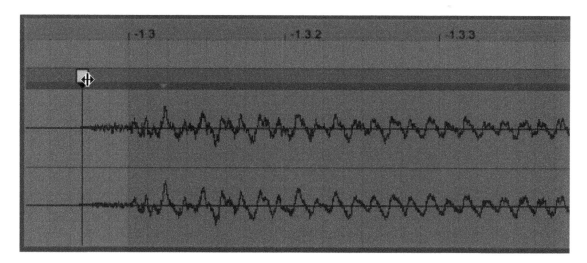

Figure 13.33 Insert and place a new Warp Marker at the start of the song.

4. Once the new Warp Marker is inserted at the beginning, select "Set 1.1.1 Here" followed by "Warp from Here". Now everything is lined up again and it plays back perfectly in sync!

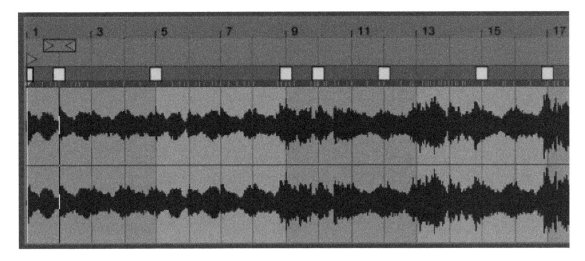

Figure 13.34 Select "Set 1.1.1 Here" then choose "Warp From Here". The correct Warp Markers should appear.

13.3.5 Adjusting Timing and Quantizing

Once upon a time there was little you could do once a performance take had been recorded to a track in the studio. The same problem existed for premade audio loop libraries, or any recording for that matter. Sure, you could cut up an audio file and tweak the timing to some extent; but the fact remained that if you wanted total flexibility with timing or correcting timing errors you had to work with MIDI. As you know, that is a thing of the past. It's now possible to edit audio recordings with the same flexibility as MIDI. That's a revolution!

There are two ways to manipulate the timing of your audio files. You can manually move the Warp Markers as we have already done, or you can let Live quantize your audio to the grid for you. Moving the Warp Markers is a great way to fix timing errors and is as easy as *grabbing* a Pseudo Warp Marker handle or Warp Marker and moving it to the correct location. Quantizing an audio sample will adjust and warp every Transient Mark of the sample into a Warp Marker, placing them on the respective grid lines based on the resolution of the Quantize Settings. To do this, select an audio loop for this demonstration. For the following example an acoustic drum loop that is not perfectly played is used. Select your clip to bring up Clip

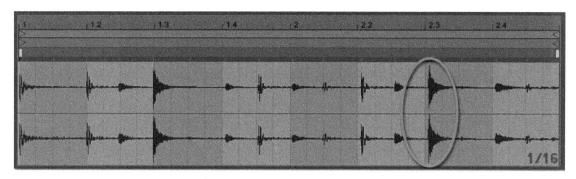

Figure 13.35 Downbeat out of sync due to drift.

View and make sure the Sample Editor is in view. From the Sample Editor you should see your audio file's waveform and its timing or tempo drift problems.

Bring up the Sample Editor's contextual menu (*right click* or *ctrl + click*) and select "Quantize Settings … " ([Shift+⌘+U] Mac/[Shift+ctrl+U] PC).

Figure 13.36 Quantize Settings Menu.

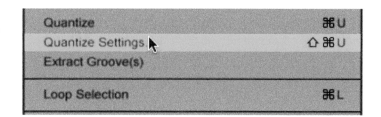

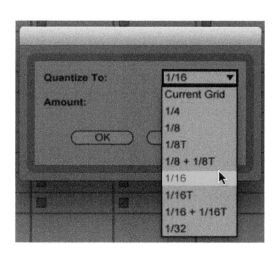

Figure 13.37 Choose a resolution based on your audio's smallest beat subdivision.

This will bring up a new menu where you can choose the desired quantization settings and the amount (quantization strength) for quantizing your sample. You could also choose "Quantize" to skip over the settings options. Keep in mind the smallest subdivision you'll need for your sample and how close to perfect you want it to be align to the grid, 100 percent being max strength. Just because you have this power of control over timing doesn't mean you should always use it. Quantization takes the human feel out of live recordings or samples, etc. It is best to use this feature only when absolutely necessary or when the genre calls for it, such as various types of electronic music where a precise robotic-like timing is desirable.

After you select your quantize resolution, press "OK" and you will see in the Sample Editor a perfectly quantized sample.

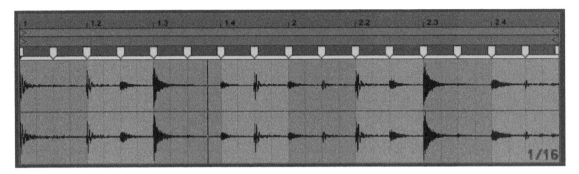

Figure 13.38 Quantized audio with Warp Markers.

Hot Tip Quantizing your audio may take a little getting used to at first. To help you wrap your head around this whole concept, keep in mind that the Grid alignment in the Sample Editor plays an important part in the quantization process. Take some time to experiment with the various Grid resolutions to quantize your audio's timing and feel; you may, in fact, be surprised with the possible results.

13.4 Musical Concepts

The continual development and improvement of time-stretching algorithms and techniques to manipulate and quantize audio seems to be a never-ending quest for software developers. With every new development in this area, we get closer and closer to effortless and undetected pitch and time manipulation of audio, just as if it were MIDI. With that in mind you now have the ability to warp audio for any number of reasons. Whatever the reason, think about how much easier it can be when you want to clean up the timing of a not so great recording session, or to put that drummer or bass right in the pocket, with the rest of the band's multi-tracks to follow suit. So many great and flexibility options with audio! Now that you have all the warping features at your disposal, you have to decide how you want to make your tracks groove. How about this? Try quantizing your audio in real-time. Say what? Read on and you'll see!

13.4.1 Produce: Quantize Audio in Real-Time

You can adjust your quantize settings easily in real-time using ([shift+⌘+U] Mac/[shift+ctrl+U] PC). This will bring up the quantize window allowing small to large adjustments in your quantizing resolutions and strength. Aside from changing the percentage of strength, one trick is to make these adjustments based on altering and changing the Grid resolution within the Sample Editor. To do this, set *Quantize To* to "Current Grid". A narrower Grid resolution will allow the Quantize setting strength to latch to the closer grid lines along the timeline. A wider grid will allow for a more general snapping of your Warp Markers once engaged. Keep this in mind when using the Quantize Warp function in the contextual menu since it can really make a difference in how this function works with your Sets. Speed up the changing of grid resolutions by utilizing key commands ([⌘+1, +2, +3 +4] Mac/[ctrl+1, +2, +3 +4] PC). This allows for a more intuitive workflow.

Loops, Slicing, and More Looping

14.1 Loops Demystified

Much of computer-based music production is centered on loops and various loop technologies; therefore, it is important to shed some light on the topic as well as dispel misunderstandings about them. Loops serve as a compositional tool and foundation for many composers, providing rhythmic or melodic material that they otherwise would not have access to or the ability to create. Well, at least the time to create anyway. Don't be fooled by the title "computer-based music". Virtually all produced music spends time "in the box", meaning that a Digital Audio Workstation (DAW) and all of its virtual instruments and effects are the backbone of studio recording and music production regardless of genre—yes, even for the analog purists (are we still having that debate?). That being said, acoustic-based singer/songwriters and performers will often use loops, whether or not they end up in the final mix or not. Many producers create their hooks in the studio by recording a live band or instrument(s) to the playback of pre-recorded loops or build loops out of live recorded material for either use in their current production and for other studio musicians to record along to. Frequently, the backing track becomes a groove that gets looped and even improved over!

Regardless of how or why you use loops, they are a fantastic resource for composers, producers, and performers alike. The most important thing is that you understand the concept of loops, what type of looping technology is available to you, and how to use them effectively. The word "loops" doesn't mean repetitive music, cheating, or unoriginal—rather, the contrary. With the current loop technologies and programs such as Live, loops can be as original and creative as any other recorded instrument or performance. In all actuality, they are malleable, flexible, and completely customizable, not to mention performed by real musicians in the first place. Think of loops as a means to an end, an instrument to create, produce, and perform with.

14.2 REX Loops

We briefly discussed REX formatted (.rx2) loops in Clip ▶ **13.1** . They are created with Propellerhead's Recycle, but REX loops can also be purchased from third-party developers as loop bundles of sorts. As a quick review, REX files are created from traditional audio files and are intended to be looped. They can be perfectly cropped to a loop/bar length, or not: it doesn't really matter. Recycle will handle both types; it will just be your job to cut and create the perfect loop out of the material you feed ReCycle. The process is fairly simple, especially considering that we've been talking a lot about Transients and Warp Markers, which is very similar to how ReCycle interprets audio samples and loops.

REX loops are created by establishing a start and endpoint within an audio file/sample to create a loop region, then by slicing the audio file/region at each transient. This cuts the audio loop into little chunks that become anchored by the slices, similar to the way in which Warp Markers anchor Transients in Live. The slices are locked to a time-based grid, allowing the audio to be adjusted to any tempo without changing the pitch or reasonable loss of sound quality. This will be apparent in the example showing a loop sliced up in ReCycle.

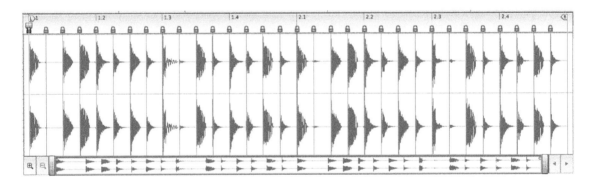

Figure 14.1 Sliced up "Rexed" audio loop in Recycle.

That's not where the ReCycle technology stops. REX files also carry a MIDI component that allows the slices to be mapped individually to the MIDI keyboard triggered in a chromatic note sequence. This allows for some really creative and flexible ways to manipulate audio loops similar to how Live slices to new MIDI ▶ **14.3** , but we'll save that for later. Now that you know what REX files are, let's look at how they translate and can be used in Live. You will find REX files on the companion website to use throughout this Scene. Download "**CPP_REXLoops**" and use them as you follow along—maybe get a bit creative along the way.

14.2.1 REX Mode

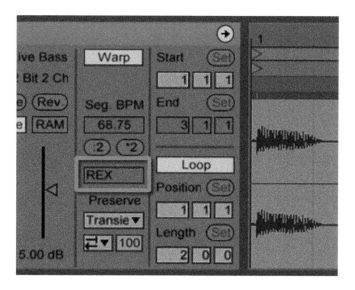

Figure 14.2 REX Mode.

Upon import of a .rx2 file (REX), Live will automatically run the clip in REX Mode. This is unique to REX files and cannot be activated manually. Live handles REX files with ease since all of the calculations, tempo, and length information has been predetermined and stored in the .rx2 file. Looking at the Clip View Sample Display/Editor you will see that Live has quickly translated and interpreted the loop perfectly with Transient Markers. If you were to open the loop in ReCycle, you would see that the Transient Markers in Live are the same as the slices in ReCycle. Just like Live's Warp feature, these loops sync perfectly to your Set's tempo and can be sped up, slowed, and transposed.

It should be pointed out that Warp Makers and any warp-associated parameters are unavailable in REX mode, other than the Granulation Resolution, so you're stuck with the REX slice programming as is. At the end of the day, this is only one dimension of working with REX files in Live. For more flexibility and power, you can map the individual slices—in this case, Transient Markers—to a chromatic MIDI note scale! Let's go over this concept next.

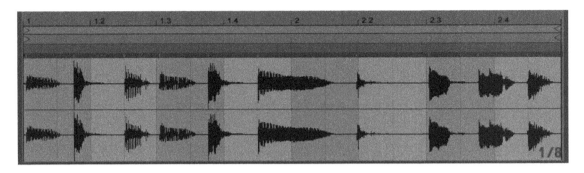

Figure 14.3 Transient Markers in Live.

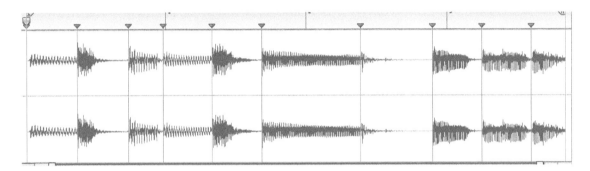

Figure 14.4 Slices placed identical to Live's Transient Marks.

14.3 Slice to New MIDI Track

As you already know, Live provides you with the ability to manipulate, quantize, and customize the transients and rhythms of audio files through Warp. Taking it a step further, there is another method for manipulating audio loops: *Slice to New MIDI Track*. With this feature you can instantly turn an audio file into controllable playable MIDI notes based on its transients. Any audio file is eligible to be sliced, including REX files. Once transients are sliced and assigned to MIDI notes, you can then re-sequence them, trigger only specific slices, or create custom rhythms. With the ability for an audio loop to be triggered via MIDI, you have the ability to unleash the true creative powers of audio loops, all integrated right into Live. In the following example you will see how flexible slicing to MIDI can be. The concept is loosely derived from ReCycle and REX technology, but

Hot Tip

*Quickly turn a **REX Mode** clip into a warped audio clip. Freeze a REX clip and then drag it to a new Clip Slot on an audio track. Now you'll be able to access its Transient Marks and warp it just like any other Live audio clip.*

288

Figure 14.5
Original sliced
MIDI clip.

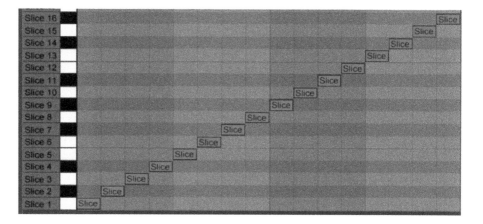

Figure 14.6
Edited MIDI slices.

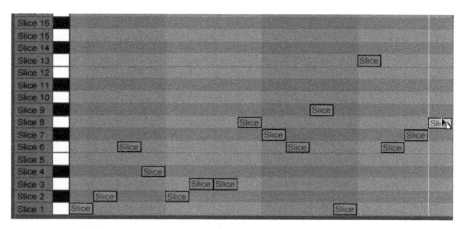

Figure 14.7
Edited MIDI slices.

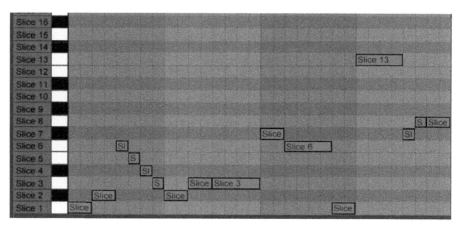

you will also find this feature in other popular software such as Propellerhead's Reason, Spectrasonic's Omnisphere, Logic's Ultrabeat, Native Instruments' Kontakt, and more—that's how important loops are to music production!

14.3.1 Audio Loop Slicing

Fortunately, you don't have to rely on ReCycle or any other third-party programs in order to slice up your loops to MIDI. Live makes it possible to turn any audio loop into the equivalent of a REX-type loop and with great flexibility. Using a traditional non-REX audio loop (.wav, .aiff, .mp3, etc.) let's slice it to a new MIDI track. To begin, locate an audio loop on your computer. This is where adding a custom folder to Places becomes helpful:

1. Import your audio loop into an empty Clip Slot or as a new audio track in the Session View.

2. Launch your new clip to ensure that your audio loop is warped correctly. This means that the best Warp Mode has been chosen and that the Transient/Warp Markers are positioned correctly.

3. Bring up the clip's or Sample Editor's contextual menu (*right click or ctrl + click*) and select "Slice to New MIDI Track". This will open up the Slicing options for the audio loop.

4. Choose a "Slicing Preset". Live has provided a number of slicing presets with some very cool effects. For now just choose "Built-in", then select how you want each beat to be sliced. "Create one slice per … "; use "Transient", then *click* "OK". This will create a new MIDI track and Drum Rack containing all of the audio file's slices based on the Transient Markers assigned to the audio loop. Launch to ▶Scene16 for more on Device Racks.

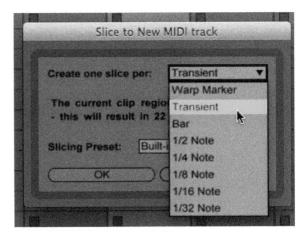

Figure 14.8
Slicing options.

5. Launch the new MIDI clip that was created. It should sound virtually the same as the warped audio loop you imported at the start. Remember to mute your original track source that you sliced from. It's easy to forget sometimes and you'll hear both tracks playing at once if you're not careful.

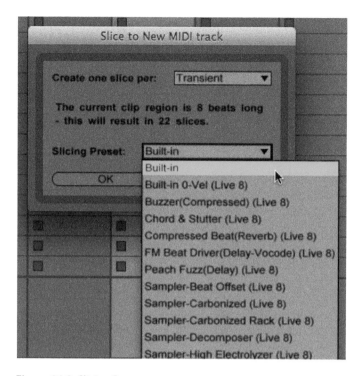

Figure 14.9 Slicing Presets.

That's it! You've successfully sliced your audio loop to MIDI and can now manipulate it just like a MIDI clip and beyond. If you look at the MIDI clip, you will see that Live sliced it based on the preset you chose. For example, if you choose to use a beat value as your slice resolution, then you will see that slices converted to MIDI notes in the MIDI Note Editor are placed at that resolution, such as a note on every 16th note. Don't forget that any audio file can be sliced and re-sequenced. As a tip, depending upon the nature of the loop and your plans for manipulation, you might decide to pre-warp using custom Warp Markers, quantize, or use groove to adjust the timing before you slice to MIDI. This can make it easier to re-sequence some loops.

Now's a good time to try slicing loops on your own and experiment with each different Slicing Preset. It can be a lot of fun! For more fun, check out the *Convert Audio To* feature for more ways to exploit audio loops! Launch to ▶ 8.3.2 .

14.3.2 REX Loop Slicing

Slicing a REX loop to MIDI is a no different than standard audio loops. It's a simple task with a ton of benefits. All you need is a REX formatted file, which can be obtained from the companion website in .rx2 format. Download and locate the REX loops. Now let's begin to slice and dice:

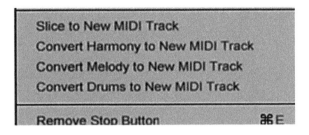

Figure 14.10 Slice to New MIDI command.

1. Import your .rx2 loop into the Session View. You can do this by *double clicking* on it in the Browser if you've added a folder or *drag* it in from your Finder/Explorer.

2. Bring up the Sample Editor's contextual menu (*right click or ctrl + click*) and select "Slice to New MIDI Track".

3. This will bring up a new menu telling you how Live will slice the REX file. Live will slice your REX file based on the slice information that was translated from the file to REX mode. Slices will be a derivative of the loop's length in beats.

4. Now choose a Slicing Preset. To keep it simple, just choose "Built-in". This will create a new MIDI track and Drum Rack containing all of your slices. (Figure 4.12.)

Figure 14.11 Menu shows how many slices and provides a Slicing Preset chooser.

5. Launch your new MIDI clip. It should sound virtually the same as the REX loop clip.

6. Unfold the new MIDI track by *clicking* the "Chain Mixer Fold button" from the track's title bar. With the I/O section and Mixer in view, you will be able to see how Live has sliced and routed the slices to each MIDI note using *chains* (Rack Chains) along with associated mixing parameters. (Figure 4.13.)

7. Each one has been mapped starting at MIDI note C1 and ascending chromatically until the last slice. Select your new MIDI clip generated by the slice command and you will see how the MIDI notes have been sequenced chromatically. Also notice that the note positions and lengths are relative to the slice/transient positions and lengths.

Figure 4.12
(facing page, top)
New MIDI track and Drum Rack containing the new MIDI slices assigned to its drum pads.

Figure 4.13
(facing page, bottom) MIDI slice (note) Rack Chains.

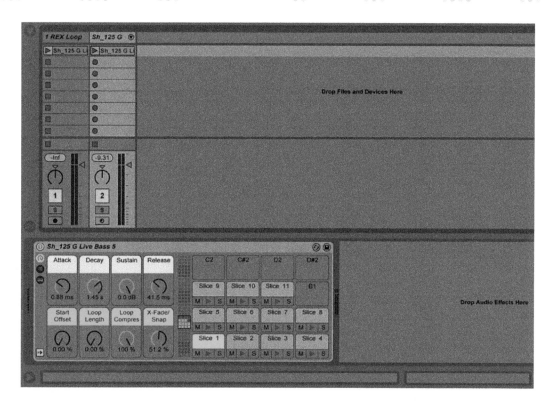

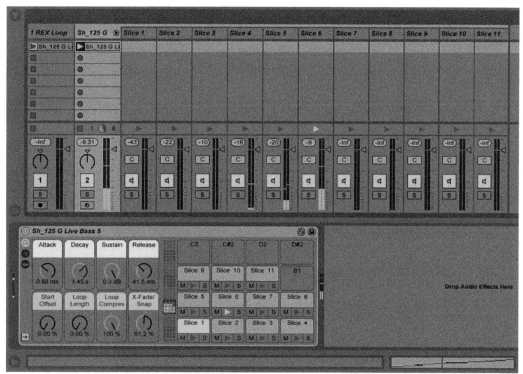

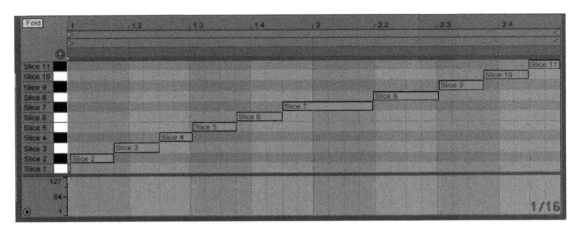

Figure 14.14 Chromatic note slices in the MIDI Note Editor.

That's it. You have successfully sliced your REX file to a new MIDI track. Of course, there is much more to learn about tweaking your new Drum Rack Device and all of its Macros. Launch to ▶ **16.3** for more on Drum Racks. For now, understand that Live creates a Drum Rack regardless of the loop content—for example, a melodic instrument loop versus drum loop. Therefore, your REX file MIDI slices will be triggered via Drum Rack virtual drum Pads.

14.3.3 Re-sequencing

Now it is time to begin customizing your loops, which is why we sliced them to MIDI tracks in the first place. It is all about the ability to liberate yourself from the original patterns of your loops. If you haven't noticed already, editing the loop's MIDI note durations, positions, and velocities affect the once cemented audio loops. To re-sequence your loops, go to the MIDI clip's Note Editor and alter the MIDI notes as you wish, as shown in the examples in Clip ▶ **14.3** . You might change the beats around for altering the feel. You could also delete or repeat notes to add fills and variations.

Remember that Live allows you to do all this in real-time, while everything is playing and looping. Customizing the sequence breathes freshness, eliminates repetition, and allows you to manipulate the audio as MIDI. Think about what you can now do if you start adding MIDI effects or slice the loops up with a creative preset! Or better yet, how about slicing a solo vocal or guitar phrase? There is great potential here for complete control over your audio.

> **Hot Tip**
>
> Create that cool **buzzing/stutter effect!** Use various Grid resolutions when editing newly sliced Audio-to-MIDI notes in the MIDI Note Editor. With a tighter grid (higher resolution), try duplicating several notes consecutively and vary velocities for groups of notes. You'll quickly see that mixing up the grid resolutions for adding and subtracting notes in the Note Editor opens up some very creative possibilities.

Hot Tip Try changing the Grid to Triplet Grid ([⌘+3] Mac/[ctrl+3] PC). This gives you the advantage of adjusting notes to a new start point on the Grid and syncopating notes for a completely different feel.

14.4 Working with Loops

You've just seen what is possible with audio loops, so let's look at how to work with them in Live. Since loops can be played in the Session View and Arrangement View, we'll go over how they are used in both and in what ways they differ.

Session clips essentially function as loops. Of course, they don't have to, but for discussion sake let's stick to that assumption. They repeat infinitely unless you do something to stop them. Well, that's great, but you wouldn't want them to loop without any variation over time. That's boring and uninspired. For that reason, it is necessary to find ways to vary them. How do you do that? On a basic level, while building out your Set and musical ideas, you most likely will copy clips and paste them into consecutive slots, one after the other to create multiple scenes. Why not use this as an opportunity to add variation to your clips by altering each duplicate to be a variation of the previous loop clip? In this way, you can alternate between them as you wish, keeping things interesting. One especially daring way to bring variety to your duplicated Session clips is to create *Follow Action groups*. This allows you to predesign how many times a loop will cycle before automatically executing an action of sorts, such as advancing to the next clip, or randomly jumping around. This automated process resembles that of a fixed timeline approach that you might have created in Arrangement View, but instead used Session View. The goal here is to create variations over time to keep your loops fresh. Each clip in the Follow Action group could be a segment of the previous clip, an intro, break, fill, or simply an effectual variation of the original. Launch to ▶ 7.10.1 for more on Follow Actions. If you're still looking for ways to mix things up, another consideration for audio loops is to slice them to MIDI, as talked about earlier. Try re-sequencing and adding MIDI Effects to create musical variety to potentially static loops.

14.5 Looping in the Arrangement View

The beauty of clips is that each one is independent of the other and for that reason loops can be changed and customized so that your music stays fresh and your workflow is never interrupted. Therefore, the important concept of loop

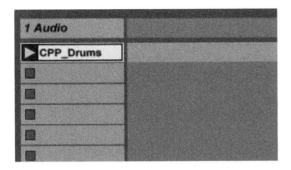

Figure 14.15 Session clip launched and playing back.

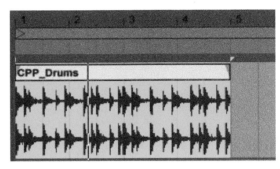

Figure 14.16 Arrangement clip playing back (Linear).

manipulations and variations should most definitely be carried over to your arrangements in the Arrangement View.

Here in the Arrangement View, you can tweak and customize your clips just like you did in the Session View, but now you have the ability to edit them as whole or as partial clips, varying them along the timeline. This then brings us to some physical operations that must be familiarized in order to create and vary loops in the Arrangement View. Most of this information you may be somewhat familiar with as it relates to the actual process of looping and editing clips.

Working in the Session View should be incredibly creative and a ton of fun, but you'll eventually work in the Arrangement View to edit, enhance, mix, and remix your performances. Part of this process involves looping and manipulating your clips. There are a few different approaches to looping clips in the Arrangement. First of all, you can copy and paste clips one after another, or use the duplicate command ([⌘+D] Mac/[ctrl+D] PC), resulting in the same outcome.

Figure 14.17 Duplicate clips along the timeline.

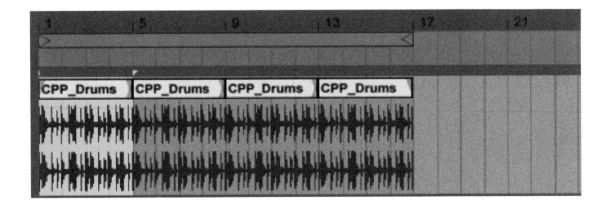

Once you have a number of loops laid out along the timeline, you can make subtle changes to each clip to help evolve your music over time. When, where, and how much evolution will be up to you and the structure your song dictates. Use the traditional keyboard shortcuts to copy and paste or to duplicate. Another way to loop your clips in the Arrangement View is to extend the clip from either edge by *dragging* its length out for as far as you want it to loop. When the Loop Switch is active, Live will automatically draw out the clip as a single looped clip without breaks. What is unique about this is that the clip is still interpreted as a single clip even though it has been extended beyond its original length. Essentially, Live references the original clip, as seen in Clip View, but then repeats it in succession for as long as you extend it. This provides a quick and efficient way to extend a loop over a selected timespan while minimizing added strain on your computer hard drive or system speed, as Live is not actually building, consolidating, or recording new physical regions of audio to disk. There are a few innate limitations to this approach in that any adjustments or changes to the clip will affect each loop cycle of the looping clip. For this reason, event variations should be made by duplicating, copying, or splitting a clip(s). In addition, consider using Clip Envelope automation for altering each duplicate, copied, or split-up clip. Launch to ▶ **10.3** for a refresher on Clip Envelopes and automation.

14.6 Loops with Unlinked Clip Envelopes

Creative variations can be achieved with even a single looped clip by using *Unlinked Clip Envelopes*. This creates progressive variation over each loop cycle of a clip, which can be customized to taste. The concept is that a clip can loop indefinitely while changing over time. This can be done with any clip in both the Session and Arrangement View. The difference between this method and duplicating/copying clips is that envelope-based variations are made within a single looping clip as opposed to multiple variations of a clip.

Take a moment to walk through this exercise and see how looping with Unlinked Clip Envelopes is done. You can use one of the audio loops from the companion website or use your own for this:

1. Using a clip that is two bars long, open its Envelope Box/Editor from Clip View.

2. Choose "Volume Modulation" from the clip Control Chooser. Notice the pink dashed envelope line at the top of the Sample Editor that you'll be working with.

3. **Unlink Clip Envelope:** *click* on the yellow "Linked" button at the bottom left of the Envelopes Box to switch to "Unlinked". The button should turn orange. (Figure 14.18.)

4. With Volume listed in the Control chooser, extend the Loop Brace in the Envelope Editor to four bars (1 to 4) or type in "4" in the *Loop Length* input field at the bottom right of the Envelope Box, if not already. Make sure to zoom out to see all four bars. (Figure 14.19.)

5. *Click* on the envelope at bar 3 to insert a breakpoint. This will anchor the volume automation value at the end of the second bar.

6. Create another breakpoint at the end of bar 4 (start of bar 5), then *click + hold + drag* this breakpoint downward to create a gradual fadeout with the envelope. You can also make this a sharp cut in volume by highlighting the entire range from the end of bar 2 to the end of bar 4 (bars 3 to 4) and *dragging* the envelope downward. (Figure 14.20.)

7. Play back your clip and listen to the results.

Now that you have completed this exercise, check your Set against the example Set. Download "**CPP_14–6_Unlinked-ClipEnvelopes Project**" and see if they sound the same.

There are many uses for this feature. On a basic level, it allows you to use a single looped clip and have it driven by unlinked envelope loop automation in order to customize its playback over time. Use it with any of your clips. Another interesting way to take advantage of unlinked envelope automation is to split a longer MIDI clip sequence in the Arrangement View into multiple clip segments for affecting each clip segment differently. Just split a clip as many times as you like at various points where you want to alter the automation effect or event change. You can then apply a different envelope effect to each new clip that is created. Make an insertion with the mouse pointer where you want to split the clip into a new clip, then *press* ([⌘+E] Mac/[ctrl+E] PC) or choose "Split" from the Edit Menu.

14.7 Looping Concepts

Hopefully by now it has become obvious that loops are very powerful tools for a musician. Love them or hate them, they are here to stay and there is no reason why a digital composer, producer, or performer should go without. Regardless of whether they end up in a final mix or not, they're an invaluable resource, especially since they are so malleable in Live. At the least, a digital composer will and should

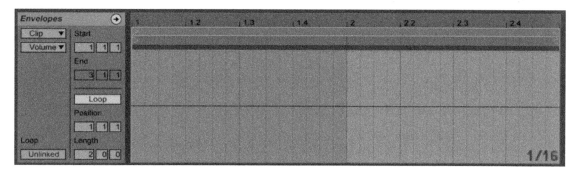

Figure 14.18 Unlink Clip Envelope.

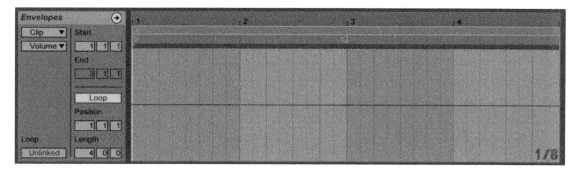

Figure 14.19 Unlinked Clip Envelope.

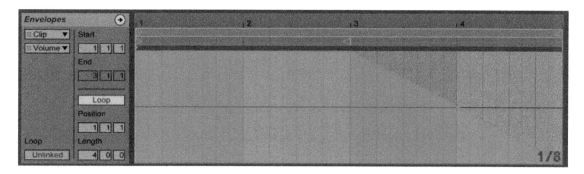

Figure 14.20 Unlinked Envelope with fade out starting at bar 3.

at some point make their own if they don't use a third-party loop library. Look at loops as a challenge. Some of the greatest digital composers are those who not only have the resources, but can also create unique sequences from the simplest of loops.

14.7.1 Create: Custom Loop Library

Creating your own custom library of loops is easy and satisfying. Once you begin to fine-tune your skills of slicing your loops with the Slice to New MIDI Track feature, you'll probably want to start archiving or saving your work. You can keep these as sliced MIDI clips or resave them as audio clips. They both have their advantages; but for a moment consider converting your newly sliced and edited MIDI clips back to audio to save in the Library for future use. Here's how.

Create a new audio track next to your sliced MIDI track. Select the Freeze Track function for the Sliced MIDI track, then *drag* the frozen Sliced MIDI clip to your new empty audio track's Clip Slots directly to the right. Presto, there you have it,

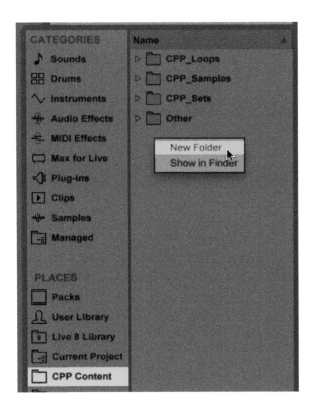

Figure 14.21
Create a new folder for your custom clips. Right-click or ctrl + click on a current folder or in the area just outside of a folder to open the contextual menu.

a new custom audio clip (preset) that is looped and contains all of your hard work. Now it's ready to be saved and archived in the User Library as a Live Clip. *Drag* it your User Library folder or any added folder or elsewhere on your computer.

This will keep things organized. Follow this same procedure for saving your sliced MIDI clip as a preset (Live Clip) instead of converting it back to audio. Use this functionality to build up your arsenal of custom sliced loops. Launch to ▶ **15.2.4** for more on custom presets.

SCENE 15

Instruments and Effects

15.1 Introduction to Live Devices

Software instruments and effects are the backbone of the Digital Audio Workstation (DAW). Whereas the project studio was once crowded with racks and racks of outboard gear, hardware samplers, synth keyboards, sound modules, reverb units, and drum machines, it's more likely now that your whole production studio fits in your backpack or at the most takes up a small corner in your bedroom, just you, your computer, an audio interface, and your MIDI controller. By no means should you abandon all analog devices or reduce your project studio down to a shoulder bag, but trends have led us to the point where there are many amazing software-based sample libraries, virtual instruments, and every imaginable digital signal processor (DSP)-powered effect plug-in you can imagine. Some of you have never even handled outboard gear except at the used and discounted section of your local pro audio store. That being said, don't go running out and trading in all your hardware—not that anyone would take it; but you should consider buying at least some more RAM and a large capacity hard drive! While you're at it, don't forget to pick up or upgrade to Live 9 Suite and download a few Packs too!

All right, enough of the philosophy. A good DAW is the sum of its parts, so what good is it if you don't have anything to make music with? It's paramount that you have a useful toolbox of virtual instruments and effect plug-ins and Live is in no short supply of these, especially if you own Suite. As you have seen throughout the many examples and exercises so far, Live Devices (instruments and effects) are always accessible and never require your music and workflow to stop. Now all you have to do is master them.

15.2 Working with Live Devices

From the Live Browser you can access all of your Live *Instruments*, *Audio Effects*, and *MIDI Effects*, which include a nice variety of factory presets ▶ 2.4 .

Figure 15.1
Live Devices.

Between Live 9 and Live 9 Suite, all of the essential instruments and effects you'll need to create, produce, and perform are right there in front of you. Keep in mind that Suite is Ableton's premier package of instruments and effects bundled with Live 9. If you haven't already, you will eventually want to upgrade to Suite. For a detailed overview of Suite, launch to ▶ **Web** . In the meantime, check out Ableton's website and grab any free Packs you can find! It never hurts to have more free samples and presets (www.ableton.com/en/packs). Review installing Packs ▶ **3.4** .

15.2.1 Overview of Devices

Here are the basic guidelines for working with Live instruments and effects:

Instruments: receive MIDI note information through MIDI tracks and convert that data to audio, which outputs via the MIDI track that the instrument is placed on. For this reason, they cannot be inserted on audio tracks, Returns, or the Master track. Instruments can only be inserted on MIDI tracks ▶ **3.4** .

MIDI Effects: work exclusively with MIDI notes and can only be inserted on a MIDI track. They generate some creative performance-based manipulations, affecting how MIDI notes are interpreted and distributed prior to being converted into audio signals. They are inserted first in a MIDI signal chain, before an instrument in the chain. MIDI effects are inspiring devices that literally manipulate and transform your music into dynamically complex and rich performances—a great way to start your day! ▶ **15.4**

Audio Effects: can be hosted by any track type, even MIDI tracks, so long as they are inserted after the instrument ▶ **15.5** . Located under Categories in the Browser are tons of unique Audio Effects, from the classic tracking, mixing, and mastering processors to wildly manipulative effects. Combine these with Device Racks and the possibilities are endless ▶ **16.4.2** .

There are three basic ways to insert a device into a track as long as the track type supports the type of device being inserted:

1. *Double click* on the device name in the Browser, which creates a new track.
2. Pre-select a track, then *double click* on a device or *press* the return/enter key on your computer keyboard. The device will be inserted on that track.
3. *Drag* and *drop* the device to a track, device Drop Area, or to the Device View. Use any one of these methods in both the Session and Arrangement View.

15.2.2 Device View

Figure 15.2 Shown here is the Operator instrument.

Device View is where you'll interact with an instrument and effects user interfaces. As you've seen already, Device View shares Detail View with Clip View, at the bottom of the main Live screen. Toggle back and forth with the Device View Selector or *double click* on a track's title bar to bring it up. Feel free to insert devices directly to Device View by *dragging* and *dropping*. There are many ways to insert devices, so don't feel obligated to do it in any one way. To remove a device, *click* on its Title Bar and then *press* the delete/backspace key on your computer keyboard.

> **Hot Tip** *Use the keyboard shortcut ([shift +Tab] Mac/[shift +Tab] PC) to quickly alternate between Device View and the Clip View.*

Since Live handles clutter and space exceptionally well, make more room on your screen by hiding the entire Device View, which as a whole is called Detail View. Just *click* on the *Show/Hide Detail View* at the bottom right of the main Live screen and it will be tucked and folded away.

Figure 15.3 Show/Hide Detail View and Device View Selector.

Every device has a *Device Activator*, *Hot-Swap*, and *Save Preset* Button on its Title Bar. To temporarily bypass a device, *click* on its Device Activator Button 🔘 in the upper left-hand corner of the interface (yellow = On). On the opposite side of the Title Bar to the far right is the Hot-Swap Button 🔄. This is used to quickly audition effects in the same way you would for all Hot-Swap Buttons. Next to Hot-Swap is Save Preset 💾. You can save your device settings as a preset in the User Library. Some devices will have an additional *Parameter Fold/Unfold Button* adjacent to the Device Activator. This will be on a per device basis.

On the left and right edges of every device is a Level Meter (input and output). These will display the active signal flow (left to right), lighting up as MIDI or audio signals pass through a device or between devices in a chain (from device to device). They also light up when a device is receiving an input signal (audio or MIDI). Level Meters are very helpful for identifying if and when a signal is reaching a device in the chain. Note that audio and MIDI signals are each represented through a different type of Level Meter indicating the difference in signal.

Figure 15.4 General Device Layout.

Figure 15.5 Audio effect device Level Meter indicating input and output.

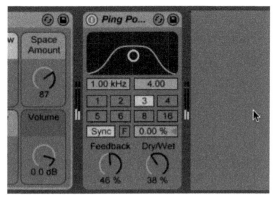

Figure 15.6 Instrument device Level Meter indicating input.

Depending upon the device, there may also be a *Sidechain Toggle* located next to the Device Activator Button (i.e., Compressor). Not all devices have built-in sidechain controls. *Click* the toggle to unfold a device's sidechain parameters/controls when applicable. From there you will be able to activate a sidechain, choose an input source, and set specific parameters. Launch to ▶ **15.7.3** for more on Sidechaining.

15.2.3 Device Presets

Presets store device parameters and settings. They are valuable tools for everyone, even if you think otherwise. Someone spent a lot of time building and refining them, so you might as well see what they have to offer. You'll find that they can be a great starting point and time saver. Presets are located in the device's respective folders. To load a preset, *double click* or *drag* and *drop*; it's all the same. One of the best ways to navigate presets is to load up any one to begin, and then *click* on the Hot-Swap button to start trying them all out. This is even a great way to add any instrument. Sometimes you may not feel like scrolling through the instruments or won't remember where your device came from. All you have to do is *click* the Hot-Swap button and you're instantly taken to the exact folder location of the device you are currently using. Use your mouse or computer arrow keys to toggle through presets, using the return/enter key to load a preset. If you have a scrollable mouse, then you can move the Browser content up and down that way.

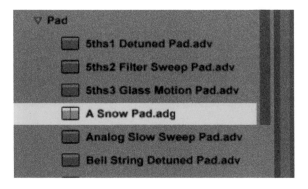

Figure 15.7 Device Presets.

15.2.4 Customizing Presets

There are a lot of great Live device presets to get you started making music, but what is really exciting is the intuitive presets functions for loading, customizing, and saving presets. One might assume that presets have limited functionality; simply load a preset and then hit the save button if you want to recall your favorite setting later. While this is true to some extent, customizing and saving presets the "Ableton way" goes quite a bit further than the basics. In Live you can *drag* your device to the User Library, giving your custom device a dedicated home away from the factory presets in the Browser. This is convenient because most of you are used to the standard save preset command where your preset becomes a lost name in a dropdown menu or saved to some special hard-to-find location. In Live it's all about *drag* and *drop* to a location. That's all a standard feature for the dynamic Live Browser and the User Library is designed to retain all of your custom settings as presets and it can do it in real-time.

Hot Tip

Create **custom device presets**. Click the Save Preset button 💾 or drag a device directly to the Browser. This way you will be able to recall the exact settings for any Live Set!

Create an "**all-in-one clip preset**". Drag a clip to Places to save its performance data and its associated device. This way you will be able to recall the clip's performance, device, and effects by dragging the **Live Clip** into any Set.

There are a few options for saving presets. You can *click* on a device's *Save Preset* button to save its current settings, or you can *drag* the device from its title bar directly into the User Library. Either way, the custom preset is stored in the Browser Places as a recallable preset for all of your projects. You will be prompted to

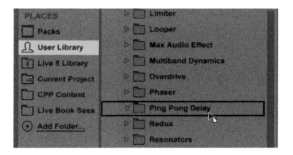

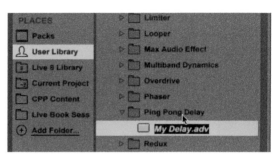

Figure 15.8
Drag your device to the User Library to save as a preset.

Figure 15.9
Type in a custom name.

name your new preset. If for some reason you don't want to save the preset, *press* the Esc key on your computer keyboard and the save will be cancelled as long as you are still at the naming phase. Later on if you decide you don't want to keep your preset, select it and *press* the delete/backspace key to move it to the trash.

But wait, there's more!

You can also *drag* your clips to the User Library or Current Project—or to any location on your computer for that matter—to save a device and the clip data as an all-in-one preset. Clips retain device and performance data when stored in any location, including in other projects; the official name of the clips is *Live Clips*. In this way you can quickly recall your favorite device, a performance, or both from

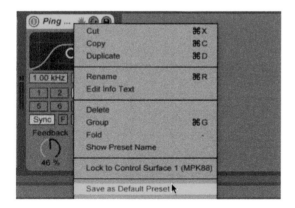

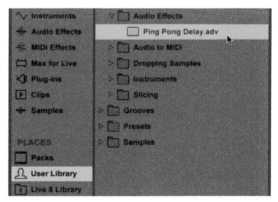

Figure 15.10 Save as Default Preset from a device contextual menu.

Figure 15.11 Navigate between content folders by hovering over their individual icon while click + holding a preset with the mouse.

the User Library or navigate to another Set using Places in the Browser. Just like that you can recall the device, effects, and your clip performance data for any Project. If you would like to save your current device settings as the default setting for that device, then *right click or ctrl + click* on the device's Title Bar and select "Save as Default Preset". This will store your custom settings as the default for the selected device in the User Library Defaults folder. From then on, every time you load that particular instrument, it will load with your title and custom settings until you delete it from your Defaults, then it will revert back the Core Library default. This works for all of your device types. Note, when *dragging* various items to or between folders, you can navigate between each different content area by *hovering* over each individual folder while *clicking + holding* the item with the mouse.

Additionally, under the Defaults there are a number of additional Default Preset folders. Each one allows you to determine how Live should behave related to the default type. For example, "Dropping Samples" allows you to dictate how Live should behave when dropping samples onto the Device View or a Drum Rack. All you need to do is create a new empty Simpler/Sampler on a MIDI track, customize its settings to your desired default, then *drag* it to the "On Device View", "On Drum Rack", "On Track View". With this set, the next time you *drop* a sample in either the Device View or on a Drum Rack, the Simpler/Sampler device will load with your default settings. For more on Simpler, launch to ▶ **15.3.2** .

15.3 Instrument Basics

As mentioned earlier, all versions come with a core set of instruments and presets. The difference is that Suite includes all of Ableton's software instruments. For those of you without Suite, be aware that Live is limited to only a few instruments. That being said, we'll focus on just a few specific Live 9 instruments: *Impulse, Simpler,* and *Sampler* (Suite). For more on the Suite instruments, launch to ▶ **Web** . You will find Live's instruments located in the Browser under *Categories>Instruments*. They are part of the Core Library pack that installed with your version of Live. These are core instruments of Live that will get you up and running in no time.

Impulse is an eight-note basic sample player for all types of sample content, but it is primarily used for triggering and playing back sampled drums and percussion, along with other one-shot audio files. Simpler is a standard sample player for all types of sample content using the entire keyboard note map, but is slimmed down in features. It's the stepchild of Sampler, which is Live's full-fledged sampler for all types of sample content. These devices are the main devices you will need to begin making music without investing in third-party virtual instruments. They all come with presets, sample-based instrument patches that include acoustic instruments, synths, drums, and much more, so feel free to start exploring their capabilities as soon as you can. Although these are useful instruments in their own right, they will be a whole lot cooler when used inside Racks. In fact, most, if not all, of their presets are stored as Racks.

Since Impulse and Simpler are included in Live 9 Intro and Live 9 Standard, as well as make up the core of many of the Rack presets, we'll focus on them first to eliminate some of the more complex variables while learning.

15.3.1 Impulse

One of the most essential elements of music is rhythm. Without it, music cannot exist. Live includes many ways to create percussive rhythms and beats using MIDI-driven drum samplers and synths delivered as Drum Racks, Instrument Racks, and as standalone (non-racked) instruments. Here the focus will be on the standalone device, Impulse. We'll get to the much anticipated Racks later ▶Scene16.

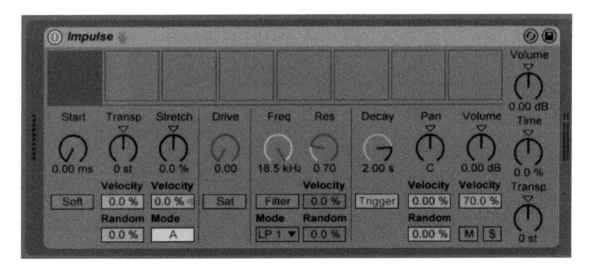

Figure 15.12 Impulse User Interface.

Impulse is located under *Categories>Instruments*. *Double click* or *drag* the device to a MIDI track, Device View, or Device Drop area in your Set to instantiate Impulse. This automatically brings up the Impulse user interface in Device View. Impulse is about as easy to use as any device can be.

User Interface

Impulse is divided into two sections: Sample Slots and Parameters. There are eight *sample slots*, each of which can be filtered and modulated by various parameters.

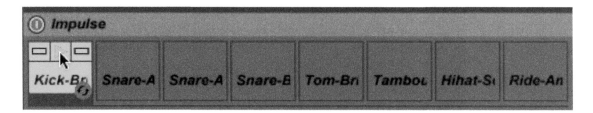

To create a totally original instrument, *drag* a sample from anywhere on your computer to an Impulse sample slot and you're ready to create a beat. No sample? No problem! There are a few samples that come with Live 9, which are totally customizable using Impulse's modulation parameters. You can also use the samples included on the companion website. Test our various samples until you find the one you like and Hot-Swap them on the fly. Each sample slot is automatically mapped to MIDI notes C3 to C4 or computer keys "A to K". They can also be triggered with the mouse if you *click* on the slot. An arrow will appear indicating where to trigger the sample. To the right and left of the arrow are the mute and solo buttons. Assign your MIDI controller's trigger pads to each slot if you've got one! Overall, these features make it very convenient to trigger Impulse via MIDI, computer keys, or the mouse. Unique to Sample Slot is a Link button that appears in the lower left corner of the Impulse interface when slot 8 is selected. This button links slot 8 with slot 7, allowing each the ability to mute the other's (7 or 8's) playback when one is triggered after the other. The main purpose of this is to emulate how closed hi-hats silence open hi-hats.

Impulse's available parameters provide some quick, useful ways to shape the sound of your sample kits. Each sample slot can be individually modified with its own set of parameters located directly below the sample slots. *Click* on the slot to select the sample, and then set its unique parameters as you like. You can also assign envelope automation to each parameter via the clip envelope editor. The following is a brief description of each parameter.

Below the sample slots starting on the left side of the user interface are the Sample Parameters. Each slot's parameters are adjustable. Select a slot to access its individual parameters. *Start* determines where in the audio sample playback will begin when an incoming MIDI note triggers it. *Soft* creates a slight fade-in on the attack phase of the sample. *Transp* controls the transposition (pitch) of the sample. *Random*

Figure 15.13
Impulse: eight customizable Sample Slots.

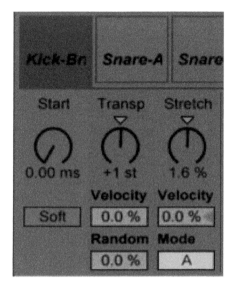

Figure 15.14 Impulse: Sample Parameters for each Sample Slot.

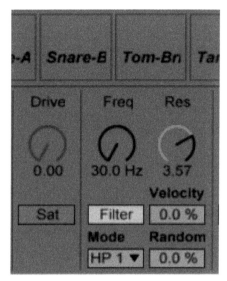

Figure 15.15 Impulse: Filter Parameters for each Sample Slot.

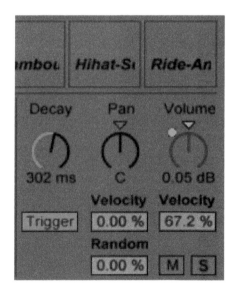

Figure 15.16

Impulse: Envelope Parameters for each Sample Slot.

controls the percentage of randomized modulation that can be applied to the transposition in order to create some interesting effects. The *Stretch* knob controls the amount of time-stretching that can be applied to the sample. The type of stretching is based on the selected *Mode*. Stretch Mode A is designed for low frequency samples and B for high frequency samples. Stretch/Transposition can be modulated by incoming note velocities, which are controlled by the *Velocity* sliders. Next to Stretch is *Saturation/Drive*. This effect simulates the sonic characteristics of analog tubes and distortion, designed to thicken the sample's sound, in essence making it sound warm and more analog-like.

The center section of the interface contains the Filter Parameters. *Freq/Filter* activates and controls the frequency filter cutoff for the selected sample. From the *Mode* chooser, select between different filter types, Low Pass, Hi Pass, and Notch. *Res* controls the resonance amount of the selected filter type. The *Velocity* slider controls the amount of modulation applied to the resonance filter parameter based on a percentage of the incoming MIDI note velocities. *Random* controls the percentage of randomized modulation that can be applied to the Frequency in order to create some interesting timbral effects every time the sample is played.

On the right-hand side are the Envelope Parameters and sample controls. *Decay* controls the length of the sample's fade out in seconds and can be assigned one of two modes. *Trigger mode* applies the decay envelope immediately with the onset of the sample and applies the set decay amount. *Gate* mode suspends the decay envelope until the note is released (note off message). Next to decay is *Pan*, which adjusts the sample's left-to-right location within the stereo image. *Velocity* controls the amount of modulation that can be applied to the pan parameter by a percentage of incoming velocities. *Random* controls the percentage of randomized modulation that can be applied to the pan amount in order to create some interesting imaging effects. *Volume* controls the amplitude level of the sample and can be modulated with the *Velocity* slider. This controls the amount of modulation that can be applied to the sample's volume parameter by a percentage of incoming MIDI note velocities. *M/S* buttons mute/solo the sample.

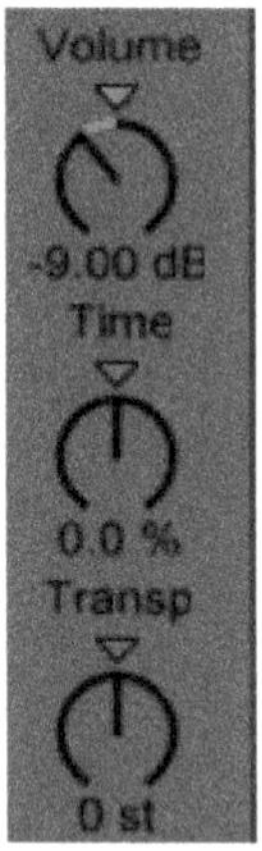

The Global Controls on the far right of the interface affect the entire instrument. *Volume* sets the overall amplitude level of the Impulse instrument. *Time* controls the overall time-stretching that can be applied to the Impulse instrument and *Transp* controls the overall transposition factor of the Impulse instrument.

Once you get the hang of creating your own Impulse instrument, then you can explore the presets that come with it. Each preset usually functions within an Instrument Rack. For more information on Racks, launch to ▶**Scene16**. All in all, Impulse provides you with the ability to program rhythmic patterns and percussive layers into your music.

Figure 15.17
Impulse: Global Controls.

Rerouting Impulse Individual Outputs

Traditionally, a device's output is a summation of its entire internal sound engine or sample set. In this case, the output is directly routed from the instrument, chain of devices, or track output to Live's main outs. With Impulse, each of its eight sample slots can be individually routed to another audio track instead of being summed altogether on its own track. In this way, the sample can be broken off from the track output and essentially sent (bussed) to another track. This allows you to isolate the sample—let's say, the snare—of a kit from the internal or track mix and add audio effects and/or mix it separately from the rest. Let's take a quick look at how to reroute the internal mix of Impulse. This can be done in the Session or Arrangement View.

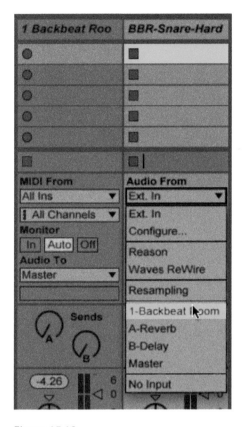

Figure 15.18
Audio From sent
to the Impulse
instrument track.

1. Load Impulse preset "Backbeat Room" or a drum kit preset of your choice into your Set.

2. Insert a new audio track and name it "BBR-Snare-Hard" or similar.

3. Set the new audio track's "Audio From" to the instrument track "Backbeat Room".

4. Set Input Channel to " … /Snare-AmbientLoudPunch/ … Hard-Impulse". (Figure 15.19.)

5. Set the new audio track's Monitor section to "In".

6. Arm the Impulse track and then play it.

Now when you play your drum kit you should still hear all of the drums, except the snare you assigned will play (output) through the new audio track and not through the Impulse track. This is why we set the audio track to monitor input. If it were not set to "In", then you wouldn't hear the snare at all. The only other way to hear it would be to record-enable the audio track, thus automatically monitoring input during clip playback, or record-enable both tracks to play and monitor input ((⌘ Mac/[ctrl] PC) + *click* on the track arm to record-enable multiple tracks). Either way, having the dedicated Snare track in this

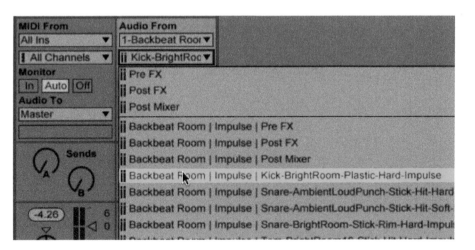

Figure 15.19 Input Channel set to Snare … Sample Slot.

example allows you to insert an audio effect device directly on the "BBR-Snare Hard" track, for instance.

15.3.2 Simpler

Before you get too excited, you should know that Simpler is not a full-blown all-inclusive software sampler; rather, it covers only the basic functions and synth parameters of a software sampler. That being said, its preset folder is minimal. Try downloading the "Solid Sounds" Pack by Loopmasters from Ableton's website. Unfortunately, to make complex multi-sample instruments, you will need to purchase *Sampler* or upgrade to Live 9 Suite, but Simpler can play multi-sample presets that have been premade by third parties in Sampler and have been converted to a Simpler preset. This is possible because the two devices are identical under the hood. The main limitation is in parameter access. Nevertheless, Simpler is designed for quickly importing a sample—automatically mapped across the keyboard map (key range)—and processing it through various parameters and then playing it back as an instrument. Samples can be *dropped* to any part of Simpler's interface and can be either played as a one-shot or looped instrument. The good thing is that you can convert your custom Simpler presets to Sampler Presets or vice versa. To do this, select the "Simpler —> Sampler" command from the Simpler's title bar contextual menu. Launch to ▶ 15.3.3 for more on Sampler.

Simpler is located under *Categories>Instruments. Double click* or *drag* the device to a MIDI track or Drop Area in your Set to instantiate it. This will automatically create an instance of Simpler and bring up its user interface into Device View.

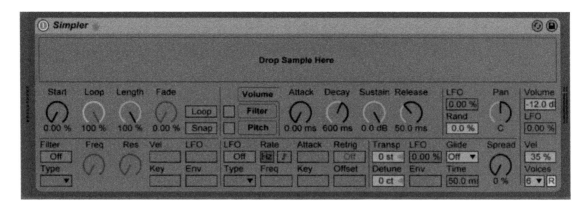

Figure 15.20 Simpler User Interface.

Simpler presets can be saved or Hot-Swapped at any time, just like any other device. You can also swap the entire device out. *Click* the Hot-Swap or Save button in the upper-right corner of Simpler in the Device View.

User Interface

Simpler is divided into two main sections: the Sample Display and Parameters. Parameters are broken down into sub-sections.

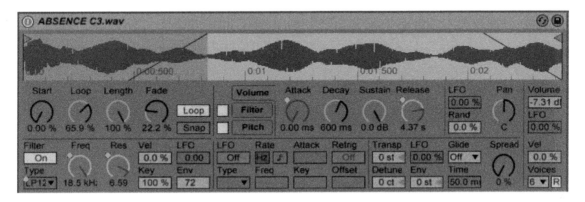

Figure 15.21 Simpler: Sample Display.

The Sample Display serves as a visual aid for setting the sample start, loop region, sample length, and loop crossfade. To zoom in/out in the sample display, place your mouse pointer over the display and *click + drag* the display, as you would in any other zoom-based display in Live. The rest of the interface is set aside for parameters, modulation, and performance properties. Going left to right on the interface, let's look at each set of parameters and modulation functions.

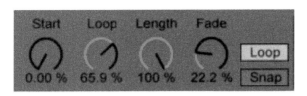

Figure 15.22 Simpler: Sample Controls.

On the upper left below the Sample Display are the Sample Controls. These are used to set how the Simpler will play back the raw sample. *Start* determines where in the sample playback will begin when triggered. This is useful for a number of reasons. For instance, you might need a sample to start a second or two into the clip due to the silence that precedes the attack transient. Adjusting the start point to the beginning of the attack will do just that. *Loop* knob (*Loop Length*) controls the length of an active loop selection within the sample. If you want your sample to sustain infinitely while a note is *pressed* then you will want to enable the *Loop Switch*, then set a loop length.

Length controls the length of the sample that Simpler will play. *Fade* controls the amount of crossfade applied to an active loop selection. Since a sample's loop region is often noticeable, you will likely want to disguise this effect. Fade allows for a smoothing of these audible transitions from one end of the loop to the other. *Snap* sets loop and length selections to the nearest zero-crossings in the sample in order to avoid pops or glitches when the sample loops.

Figure 15.23 Simpler: Filter Parameters.

Filter Parameters provide the ability to shape the timbre of a sample when Filter is "On". Use them to create a unique-sounding instrument. Enable Filter and choose the *Type* of filter to be applied: LP (low pass), BP (band pass), HP (high pass), or Notch. *Freq* sets the filter cutoff frequency and *Res* sets the resonance. *Vel* controls how much the incoming MIDI veloci-

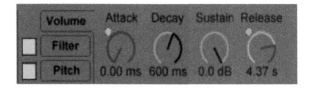

Figure 15.24 Simpler: Envelope Section.

ties affect the filter's cutoff frequency and *Key* controls how much the pitch of incoming MIDI notes affects the filter's cutoff frequency. The *LFO* (low frequency oscillator) slider sets the amount of modulation depth to affect the filter cutoff frequency. *Env* (Filter Envelope Amount) sets the amount of modulation depth that the filter envelope has on the filter cutoff frequency.

The Envelope section contains three unique envelopes and **A**ttack **D**ecay **S**ustain **R**elease knobs for modulating the sample. Use the *Envelope Chooser* (Volume, Filter, Pitch) to toggle between ADSR (attack-decay-sustain-release) settings for each. The volume envelope manages the amplitude of the sample over time. The filter envelope manages the filter's modulation effect on the sample's playback over time. The filter must be "On" to use the filter envelope. Enable the filter envelope by *clicking* the square button next to the filter Envelope Chooser. When "On", you will be able to scale the depth of modulation over the filter's cutoff frequency. This slider is located in the filter section. The pitch envelope manages the shape of modulation over the sample's pitch. *Clicking* on the square button next to the Pitch Envelope Chooser will activate it. When activated, you can also access the pitch envelope's modulation amount.

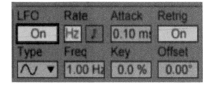

Figure 15.25
Simpler: Modulation Section.

The Modulation section is built around a low frequency oscillator (LFO)—an essential tool for any software instrument. There are a number of parameters available for the LFO. Choose the type

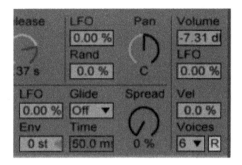

Figure 15.26
Simpler: Global Controls.

(waveshape), frequency (rate), attack, how the pitch of MIDI notes influences the rate, and how/when it triggers. The LFO can be routed to the filter, envelope, pan, and global sections via the LFO amount in each section. This is how you can really get creative, by modulating instrument parameters.

The Global Controls affect the entire instrument. From the *Glide* chooser you can set the instrument to global *Glide* or *Portamento* Mode. Both modes determine how notes transition from one to the next. *Time* (Glide Time) sets the speed of the transitions from note to note. Glide Mode is used for adding a smooth transition between different notes/pitches, creating a monophonic (one voice) instrument. When the next note is *pressed*, the transition sounds like a slide (glide) between the two notes based on the Time setting. This works especially well for electronic leads and basses. Portamento Mode is used in the same way, but functions polyphonically (multiple voices), meaning that multiple notes can be played and overlapped while transitioning. The *Pan* knob adjusts the instrument's perceived location in the stereo image. *Spread* adds a fattening chorused effect by adding a duplicate voice that is slightly detuned—a chorused synth-like effect. The overall Volume and Pan of Simpler can be controlled from the Volume amount slider and Global Pan knob. Both parameters can be modulated by the LFO. In addition, Volume can also be modulated by incoming note velocities (*Vel*) located below the LFO slider. The *Voices* chooser sets how many voices Simpler will use (polyphony). The "R" next to Voices is the *Retrigger* button. Activate this to help manage voices that sustain or overlap when trying to conserve CPU (central processing unit). When active, sustained notes will be retriggered instead of overlapping with new or additional notes that are subsequently played, which otherwise would add to the voice count and CPU load.

15.3.3 Sampler

No DAW is complete without a fully integrated software sampler for sample playback and for building or hosting sample libraries. This is exactly what Sampler is. Think of Sampler as Simpler on steroids. Lots of steroids! Sampler can be a simple single-sample instrument or a large format multi-sampled instrument. It's ideal for the creating and custom programming of user-based instruments as well. In addition to sample playback, Sampler also makes use of common FM/AM synthesis features for sound design concepts. Unfortunately, Sampler is not included

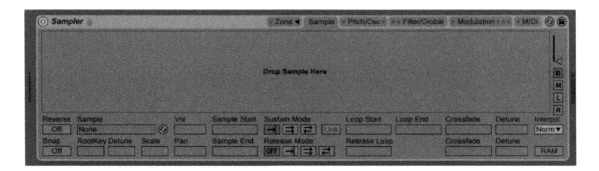

Figure 15.27
Sampler.

with the purchase of Live 9 Intro or Standard; but the full version can be purchased separately for those who do not own Suite. If you don't have Suite or Sampler, then you're going to have to use Simpler.

If you are transitioning from another DAW to Live and have instruments you'd like to bring with you, Sampler can help you out. If you're dealing with a third-party virtual instrument plug-in, then you can just use them as plug-ins—no need for Sampler. If instead your favorite instruments are exclusive to the DAW you were using, or you prefer to use Sampler's interface, you have an option. Sampler can import the following formatted instruments: Kontkat (.nki—Native Instruments, non-encrypted), EXS (.exs—Logic), SoundFont (.sf2), GigaStudio (.gig—Tascam), and AKAI -S1000, -S3000 third-party sample libraries. Note that some third-party sample programming features and parameters are not supported in Sampler due to their proprietary design; therefore, you may not get the same result that you had in their native software. Before we dissect Sampler, keep in mind that this is an overview. There is no way to discuss and demonstrate every feature of Sampler.

User Interface

Sampler is designed around a multi-page layout accessed by tabs located at the top of the interface. Each tab offers its own set of customizable parameters as follows.

The *Sample Tab* is a graphic sample editor accompanied by all of the necessary sample playback parameters. Among others, this includes sample start/end, loop, and crossfade settings.

The *Zone Tab* is where sample programming happens. Here each sample's playback range and root (fundamental pitch) is graphically managed. From this view, samples are assigned to keys (*Key Zone Editor*), velocities (*Velocity Zone Editor*), and sample select ranges (*Sample Select Editor*). Each of these mappings parameters determines when and what will trigger a sample(s) playback. Each sample, represented by a

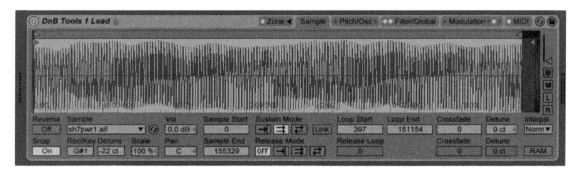

Figure 15.28 Sampler: Sample Tab.

green horizontal line-bar, will be assigned to a key or across multiple keys within the Key Zone Editor referenced along the *Key Zone Ruler* at the top. A Zone represents a specific sample that defines the MIDI note or note range it will respond to when a key is played (incoming MIDI data is received). Each sample zone will have a "root key" (*R*) note displayed within the editor. This is where the fundamental pitch should be assigned. From there the sample is transposed up or down based on how far the zone range is extended relative to the root assignment. You can lengthen or shorten the zone by *dragging* either each edge of the zone to the desired position along the key range. If you are working with multiple zones (multiple samples), you can edit them as a group by *selecting* the zone names in the *Sample Layer List* on the left-hand side of the editor.

Figure 15.29
Sampler: Zone Tab.

Hot Tip You can adjust the viewable display within the Zone editor by right clicking or ctrl + clicking and selecting "small", "medium," or "large" in the menu. This is great when working with Zone Fades and multiple zone ranges.

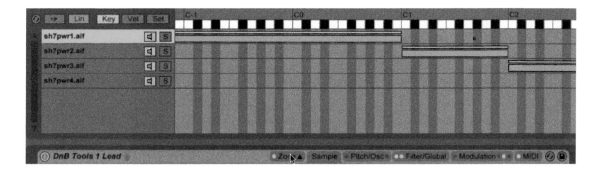

There is a Zone Editor for each MIDI keyboard note (Key), velocity (Vel) and Sample Select (Sel). A multi-sampled instrument will consist of many zones/layers—for example, a piano may have 300+ zones, one zone for each note at multiple velocity values (hard, medium, soft, etc.). In addition, just above each zone is the *Fade Range. Dragging* this from its edges will allow you to set a Fade Range to scale (fade) the volume of the Zone in the area that the fade is applied to. This is useful for overlapping two Zones so that transitions between samples are seamless. If you are, in fact, programming multiple sample layers you can assign them each to be triggered by velocity range, in the Velocity Editor. This will allow for samples to be triggered only when the MIDI input velocity meets the designated value, 0 to 127. An example of this is when one group of samples represents the loud or forte dynamic, while the second layer represents the soft or piano dynamic. Each layer then represents a different dynamic of the same instrument. In a similar way, zones can be assigned to the *Sample Select* feature. Using the *Sample Select Ruler*, samples can be assigned to 0 to 127 individual positions. This is an interesting feature that allows for a whole layer, or group if you will, to be included or excluded from playback. In this way, you could create a multi-timbre instrument, where each multilayer makes up a whole different sampled instrument. You could then embed multiple instruments on the key range, then toggle select them on-the-fly. Select positions are not controlled by MIDI input, so in order to choose different values the selector position will need to be moved manually with the mouse or assigned to a macro, key map, or MIDI map. The concept of embedding instruments within Sampler is very cool; but a more efficient way to do this is through the instrument racking process, which allows you to layer multiple Samplers within one playable device. On top of that, you can incorporate the Select feature within Racks! For more on Rack Devices, launch to ▶ **16.4.1** .

The *Pitch/Oscillator* tab reveals the FM/AM oscillator and Pitch envelope parameters. Enable either one to modulate and shape the instrument's playback characteristics. The Oscillator influences frequency and amplitude modulation. With this comes a host of parameter controls for fine-tuning its effect over the instrument. Of these, notice the oscillator waveform chooser. Using various waveforms (*Type)* and applying high intensity values (*Volume*), along with *Fixed Frequency* or *Course Tuning* will result in some uniquely creative sounds. The Pitch Envelope alters the instrument pitch over time, based on the shape of the envelope. Feel free to activate the envelope and adjust its shape and parameters. This is commonly used to affect the attack transient of an instrument of a quick pitch bend effect. At the bottom right of the Pitch/Osc section are a few global parameters, including *Spread* (adds detuned voices to enhance the stereo character/image), *Transpose, Detune, Zone Shift* (transposes MIDI notes in the Key Zone Editor only), *Glide* (pitch slide

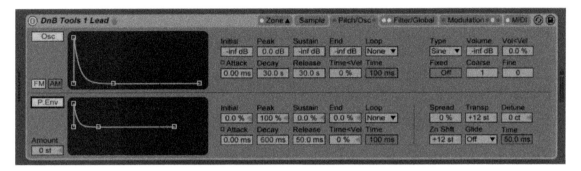

Figure 15.30 Sampler: Pitch/Oscillator Tab.

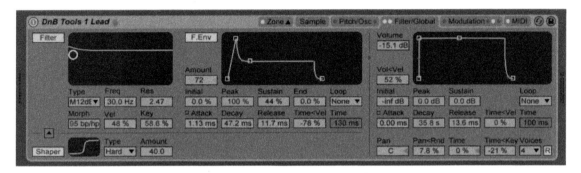

Figure 15.31 Sampler: Filter/Global Tab.

between notes), *and Glide Time*. These are all fairly self-explanatory; just be aware that they are located under this tab.

The *Filter/Global* tab includes a multimode filter with adjustable frequency and resonance parameters and filter envelope. Under *Type*, choose from a number of classic filters, such as high pass, low pass, and band pass. Each type can be customized using the black window. Simply *drag* the yellow circle around to change the frequency and resonance values. The same goes for the *Filter Envelope*. Just below the Filter section is the *Shaper* (Waveshaper). This is a nice little feature for altering the timbre of the filters of the instrument. Try using the various types from *Type* chooser and adjust the *Amount* to apply more or less intensity. The effect tends to sound like saturation or bit crunch depending upon which type—*soft*, *hard*, *sine*, or *4bit*—is applied. The Global section handles the overall volume envelope (amplitude) and various sample settings.

The *Modulation* tab manages the auxiliary envelope and three LFOs for modulating a multitude of Sampler's internal parameters (i.e., pitch, volume, filter frequency,

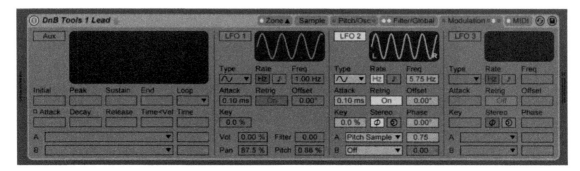

Figure 15.32 Sampler: Modulation Tab.

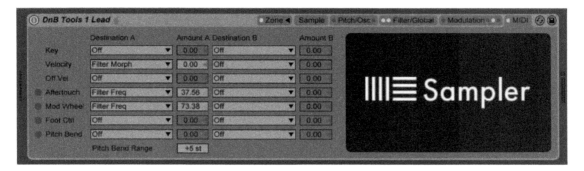

Figure 15.33 Sampler: MIDI Tab.

etc.). After spending a few minutes experimenting with LFO modulation assignments, you will be hooked. This is probably one of Sampler's best features, especially if you're at all interested in electronic music! Common applications of the LFOs are to assign modulation to volume, pan, filters, pitch, and a number of subcategories. These are just a few suggestions. Try different LFO types and rates and start applying them to your instrument. Their amount can even be controlled by MIDI note/pitch input via *Key*.

The *MIDI* tab is where Sampler's internal parameters can be assigned to standard MIDI controllers. Each MIDI controller type can be assigned to two modulation destinations.

Sampler has an enormous number of features that will keep you busy for hours on end, the result of which can be unimaginable. Definitely consider giving your instruments performance control by assigning MIDI control over the various parameters of Sampler. This will allow you to create more expressive and dynamic instruments.

15.4 MIDI Effects

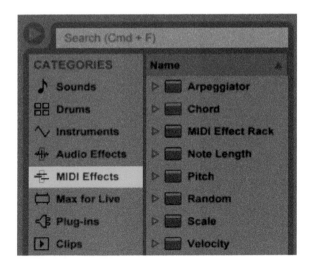

Figure 15.34
Live MIDI Effects.

Located in the Live Browser you will find some pretty useful MIDI effects. These are not to be confused with audio effects, which we'll talk about in the following section. First off, MIDI effects only function on MIDI tracks and only process MIDI signals. Since MIDI data is made up purely of digital instructions, it is very easy to manipulate that data before it is converted into an audio signal by an instrument. This is what happens with Live's MIDI effects. When added to an instrument or MIDI track, MIDI effects intercept the incoming MIDI notes and redistribute them based on the particular effect added. For example, if you *play* and *hold* a note with a basic instrument (without an effect), you will hear a single attack and then a sustain. Now, if you insert an Arpeggiator, then *play* and *hold* two or more notes, the Arpeggiator will create a repeating pattern based on the held notes—multiple attacks and repeated pattern in place of the sustain. Thus, the Arpeggiator captures the incoming data and reprocesses it into a repeated pattern. You can also look at this as a chain of events. MIDI notes enter first into the MIDI effect device, processed, and then are passed on into the instrument. From there they trigger a sound culminating as the resulting audio output.

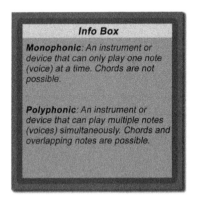

Info Box

Monophonic: An instrument or device that can only play one note (voice) at a time. Chords are not possible.

Polyphonic: An instrument or device that can play multiple notes (voices) simultaneously. Chords and overlapping notes are possible.

Live's MIDI effects include *Arpeggiator*, *Chord*, *Note Length*, *Pitch*, *Random*, *Scale*, *Velocity*, and *MIDI Effect Rack*. To learn more about Effects Racks, launch to ▶**Scene16**.

To insert a MIDI effect device or preset, select a MIDI track destination and then navigate through the MIDI Effects folders in Categories. Once you find one you like, *double click* the effect in the Browser, or *drag* the MIDI effect directly to the MIDI track or to its Device View. Live will automatically insert the effect before the instrument, as you will see displayed in Device View. You can add multiple MIDI effects, including multiples of the same effect into one instrument track if you wish. If there are no MIDI tracks currently in your Set, you can load a MIDI effect to your Set to auto create a new MIDI track with the inserted effect only. The best way to become familiar with all of the MIDI effects is to load up an instrument and a MIDI clip, then start *dropping* and hot-swapping effects to hear what happens. This will no doubt create some very interesting results.

We have already described the Arpeggiator, so here is a brief description of the rest of the MIDI effects. Each is user definable and comes with presets:

Chord—creates chords out of incoming pitches, up to six additional notes added to the original note.

Note Length—alters the length of incoming notes or to trigger a note upon a note release as opposed to a note on message (*pressing a note*).

Figure 15.35
Arpeggiator MIDI effect in Device View.

Pitch—transposes incoming notes.

Random—as its name implies, it randomizes and changes which notes are actually output/played when they are inputted.

Scale—forces incoming notes to fall into a defined scale so that pitches outside of the scale map will be altered to the defined scale degree equivalent. For example, a major scale degree could be forced to sound its equivalent minor degree when played. Scale can also be used to transpose incoming notes.

Velocity—reassigns incoming note velocities to different velocity values based on various parameters.

15.5 Audio Effects

Live 9 comes bundled with some very flexible audio effect devices for creative production and performance purposes, but three specific effects are exclusive to Suite: *Amp*, *Cabinet*, and *Corpus*. If you are using Live 9 Intro, you've got some good ones to start with, but there are quite a few more that you will not have. Time to upgrade? These devices can be found in the Live Browser Categories. Audio Effects are used to process audio signals and therefore can be added to any audio track, Return track, or the Master track. That being said, they can also be *dropped* into a MIDI track as long as there is an instrument inserted on the track and they are inserted after the instrument at its output. Simply put, when an instrument is inserted into a MIDI track it is no longer strictly MIDI. Instead, it is more of a hybrid

Figure 15.36
Live Audio Effects.

MIDI/audio track like an instrument track that you commonly find in other DAWs. In this way, the track can handle both MIDI and audio: the MIDI instrument translates a MIDI input into an audio output. Therefore, this audio output can be processed in the same way an audio track output can be processed. More on signal flow in the following section.

Audio effects are loaded from the Browser just like MIDI effects. Select a track, then *double click* or *drag* the effect into an existing track or to its Device View. If you *drag* an audio effect into the Session View Drop Area, a new audio track will be created. To load a preset, navigate to an effect's folder in the Browser. You can add multiple effects, including multiple instances of the same effect on a single instrument. This can create some very interesting results. Don't forget that audio effects can also be inserted on a Return or the Master track.

The user interface of an audio effect varies from device to device depending upon the type; but many of them incorporate some type of interactive controls, such as a X–Y controls for shaping effects in real-time. Imagine the power and creative control when utilizing MIDI Mapping for the X and Y. Assign X and Y to a different controller assignment to remote control each axis independently. Assign them to *Frequency* and *Feedback* controls and you will be a creative genius! More on how to MIDI Map here ▶ **17.2** .

Figure 15.37
X-Y controls for manual real-time sound shaping.

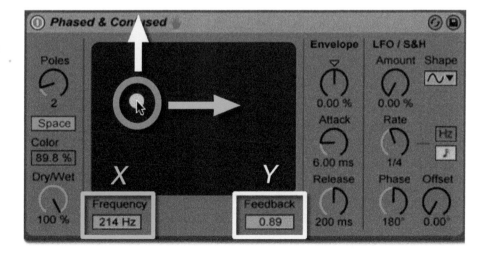

> **Hot Tip** Assign Frequency and Feedback controls to a MIDI Map assignment to split up control of the X–Y, up/down, and left/right for real-time manipulations.

Now that you have been enticed with a peek at some creative control, let's take a brief look at a few of the audio effect devices available for use in Live 9 and Suite.

15.5.1 Corpus

Corpus is an audio effect that uses physical modeling to recreate the acoustic characteristic of resonant objects. It is derived from the resonator section of Collision as a standalone device. Corpus is only available with Suite or by purchasing Collision. Launch to ▶ **Web** for information on Collision, which will subsequently cover the basic concepts of Corpus. As an audio effect, you can apply these resonators to the sound of any instrument or audio track. For example, Corpus could be applied to percussion-based sounds as a reverb-type effect, or on the Master track as a spatial effect for a Set's mix.

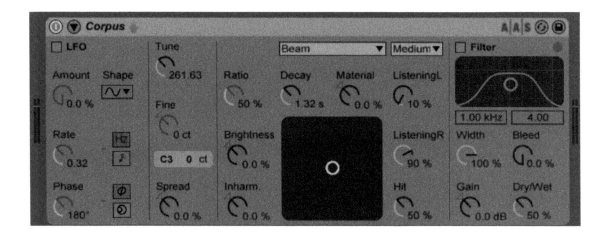

Some of Corpus's unique features include an LFO that directly modulates the resonator frequency, a built-in limiter for peak-level control, and an X–Y controller/ display area that allows for control over a multitude of parameters in real-time. To top it all off, it has a sidechain input for additional control via signal sources from other tracks in your Live Set. See sidechaining, Clip ▶ **15.7.3**.

Figure 15.38
Corpus.

15.5.2 Frequency Shifter

With Frequency Shifter you'll be able to create some out-of-this-world sounds from ordinary audio signals. Frequency Shifter combines LFO modulation with two selectable modes of processing: Frequency Shifting or Ring Modulation. Frequency Shifting mode shifts the frequencies of the incoming audio signal up or down based on the frequency shift amount assigned at the *Frequency* knob. You can control these shifts of frequencies with either coarse or fine-tuned values. Simply

put, the amount of shift is added or subtracted from the input signal, resulting in a summation of the input and tuning amount. Ring Modulation mode adds and subtracts the applied frequency amount to and from your audio's input frequencies simultaneously, resulting in multiple frequency bands added or combined with the input source— essentially frequencies above and below the original. Both modes can result in some serious sci-fi style sounds. These sounds result in flange or phasing effects similar to tremolo, as well as intense metallic crystallized effects. Frequency Shifter also allows you to add LFO modulation for manipulating the stereo image and phase of the effect for even more interesting effect over incoming audio signals.

Figure 15.39
Frequency Shifter.

Here are a few ways to apply Frequency Shifter:

1. Import a drum or rhythmic guitar audio loop into your Set, then Insert Frequency Shifter onto the same track. Now, start by adjusting its *Fine* tune frequency amount to produce some cool phasing and pitch transposition effects. From there, experiment with the *Frequency* knob (*Coarse* tuning) to create additional shifts.

2. For a completely different result, switch the Mode button to Ring for some abstract modulation effects. Combine this with the LFO parameter to add panning and sweeping effects. Switching the Rate button from Hertz to Tempo Sync will give you a wide variety of sub-divisions to choose from in conjunction with the selected LFO amount. Also try adjusting the Dry/Wet mix ratio, or even automate it to hear the difference between both the dry and processed signal.

3. Another useful function of Frequency Shifter is the activation of Ring Mode and *Drive*. This introduces a completely different effect when the Dry/Wet knob

is turned all the way down to "0%". The result is pure unadulterated distortion depending upon how high the Drive dB (decibel) level is set (located directly below the Drive button). Note that you will have to balance down the overall output of your track as you add more Drive signal.

15.5.3 Glue Compressor

Everyone needs a classic bus compressor for mixing and mastering their songs. *Glue Compressor* is Live's 1980s analog mixing console-inspired compressor primarily designed for submixing multiple tracks or for the Master mix. The Glue compressor is very straightforward and has a simple design, as most bus compressors do. The main parameter controls are *Threshold*, which determines the point at where compression begins when signals rise above the specified dB level. The compression itself is then applied based on the *Ratio*. *Makeup* adds gain back to the signal to compensate for the attenuation (reduction) in output by the threshold. It is the balancing of these two knobs/parameters that results in the "compressed" sound we are all accustomed to—the slamming and pumping of the output. A general tip is that the compressor should, when applied, not reduce or increase the overall perceived output level of the track(s). The other parameters, *Attack*, *Release*, and *Ratio* determine the behavior of the compressor as it engages upon audio signals. Use the *Range* knob to control the amount of compression applied overall. You can also use the *Dry/Wet* knob to blend the compress and uncompressed signals.

Figure 15.40 Glue Compressor.

Note: Live also offers a standard *Compressor*, which is definitely worth checking out. It has been optimized for parallel compression and expansion—among other things—and includes a nice graphic display. Many of the concepts above apply to it as well.

15.5.4 Limiter

The *Limiter* allows for mastering quality signal processing that maximizes output—like a compressor—by limiting dynamic range and reducing the overall audio output to a designated level. The Limiter is mainly used to control the overall level of your project and to prevent clipping (overloading), essentially allowing for an increase in level without exceeding 0 dB—the point at which distortion occurs. Variable control over the release time provides great flexibility for fine-tuning your output signal. In addition, the stereo switch is an interesting feature that gives you the option to separate the stereo signal for the independent limiting of each channel (L/R).

Figure 15.41
Limiter.

Other onboard features that are useful for mastering include the *Lookahead Time chooser*. This tells the Limiter how quickly to respond to and attenuate the transient spikes and peaks in the audio. This is a particularly popular feature in the world of mastering because it allows you to increase the output level of your entire Set while ensuring that output doesn't overload the Master. It won't take a whole lot of adjustments to hear the Limiter's effects on your audio. Even the slightest attenuation will be noticeable. Start out conservatively, gradually adding gain, and you'll quickly hear how it starts to affect output. Keep in mind that the Limiter can be used at the output of any individual device in order to control its output as opposed to the Master track output. As far as parameters are concerned, use *Gain* to control (increase/decrease) the input level prior to limiting and *Ceiling*, which sets the highest possible level the Limiter will output, ultimately dictating when limiting is applied. If the level exceeds the Ceiling, limiting occurs.

15.5.5 Looper

With Looper, Live brings back the classic vintage-style tape-on-tape loop recording technique first made popular in the 1960s and 1970s, but this time as a technique for the modern age of recording as a virtual audio effect. Looper makes it very easy to not only record and loop audio, but also overdub on top of that recorded

Figure 15.42
Looper.

audio loop endlessly and, best of all, hands free—just as in the popular foot pedal effects units found today. Looper has several unique features that allow you to launch and input real-time ideas in a tight and concise manner, building endless musical passages on top of one another. You can also import audio files and clips into and export out of Looper to a track or to the Browser. It's all up to you.

At first glance you will notice that Looper includes a large visual display. Once a recording has been made, graphic blocks will appear—like a step sequencer—representing the bars and beats of your recorded performance based on the subdivision of the beat. Just below this to the left are four Transport buttons: *Record*, *Overdub*, *Play*, and *Stop*. Just below these buttons is the *Multi-Purpose Transport Button*. This is an all-in-one button designed to work with a foot pedal, although not required. You could use any controller assigned to it. The point is that it allows you to use one singular control in order to activate Looper's mode in order to create on-the-fly uninterrupted. Let's walk through this, then we'll come back to the other features on the Looper Interface. Keep in mind that this is just one way to use Looper. There are a number of settings and behaviors that you will want to experiment with on your own.

Looper in action …

A fun way to use Looper is to incorporate a footswitch (foot pedal). This allows for hands-free control, keeping the creative juices flowing while jamming with Looper. By using Live's MIDI Mapping feature you can easily assign a footswitch control to the Multi-Purpose Transport Button. In this way, activate the playback of your Set from the global transport (Control Bar Play button); then simply *stomp* the footswitch when you are ready to record and then start recording your initial musical idea, such as a vocal or guitar riff. *Stomp* the footswitch again to switch from record to overdub mode, then lay down each subsequent musical layer in cycles (loops), like vocal harmonies or guitar doubles. When you're done overdubbing, *stomp* the footswitch again to switch out of overdub and into playback. Looper will then cycle over and over again, playing back your recorded overdub performances. *Stomp* twice on the pedal to stop playback or you can select stop with the spacebar. If you don't like what you've recorded or want to undo something, use the *Undo* and *Clear* buttons.

This is some pretty cool stuff and if you don't have a footswitch, then just use a MIDI controller or even the mouse; but wait, there's more. After you have recorded into Looper, you can then *drag* your looped performance directly into a track using the "Drag Me!" feature on Looper's interface, thus turning it into an audio clip. This example was the simplified practical demonstration of using Looper. Now back to the technical elements concepts.

Looper handles tempo in two different ways. With your Set playing back, you will record into Looper with the Metronome or with other tracks just as you would in an empty Clip Slot. If your Set is not playing and you record into Looper, it will calculate the tempo of your loop based on the length of your first recorded pass. Live will then change the tempo of your Set to globally sync all your clips to Looper's tempo, the tempo it determined based on your first recording. If you change your Set's tempo at this point, Looper will playback faster/slower, altering the pitch of the recording. The same goes for audio exported form Looper because the audio will be set to Re-Pitch when converted to an audio clip. Of course, you can change the clip(s) to any other Warp Mode.

After all of this, you'll quickly see that you have total control over Looper's ability to sync to your Set's tempo. Note that Live's transport can be set so that as soon as you punch out of recording in Looper, the playback commences from that point forward. Change this parameter if needed, but it's important to know that while your Set is playing, Looper is simply playing back just like an audio clip itself while you are using the audio directly stored in it. Use the *Song Control Chooser* to customize the relationship between Looper's transport and Live's global transport. By selecting "none" in the chooser, Looper's transport will not affect Live's global Transport in any way. If "Start Song" is selected, the global Transport will begin playing as soon as playback or overdub starts in Looper. The other option is "Start and Stop Song", which completely slaves global transport to Looper.

There are a few other features and controls that should be mentioned at this point. Preset the number of bars to be recorded into Looper established with the *Record Length* chooser. The *Speed* and *Reverse* function control the speed at which (as

Figure 15.43
Looper: If you select "Start Song" your Set will begin playing as soon as you engage playback or overdubbing in Looper.

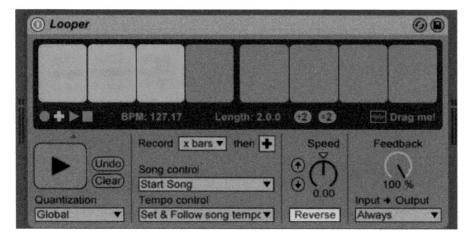

Figure 15.44
Looper: Record directly into Looper with the Reverse setting engaged to create a more radical sounds.

well as direction in which) audio is played back after it is recorded in Looper. You can actually record directly in Looper with the *Reverse* setting engaged to create a pretty radical sound.

After overdubbing a new loop you can change total length of the loop with the "divide 2" and "times 2" selector just below the bars and beats display segments and to the left of the "Drag Me!" function. For example, this works great for situations when you have a two-bar original loop and you wish to add several four-bar overdub loops on top.

Several options are available for monitoring Looper's direct Input and Output that let you decide whether or not the input signal can be heard at all times during

Figure 15.45
Looper: Record
Length.

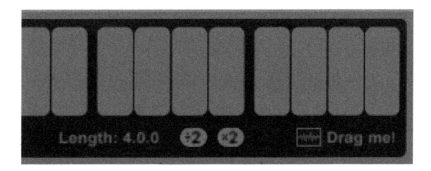

Figure 15.46
Looper: Input-
Output Chooser.

recording and playback. You'll find this chooser for Input → Output options at the bottom right of Looper.

Looper takes a bit of practice; but after a few passes you'll find that it's a wonderful extension to the Live experience, especially for singer/songwriters and guitarists.

15.5.6 Multiband Dynamics

This is probably one of the best multiband compressors included in a DAW—or at least an easy-to-use and effective favorite. This mastering caliber compressor combines the power of six discrete processors into one interface. Combined, they offer a real-time *compressor* and *expander* with several innovative real-time control parameters not found in other dynamic processors. How does it work? Live's Multiband Dynamics, or MBD, as it will be referred to here, provides both upward and downward compression and expansion across three independent frequency bands.

It comes with individual attack and release controls plus crossover points for each band. You have the option for soloing each band and bypassing compression for each band, which promotes flexibility and fine-tuning. MBD features a real-time *drag* tool available within the interface. By using your mouse and pointing to a specific area within the graphic display window, you can access and alter the volume and threshold of each band, allowing for customized sound shaping effects right from within the interface. A good way to learn the basic variables of this *click* and *drag* feature is to solo a single band on the left side of the display and then start adjusting the levels and threshold which are contained in the visual block-style segments in the display for that particular band. Keep in mind that when working with the bypass and solo buttons of each of the three bands, the Mid-band controls the entire audio signal when the "Hi" and "Low" bands are bypassed.

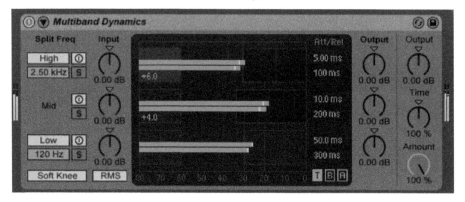

Figure 15.47 MBD - Multiband Dynamics.

You can quickly manipulate and edit a specific band's input, output, and threshold levels within the display and get quick, audible results by either *dragging* left or right on one of the block segments with your mouse to adjust the threshold levels. *Dragging* up and down on the middle of the block will adjust the signal level for that particular band.

On the right side of the display, you'll find three options for viewing the MBD's *Attack/Release* Time parameters and *Above Threshold* and *Below Threshold* values and ratios signified by the "*T*", "*B*", and "*A*" buttons. *Press* these buttons to view and edit these parameters. Each of these settings is key in determining how compression behaves, not to mention they are especially useful for precision mastering and sound sculpting. Notice that each parameter under B and A can also be controlled by *clicking + dragging* within the MBD Threshold/Ratio display.

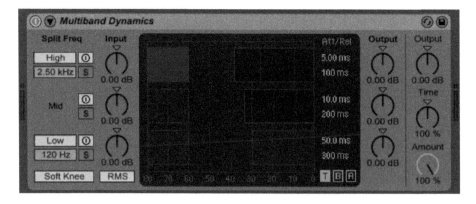

Figure 15.48 MBD: Click + drag Threshold Blocks.

Practice using these settings with a single band of audio at a time. This will help in identifying discrete characteristics of each parameter as you alter their values before you start working with all three bands at once. Speaking of all three bands, you can also use *drag* and alter across both the compression and expansion displays all at once by *holding* down the opt/alt key while *clicking + dragging* the appropriate parameter. If you wish, you can do it across all three bands at once by selecting shift + opt/alt while *clicking + dragging* the particular parameter needed.

The MBD is great for de-essing, multiband compression, and single-band processing techniques, to name a few. If you don't know what those are, just know that it makes your tracks sound better. It's also a very effective sound design tool with its ability to pull out psychoacoustic sounds in an audio file (sounds you can't hear with the naked ear). Try working with the *High Band* and *Low Band* crossover frequency settings located on the left side of the interface to see what can be accomplished or deactivate the High and Low to work in single-band processing mode. That being said, check out the presets in the Multiband Dynamics preset folder. Some of these are tailored for mastering while others are dedicated toward processing groups or individual tracks like vocals or speech. When you first load up an MBD preset, start out by adjusting the *Amount* knob and *Time Control* knob (Time Scaling). Time Control adjusts the speed of the attack and release to all the active bands. The same holds true for the Amount knob, which adjusts the overall level and intensity of both compression and expansion across all the active bands. The audible results can be minimal or very obvious. Again, using these two functions is a great way to get acquainted with this complex device. The next parameters to experiment with are the *Output Gain* knobs and the Thresholds.

Try MBD out now. Import a full stereo track you have lying around and then add MBD to the Master. You'll be pleasantly surprised!

15.5.7 Overdrive

This in-your-face distortion effect is Live's virtual guitar stomp pedal. *Overdrive* is great for adding distortion while maintaining a punchy and raw edge with great dynamics to any audio. Although the interface might look simple, it packs a powerful punch. The Overdrive interface has a few unique functions that allow you to add a subtle effect all the way to hard and heavy distortion and saturation. From the interface, you can control *Drive*, *Tone*, and *Dynamics*. Drive controls the amount of distortion and *Tone* affects the timbre of the signal post distortion for boosting the high frequency content. The Increase Drive amount yields more distortion. The *Dynamics* parameter provides control over the amount of compression applied— thus, the percentage of compression versus the original dynamic range of the signal. Try keeping it somewhere between "60 to 80%", which is a good starting point for introducing light compression into your mix. *Dry/Wet* sets the balance between the processed and unprocessed (original) signals with the exception of Tone, which always affects the sound unless Tone is set to "0%". Keep the Dry/Wet value at "50%" as a starting point and gradually increase the value along with the Tone and Drive values. Be sure to set Dry/Wet to "100%" when using Overdrive as a Send effect.

Last, but not least, is the black window at the top of the interface, which is the *Bandpass Display*. This provides direct user access and control over the bandpass filter's center frequency. Use your mouse to move the *X–Y* control (yellow circle), or the slider boxes below the Bandpass Display to manipulate the center frequency. Be sure to try this in real-time to explore the possibilities of Overdrive's broad frequency spectrum filtering effects.

Throw Overdrive onto your drum tracks and get that grungy white noise snare or bandpassed drum effect. It also works great in conjunction with other audio effects such as *Compressor* and *Dynamic Tube*. Try loading up all three of these effects as a chain within a single instrument track! You could Rack them up too!

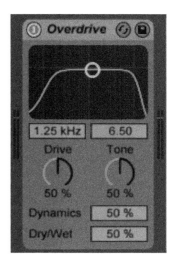

15.5.8 Vocoder

First off, there are different kinds of vocoding and therefore different methods of implementation. With that said, here is a general explanation of vocoding. The idea is that one audio signal (*carrier*) is modulated by another audio signal (*modulator*), which effectively amplifies and attenuates the frequency spectra (audio content) of the initial audio signal (*carrier*), resulting in a different sound. Putting this idea into practice, a vocoder can take a voice from a microphone (modulator), pass it through a filter bank (frequency analysis) and envelope follower (amplitude tracking), and then have that information trigger oscillators in a synthesizer (carrier). Using a synth as the *carrier* and a drumbeat as the *modulator*, the result would be something like a rhythmic synth continually influenced by the drumbeat based on whatever is played on the synth.

Figure 15.49
Overdrive.

With Live's Vocoder you have the ability to generate many vocoding effects with ease, especially since it can operate with or without an external carrier signal. Vocoder can achieve effects by tapping into its own built-in internal carrier signals or even modulate its own signal.

Figure 15.50
Vocoder.

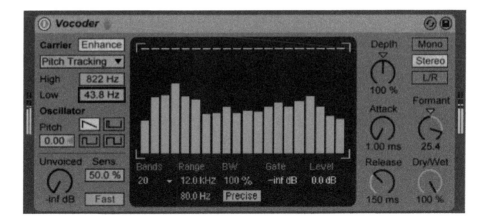

Insert Vocoder on your "modulator" track and then assign the "carrier" signal with Vocoder's *Carrier* chooser on the left side of its interface. With the intuitive routing of this sidechain input (carrier signal source) you should find Vocoder to yield a satisfying vocoding experience—which can't be said about some other Vocoder plug-ins.

Vocoder's parameters allow you to control and fine-tune variable frequency bands with format adjustments and a multitude of bandpass filters. It also offers a carrier signal source for classic Pitch Tracking, making it possible to recreate that vintage "Beastie Boys" sound effect. Amongst its parameters are *Upper* and *Lower Pitch Detection* sliders (High/Low) that control ranges over the Pitch Tracking oscillator. These detection parameters limit the frequency range that will be tracked by the oscillator, thus creating a robotic floating vocal effect similar to a mouth harp. Try putting Vocoder on a drum track and set Carrier to Pitch Tracking, then crank up the High parameter, tweak the Low parameter, and adjust the Formant to see what you can come up with. If that's not enough, you'll find four separate waveforms to choose from to manipulate the initial characteristics of the actual Pitch Tracking oscillator.

To really take Vocoder as far as possible, try using a microphone and your voice and follow these steps:

1. Using a new empty Set, insert Vocoder on an audio track, call it "Track 1". This track will be the modulator.
2. Assign your voice to be input on "Track 1" and set its Monitor input to "In" so that you can eventually audition your vocal sound after you have completed the following steps.

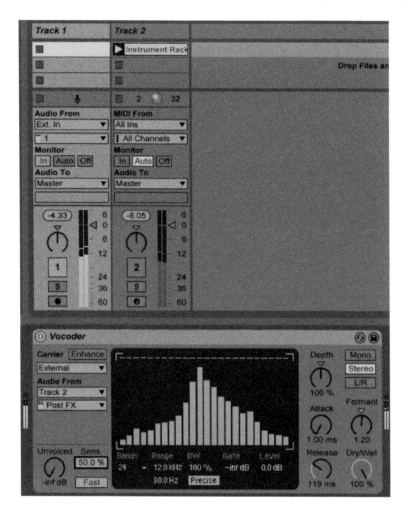

Figure 15.51
Vocoder Sidechain
in action.

3. Using a new MIDI track as "Track 2", add an instrument that has a quick attack style. Analog's "Pad"-style presets work well for this. Try "Bright1 Buzz Pad".

4. Arm "Track 2" and play some chords. After you find a few chord progressions you like, record some MIDI clips; then when you have your clips recorded, deactivate (mute) the MIDI track from its Activator button. If you need a clip, use the "SIMP-ARP" from the **CPP_MIDILoops** folder from the companion website.

5. Launch your newly recorded MIDI clip(s) and feed the MIDI "Track 2" back through Vocoder as the carrier signal by selecting MIDI "Track 2" as the External input source for "Track 1's" Vocoder sidechain input chooser.

6. Hop back on the microphone and sing or speak to get some audio activity going in "Track 1", thus into Vocoder while the chords or notes you recorded on "Track 2" (your carrier instrument) play as well.

7. You should start to hear some effected vocal sounds playing through Vocoder. Now it's time to tweak and control your new sound with the *Formant* control (shift the frequencies of the filterbank) and *Release* knob parameters (release time of the filter band's sustained amplitude). You can also try selecting filter band variations from the *Bands* chooser menu. Try something between 20 and 32 for more obvious changes in the tone and clarity.

8. By now you should notice that no matter the pitch of your actual voice, the instrument "Track 2" actually controls the real-time pitch. Pretty sweet! Who needs to sing in tune?!?!

9. Additional enhancements can be achieved by selecting different filter ranges in the higher frequency bands just to the left of the Bandwidth "BW" slider and Precise/Retro Mode button just below that. By selecting a wider BW value and choosing Retro Mode you should start to hear that familiar Vocoder sound.

The "singing synthesizer" also works great with a drum loop as the modulator. Here are some parameters to consider using for this. Try increasing the Release value on the Vocoder and you'll open up the true length of the active filter bands to hear a more sustained singing style effect. Adjusting the Attack knob will determine how fast Vocoder will respond to the dynamic changes in amplitude from the modulator signal. This is one reason drum-style clips can work well in the example above. Try adding delay and reverb along with Vocoder to create a complex device. Note that Vocoder doesn't have to be used to create a "robot voice" or "singing synth" effect, although it's a popular effect. If you want to see the Vocoder in action, take a look at "**CPP_15–5_Vocoder Project**".

15.6 Max for Live

If you've ever wanted to create your own devices in Live, you're in luck. *Max for Live* is integrated within Live 9 Suite or can be purchased as an add-on for Live that allows you to build, use, and share your own custom devices that can even look like native Live devices. If you're a Suite user, then Max for Live comes as a separate installer that includes a small collection of audio and MIDI effects and instruments. You will find these devices under Categories in the Live browser. Additional Max for Live Packs are also available from the Ableton website. Based on the graphical programming language of Max 6 from Cycling '74, Max for Live provides Live users with the opportunity to create, modify, and share new and existing Max for Live devices. This means that you can create instruments, effects, and extensions that exceed the commonplace everyday instrument designs and even reinvent your MIDI hardware controllers, converting them into unique tools for creating, producing, and performing. Max for Live makes this all possible by tapping into the limitless tools of Max through its graphical programming environment, limited only by your imagination and

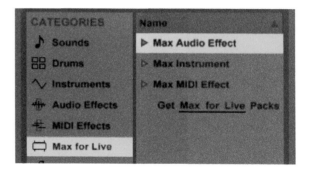

Figure 15.52
Max for Live.

knowledge of the Max language. With Max for Live, you have the ability to hack into Live's inner workings and modify just about anything in the interface and engine, as well as hardware device interaction. This is a programmer's dream.

When you first add a Max for Live device from the Browser, the Max for Live application will launch and run invisibly in the background—that is, until you *click* on the *edit* button. At that point Max will launch as a standalone application. Now let's take a look at a few of the included Max for Live devices. Keep in mind that this section does not attempt to cover all of the Max for Live devices. For a general overview of the others included in Suite, launch to ▶ **Web** .

15.6.1 LFO

Figure 15.53
Max for Live: LFO.

This Max for Live device has quickly become a favorite. Modulating device parameters is such a great way to make instruments into creative masterpieces. LFO is a very simple yet powerful tool for using waveshapes to control the physical button, knobs, and sliders of an instrument, effect, or even a track parameter on a different track. Choose from seven common waveshapes. On the LFO interface you will find an animated display of the waveshape. Below the display you can set the *Freq* (frequency rate) of the LFO to speed up or slow down the modulation effect. This can be set to sync to subdivisions of your song's tempo or to standard frequencies. Use *Depth* to moderate the amount of the effect and *Phase* to shift the waveform right or left (over time), effectively

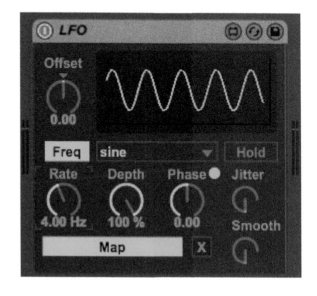

altering its start point within the waveshape. *Jitter* adds virtual noise to the shape, which creates a rugged waveshape, while *Smooth* rounds out the shape. To assign LFO to a parameter, *click* on the blue *Map* button, then on the desired parameter in another device. A really quick way to hear the effects of LFO is to map it to the volume parameter of an instrument or track. Try adding multiple instances of LFO to the same track too.

15.6.2 Envelope Follower

The *Envelope Follower* device is very similar in nature to LFO except that it tracks or traces the incoming amplitude of audio signals—any audio source, including the output of a MIDI instrument—to create an envelope to match. That information can then be used as a control over a physical parameter. On the interface is a display window that shows the shape of an incoming audio source. The *Gain* knob controls the amount of signal received. The more gain, the greater the incoming signal, thus a wider potential range for the envelope to draw. *Rise* controls the intensity of the transient shapes that are tracked by the Envelope Follower. Increasing this value will effectively smooth out the transients, essentially the attack of the envelope. *Fall* similarly controls the decay of the source material, smoothing out the release of the envelope. The Minimum and Maximum slider boxes scale the range of the effect sent to a mapped parameter. The *Delay* parameter delays the effect of the envelope; otherwise it would mirror the input source in perfect sync. By default this is set to "Time" in milliseconds. For more beat-driven audio sources, set this to "Sync". In sync mode, delays are based on subdivisions of the beat, which can create some very musical and creative effects. To make a mapping assignment, *click* on the Map button and then on the parameter you wish to control. Try this one out with auto filter by mapping Envelope Follower to Auto Filter's frequency, boost up the Q resonance, and let the magic happen. This can be even more effective than a gate sidechain effect.

Figure 15.54
Max for Live:
Envelope Follower.

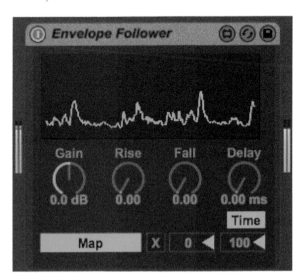

15.6.3 Mono Sequencer

This MIDI effect pattern sequencer comes in very handy for programming leads and basses without having to physically play MIDI notes or program them into clips. There are 12 programmable patterns in all that can be created, copied, or pasted between sequencers. These patterns are launched from within the device, each of which can be chosen on-the-fly. Use the *Quantize* button—similar to clip launching—so patterns always engage in tempo with your Set. Beyond just notes, the *Mono Sequencer* sequences pitch, velocity, octave, duration, and repeats. Each of these *Sequencer Data Lanes* is selected from the left side of the sequencer window. Sequences are programmed from within the display by *clicking* and *dragging* the blue bars in an upward or downward direction. Some sequencer lanes are bipolar (positive/negative values) while others use standard minimum to maximum values depending upon the parameter. *Pitch* is a good example of the bipolar feature.

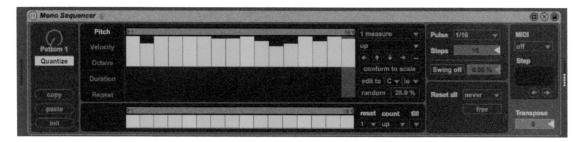

Figure 15.55 Max for Live: Mono Sequencer.

You'll find many of the sequencer parameters self-explanatory, such as *Pitch*, *Velocity*, and *Octave*, but *Duration*, *Repeat*, and *Step enable* are probably a bit less familiar. The Duration sequencer allows you to sequence the length of each step—steps represent some subdivisions of the beat based on the *Pulse* and *Steps* settings—translating into how long a note is held out for. The Repeat sequencer sets how many times a note repeats over the duration of the step to which a value has been assigned. Up to 16 repeats are possible per step. Below the Repeat button is the Step enable sequencer lane, which controls if a step is active or not. This is essentially a gate sequencer for creating rhythmic effects. Notice, too, that all sequencers also have a loop brace. Sequencer playback performance parameter controls are located to the right of the sequencer window. Each sequencer can be assigned individual playback behaviors. At the top is the *Sequence Reset chooser*. This regulates when the selected sequencer retriggers from its start. An interesting setting is MIDI key, which listens for MIDI input to rest the sequence. Below that is the *Step Behavior* chooser. This sets the direction in which the steps advance. Up means forward; left to right and down means backwards, right to left. The others settings are variation of the two. Beneath this are arrow buttons that enable you to shift the step placement of the sequence forward, backward, or flat. Step values can also be raised or lowered as well. Use the *Conform to Scale* button to quickly adjust a sequence to match a musical scale (i.e., major, minor, or modes). There are additional quick controls for scalar patterns.

On the right-hand side of the sequencer display are the Pulse and Steps controls. Here you can choose the resolution of the steps, dictating the speed as advancement occurs. Steps determine how many total steps there are in the sequence. Add swing and scale the amount applied here too. The *Reset All* section is a global reset for all sequencers. When Sync is enabled, all sequencers will reset when Live's Arrangement playback is initiated, otherwise they would continue from where they left off, hence "Free". Just to the right of these global controls is the *MIDI Input* routing section. This allows you to program steps, parameters, and select patterns using MIDI input. This makes for great on-the-fly triggering and sequencing.

15.7 Device Chains

Up to now we have focused primarily on individual devices and haven't spent much time exploring how multiple devices will interact when put together on the same track, such as combining a MIDI effect with an instrument. Before we start stringing together devices into what is called a "device chain", let's define it. A device chain in Live is the pseudo-physical connection of multiple devices linked in succession, passing audio and/or MIDI signals from device to device. Remember, devices are instruments, effects, or plug-ins, and just like any other DAW you can insert multiple devices on the same track (MIDI, audio, Return, Group, or the Master track) or in the same device rack forming a serial connection of devices. As you begin chaining devices together, you'll find there is a hierarchy to the chain. It's all about signal flow. Devices always receive and pass audio or MIDI signals in succession (serially) from left to right. This is not a new concept. In fact, it is the same process that all DAWs follow except they usually pass signal from top to bottom. A device chain can be any combination of instruments, effects, or plug-ins. You will see in the following example an Arpeggiator, synth, compressor, and a delay creating a complex instrument. This is an example of an instrument-based device chain. For a detailed description of signal flow, launch to ▶ **16.2.2** . Note in the example that the Arpeggiator and customized Compressor devices are folded within the Device View. This is a useful way to see more devices in the Device View.

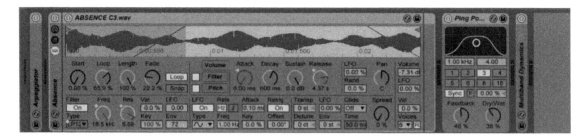

Figure 15.56 Fold devices for more viewable space in Device View.

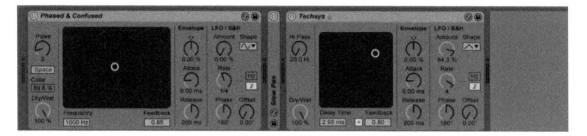

Figure 15.57 Chorus "Slow Pan" folded between Devices.

To illustrate the idea that a device chain can be created on an audio track, Return, Group track, and Master track, here is an example of a Phaser, Chorus, and Flanger effect making up an effects-based device chain (effects processor). Note that the Chorus device is folded in between the Phaser and Flanger as a space-saving decision.

15.7.1 Chaining Effects

There are two ways to go about chaining effects together. The concept is fairly simple. The key is to understand the signal flow and the difference between an insert effect and a send effect. First off, when you are building custom instruments in Live, you will generally make use of audio effects as inserts. This means that they are inserted directly into the signal path of instrument track or audio track containing some audio file source, creating a chain. You saw this when a compressor and delay were added to an instrument. In that way, inserted effects devices make up a part of the instrument and its sound. The alternative to this is to create a chain of effects on a Return track, thus creating an effects processor to be used as a Send effect, such as a reverb. In this scenario, the Return track receives a copy of an audio signal from an audio or MIDI track via track Sends—hence the name "send" effect. This signal is received and processed through the Return track's device chain—left to right in Device View—then output to the to the Master track. At the Master track the effected (wet) signal received from the Return track is blended with the original track source (dry) signal.

Device chains may also be created on the Master track where the audio signal received is the sum of all of the tracks in your Set. The Master track then processes the "Mix" through its effects chain. This is the common scenario for self-mastering your mixes. Live's Multiband Dynamics, Glue Compressor, and Limiter are great choices to experiment with for this purpose!

We have just looked at effects chain examples consisting of basic devices that can be created and chained one by one within a track. Taking the concept of chains a step further, Effect Racks can add a whole new level of integration, flexibility, and performance parameters. For more on Effect Racks, launch to ▶ 16.4.2 .

15.7.2 Interfacing with Device Chains

As you begin to build device chains you'll notice that the Device View gets filled up quickly. Since devices visibly extend left to right; it's quite possible that some will already be off the screen and out of view after only a few devices have been added. For this reason, there are a few ways to view and navigate the track display in order to maximize efficiency.

When you first add a device to a track it will appear in full view, meaning that you will see its entire user interface. As mentioned before, use the Device View Selector to scroll back and forth through the device chain.

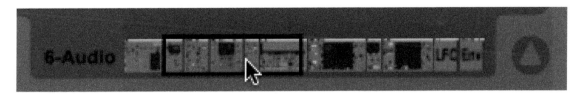

Figure 15.58 Use the Device View Selector for Scrolling Device View.

You can also navigate left and right with the keyboard arrow keys or with horizontal scrolling on your mouse if it's capable. To make room or view a single device in Device View, *double click* the Title Bar of any device in the chain and fold it to thin vertical bar, like a book on a shelf. On the spine of the device you will see the Title, Save Preset, and Hot-Swap Button. Collapsing devices creates a ton of space in the Device View for you to clear clutter and focus on the effects that you are

Figure 15.59
All Devices folded.

working with. *Double click* again to unfold the device back into full view. Collapse all of them if you really run a tight ship.

Moving effect devices around in a chain is very easy. Remember that signal flows left to right through the chain. With that in mind you can reorder your effects however you prefer. Just *click + drag* a device to the desired location in the chain.

Hot Tip When working with active clips and effect devices at the same time on the same track, you can easily toggle between detail views (Device/Clip View). If you are working in Clip View you can switch to Device View by double clicking a track's title bar or simply pressing shift + Tab on your computer keyboard. To return back to the Clip View, double click directly on the clip you wish to view or use shift + Tab again. This applies to both Session View and Arrangement View.

15.7.3 Sidechaining

Sidechaining is a popular process in music production, yet can be quite elusive when it comes to figuring it out and to master the process in a DAW. You've heard the effect in your favorite songs; now you want to do it for yourself. This all goes back to the question you've probably asked at some time or another: "How do I get that produced sound?" Well, this is one of the ways. Thankfully, Live makes this actually pretty easy to comprehend and execute.

What is sidechaining? The practice of sidechaining audio consists of taking an audio signal from one track and using it to affect (modulate) another by way of an effect device of some sort that supports a sidechain input—in essence, using one sound to change another sound. A very common use is to affect the dynamics or output of one signal by another signal.—for example, the concept of "ducking" in radio commercials, where when a voice speaks the backing music dims. Another example is the trance chord breathing-suction effect (sidechain compression). There is also the synth-pad gating effect, where sidechaining results in the carving out of rhythms within a sustained pad or chord!

Several of Live's audio effects support the sidechain input feature: *Compressor, Corpus, Gate, Auto Filter, Glue Compressor, MBD*. In all of these audio effects with the exception of Vocoder you will see a small arrow button located next to the Device Activator as shown in the Compressor example. This is the *Sidechain Toggle*.

Figure 15.60
Sidechain Toggle.

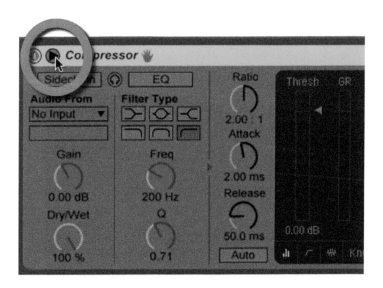

By *clicking* on the Sidechain Toggle, you'll see all of the components for the sidechain feature. Conveniently, Vocoder's sidechain input parameters are already in view and ready to work with. For this demonstration we'll reference three examples depicting the Gate audio effect. In the first example you will see Gate as a standalone

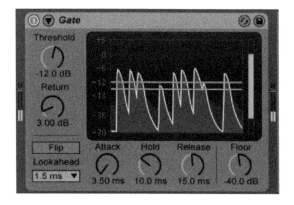

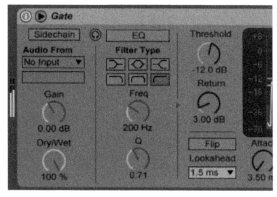

Figure 15.61
No Sidechain active.

Figure 15.62
Sidechain parameters.

effect without sidechain open or activated. The second example shows the sidechain parameters open and ready to be activated.

Last, the third example shows the sidechain feature fully activated with the "Audio From" input chooser set to a source track of audio/MIDI that is activating the Gate sidechain feature.

Once a signal is represented in the sidechain process, it's time to set the appropriate parameters to achieve the desired sound. Let's dive into a quick run-through on working with the sidechain feature using the Gate audio effect.

With Gate, it's possible to achieve a stutter effect (rapid muting effect). The audible effect is very fast silencing in the audio that makes it sound glitchy or gapped, but at a speed that doesn't result in much silence at all. To achieve this effect you'll

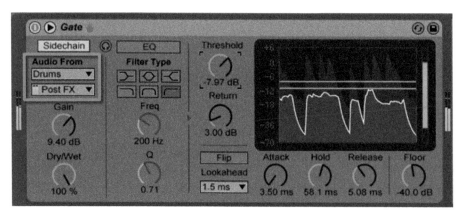

Figure 15.63
Sidechain activated with the "Audio From" input chooser set to a source track of audio /MIDI/.

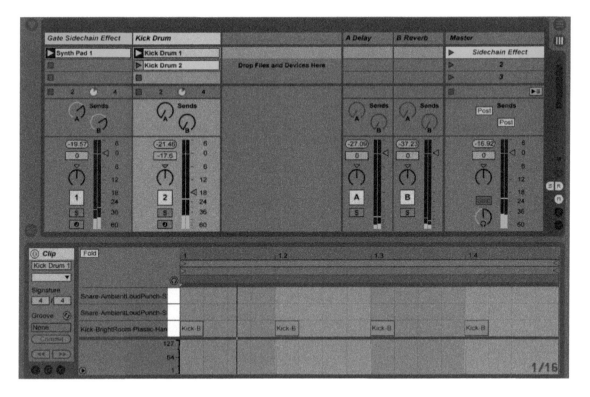

Figure 15.64 Gate sidechain effect with drums for getting a stutter effect. Kick or snare sample selected for drum source.

need two MIDI tracks, "Track 1" with a synth pad and Gate; "Track 2" with a drum kit instrument. Use presets from Live 9. "Track 1" should be a legato pad-like synth. A sustaining synth pad instrument patch will work perfectly. "Track 2" ("source track") can be any drum kit you like:

1. Select a solid kick drum or snare drum sample for your drum source "Track 2" to feed into the sidechain input chooser window. (Figure 15.65.)

2. Program MIDI notes for the kick drum to play back as a frequent beat pattern for a length of two bars at 120 BPM (beats per minute).

3. Now using your legato synth pad you selected for your Gate sidechain track, create a two-bar sustained chord loop or simple melodic phrase into a clip in the same Scene as the kick drum clip.

4. Launch your Scene with both clips playing: the synth pad "sidechain" track and the source track (your kick drum beat pattern).

You'll notice that your signal level on your Gate effect is activated. That's what you are looking for. Now control Gate's threshold parameter to begin dialing in the perfect balance between a ducking gate-style effect and the original synth pad melody or chord. If you don't hear an immediate effect after adjusting the threshold level, make sure the Sidechain button is activated in the Gate sidechain window. Generally, the higher you raise the threshold level the more obvious the gating/sidechain effect will be.

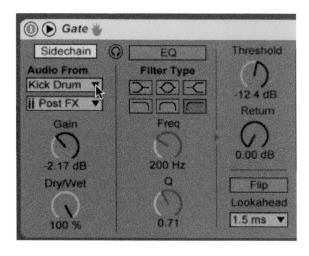

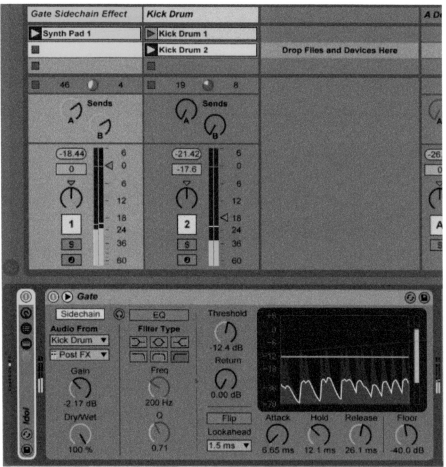

Figure 15.65
Source fed into Sidechain input chooser on Gate.

Figure 15.66
Both clips launched. Synth Pad track gated by Kick Drum.

Hot Tip Adjust the Release and Hold parameters to give a natural decay to the affected sound. While the clips are playing back and gating, fade down the drum source track volume to hear just the effect of the Gate. Pretty cool! If you want to get creative, you can also add a Ping-Pong Delay after the gate sidechain effect to animate and bring the new sound to life. Now try dropping in a full drum loop clip on another new audio track and hear the creative possibilities start to unfold.

15.8 Plug-in Devices

Plug-ins, by definition, are applications that function within another software application, enhancing the host software's functionality. In this case, these are instruments and effects that Ableton does not develop. Plug-in devices are designed by third-party developers to operate within any DAW. As you know, Live includes many well-integrated instruments and effects, but once you have exhausted all of them—and that may take a while—consider branching out and investing in some additional third-party instruments and effects. Live supports third-party audio units (AU—Mac only) and virtual studio technology (VST) formats for effect plug-ins and virtual instruments. This will allow you to work with thousands of additional unique instruments and effects using the Live Browser for seamless integration of your third-party instrument and effects, just like Live's. You can even mix and match Live's instruments and effects alongside third-party plug-ins within tracks and in Device Racks. Now, before you spend all of your hard earned cash on the hottest new plug-ins, understand that no matter what plug-ins you own, your creative ideas, arranging skills, and perception of sound are far more important. Just like any other hobby, profession, or sport, it's easy to get caught up in the technological glamour of it all. With that in check, the more resources you have, the greater the potential for realizing your music. Once you own Live 9 Suite and max out your Live resources, then stock up on some cutting-edge third-party plug-ins!

15.8.1 Plug-ins Browser

There are two types of plug-ins: Instruments and effects, both of which are accessed from the *Plug-ins* category listed under their manufacturer. Before you can use these plug-ins with Live, you'll need to enable them for use in the Live Preferences menu (*Preferences>File Folder>Plug-in Sources*). Once activated, Live will scan for all of your AU and or VST Plug-ins, making them available in the Plug-ins category for use.

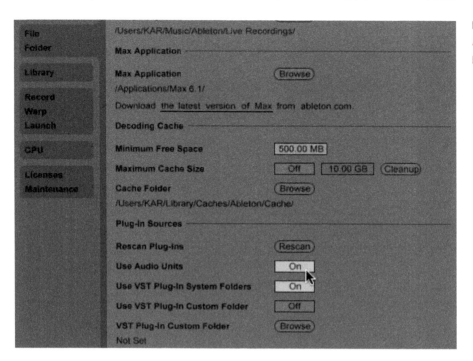

Figure 15.67
Activate AU/VST
plug-in formats.

Unfortunately, the Browser sorts these virtual instruments and audio effects together into one single folder structure, so you may have to do some digging to find what you want—but it's well worth it!

15.8.2 Third-party Instruments

Third-party instruments come in all shapes and sizes and are often called "software instruments", "virtual instruments", "soft synths", "plugs", etc. To clear up any confusion, these titles are the generic "on the street" labels; but there is a difference. First off, "virtual instrument" is a fair name, but beyond that, the rest are slightly inaccurate. Software synths or soft synths are instruments that generate sound through synthesis, a synthesizer if you will. By definition, that leaves out samplers and sample

Figure 15.68 AU/VST Plug-ins Category.

players, which are designed to store and play audio samples as opposed to generating tones through synthesis techniques. So, then, we are back to "virtual instruments". Add the word "plug-in" to the end of virtual instrument and it's settled. As you know, Live uses the term Plug-in Device. This is because they don't delineate between instruments and effects. Unfortunately, you will have to memorize what plug-ins are what types. This is not that big of a deal, but it can slow you down when you accidentally add what you think is an instrument and it turns out to be an effect, or vice versa. Thankfully there is the undo command ([⌘+z] Mac/[ctrl+z] PC) that will save your life, time and time again.

15.8.3 Third-party Effects

As for effects, they are what they are: digital signal processors. They too come in all different shapes and sizes covering an array of dynamic and time-based processors. Effects serve one of two purposes: sound shaping (e.g., dynamics, timbre, pitch, etc.) and performance enhancing (e.g., specialization, timing/tempo-based). Common sound-shaping plug-in effects include EQs, compressors, limiters, de-essers, and saturators, etc. There are some large-format third-party plug-in effects that come packaged as an all-in-one effects unit featuring multiple effects within one dedicated user interface (i.e., Izotope Ozone 5). Common performance-enhancing plug-ins are delays, chorus, flanger, modulators, and reverbs, just to name a few. Of course, many effects border the line between both sound shaping and performance enhancing. In the end, the classifications don't matter, just as long as you know what each effect does, how to use and insert it, and, most importantly, that you like the result.

15.8.4 User Interfacing and Layout

Interfacing with third-party plug-ins in Live is much more different than the way in which you were interfacing with Live's own devices. You will still load the plug-ins the same way, but the plug-in's user interface (GUI) will appear in its own unique floating window instead of in the Device View and will appear as a generic device with X–Y control in Device View. To close the plug-in's window, *click* on the small red button in the upper left of the interface window.

When the Plug-in's floating window is closed, access the plug-in via the Device View where you'll see it's the generic device interface with assignable *X–Y Control* window. This lets you assign any two specific plug-in parameters at a time to the control window for free-flowing control over the plug-in's parameters in real-time. If your plug-in has 32 or fewer automatable parameters, they will be displayed as sliders to the right side of the Control window and will be ready for assignment

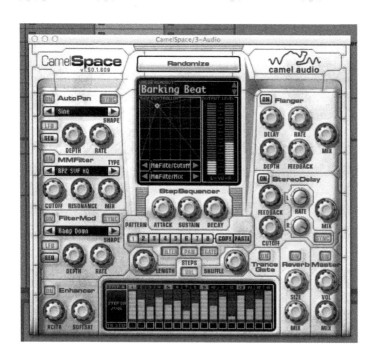

Figure 15.69
Plug-in user interface appearing in a Floating Window.

in the X–Y Choosers at the bottom of the interface.

If it has more than 32, this area will show nothing. Instead, you must manually add parameters to be displayed. On the Title bar there is an *Unfold Device Parameters Button* and just below that is the *Preset Chooser* and *Load Preset Button* when presets are available (not available on all plug-ins). Unfold Device Parameters to view the available internal parameters for the plug-in. If none are available, then *add* parameters by selecting the *Configure Button* on the right side of the Title Bar. You can also rearrange and delete parameters. *Click* "Configure" (green = On), then re-open the plug-in's main interface floating window by *clicking* on the *Plug-In Edit* "wrench" button (if closed) and *click* on a parameter on the plug-in's interface to add it to the listing of available device parameters.

Figure 15.70
Assign any two specific plug-in parameters to the X-Y control window for real-time manipulation.

Figure 15.71
Plug-in device
X-Y Control/
Parameter layout.

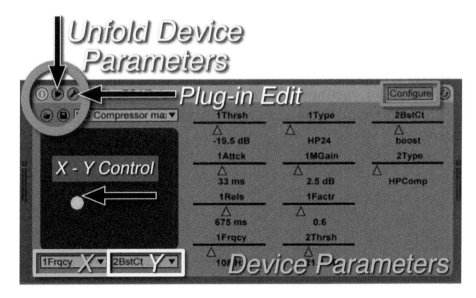

To delete, *click* on a green-lit parameter in the device and *press* the delete/backspace computer key. Not all plug-ins are compatible with the Configure feature, so don't be disappointed if you can't assign any controls or access presets from the Device View. For example, Native Instruments' Kontakt sampler is not compatible with this feature. Instead, you would assign automation through its own internal automation window.

Figure 15.72
Configure Mode.

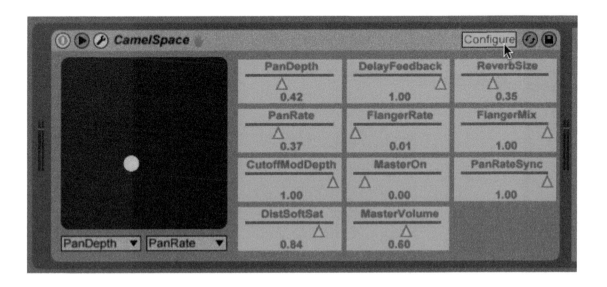

Figure 15.73 Kontakt floating window with Sample Logic sample libraries.

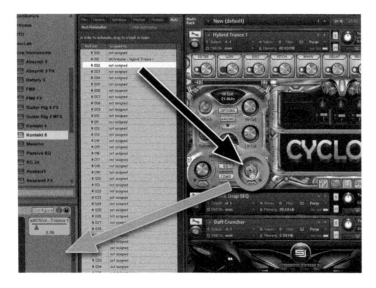

Figure 15.74 Must use the third-party automation assignments for some devices in order to make assignments to the X-Y control parameters in Live.

Once you have selected parameters of a device they will become available for MIDI/Key mapping in Live as well as assignable in the X–Y choosers. Choose parameters from the chooser that you wish to control in the Control window or map them as you see fit. Next to the Unfold Button is the *Plug-in Edit* button. *Click* this to open and close the plug-in's floating window. As for the rest of the Device View, controls remain the same.

> **Hot Tip** To control Kontakt 5's instrument parameters with the X–Y feature, or even Macros, click the Configure Mode button on the Kontakt device in Device View, then in Kontakt, navigate to the Auto tab in the Kontakt Browser. Under Host Automation, click and drag a host parameter number #001 or greater and drop it onto an instrument parameter in the Kontakt interface. Instantly that parameter will appear as a device parameter for the generic device/X–Y controller. You could then assign these controls to Macros in an Instrument Rack.

15.9 External (MIDI) Instruments

Yes, it's true: people still use external MIDI instruments to make music. The most common instruments are generally synth keyboards, virtual analog synths, drum machine/sequencers, and digital keyboard workstations. We won't go into the specific manufactures or devices here, but let's look at how they function in Live. The External Instrument Device is not included in Live Intro!

15.9.1 Routing

Live's MIDI tracks cannot only be routed internally between tracks and instruments, but they can also be sent to the external world for use with external MIDI hardware devices. A configuration such as this requires a MIDI track to send out MIDI data to an external sound device and an audio track to monitor input of the external device's audio output so you can hear the sound of your gear and record it to track. The best way to do this is to use Live's External Instrument device located under Instruments in the Browser. This instrument type is a gateway to your external MIDI devices. There is no actual instrument or sound associated with it; instead, it acts as a bridge to your external device by converting a MIDI track into a hybrid MIDI/audio track. This means that you can trigger an external MIDI device and monitor its output in Live, all on one MIDI track. This is similar to instrument tracks in other DAWs. For now we'll focus on external MIDI.

To use external instruments with Live, they must be connected to your computer via a MIDI interface or USB connection and their audio output(s) should be routed into your computer via it's audio interface in order to the audio output of the instrument through Live. When routed back into Live you can record and process this output.

How to set up the external MIDI device:

1. From the *Categories>Instruments* folder, load the External Instrument Device onto an empty MIDI track or *drag* it into the Device Drop area. This will open the Device View and you should now see a device interface with "MIDI To", "Audio From" and a "Gain" knob.

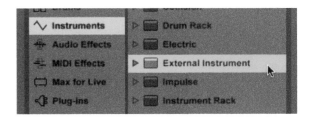

Figure 15.75 External Instrument.

2. Set "MIDI To" to the MIDI port to which your device is connected and set "Audio From" to the audio input channel the external device is connected to. This will vary depending upon the signal flow of your studio. The only way to monitor through Live is to route the audio back into the computer.

3. Set your MIDI track to receive MIDI from all inputs and all channels.

4. Arm your MIDI track and begin playing.

You should now hear your external MIDI instrument through Live. If so, then you are good to go. If you don't, make sure the Monitor section is set to "Auto" and that the MIDI input channel indicator is lighting up when you play. If you still aren't hearing anything, check all of your connections, inputs, and outputs for MIDI and audio. Last, consult your MIDI instrument's manual. Other common issues can be related to the instrument's Local Control, channel settings, volume control, etc.

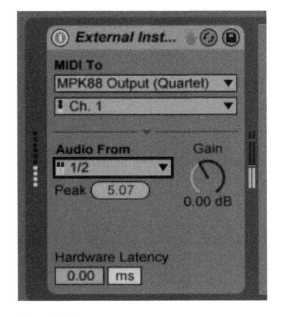

Figure 15.76
External Instrument routed to the outside world and back.

15.9.2 Rendering External MIDI

Rendering MIDI to audio is a common and important process, whether an external MIDI instruments or Live-based MIDI devices. Most of the time you will render audio offline, meaning that the process happens faster than the actual time it takes to play through the song, and you won't hear playback during the rendering process. With external audio, the process cannot be done offline because the audio must be converted from analog to digital since it lives inside the external instrument. This means the external device must physically play the notes in order for it to be rendered. The good thing is that Live handles this for you. When you render, Live will automatically determine if the content contains internal or external audio. If it does contain an external audio source it will render in real-time.

Whenever you want to render in Live you have a few choices. In ▶ **8.3.1** we talked about Freezing and converting MIDI to audio, so that is one option. A favorite! Your second option is to render/bounce your MIDI (Export Audio/Video). The third choice is to print your MIDI as audio to a track by recording the external device's output onto a new audio track in Live. The decision is all yours. Launch to ▶ **8.5** for more about exporting and printing audio and MIDI.

15.10 Working with Devices

There are a number of exciting ways to create, produce, and perform with third-party and native Live Devices. With the popularity and convenience of software-based samplers, it has become increasingly important to understand how to work with multi-instruments (multis) in your productions. If you are unfamiliar with software samplers, they are powerful virtual instruments that allow multiple instrument presets to be loaded into a single interface, like a rack. In this way the various instruments can be addressed all via one MIDI channel or via multiple unique MIDI channels. In the former, instruments are triggered using the same channel for all instances or OMNI mode. In the latter, each instrument is assigned its own MIDI channel and triggered separately. Let's take a moment to look at a plug-in device as a multi-instrument in detail.

15.10.1 Produce: Multi-instrument Plug-in

For this example you will see Kontakt 5 (plug-in) by Native Instruments. Kontakt allows you to selectively assign multiple discrete MIDI channel inputs from Live MIDI tracks to different instruments.

Here's how to do it:

1. Open a new Set and delete all of the tracks except one MIDI track.

2. Select Kontakt 5 (or your favorite multi-instrument plug-in) from the Plug-ins category and insert it on the MIDI track.

3. Create four new MIDI tracks. You'll have five MIDI tracks in all.

4. Using the multi-parameter select function (*shift + click*), highlight tracks 2 to 5, then select "1-Kontakt 5" in the "MIDI To" input chooser on track 2.

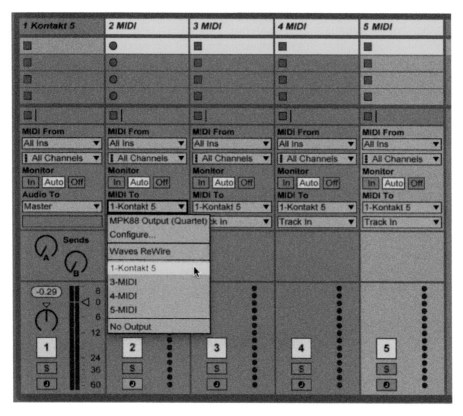

Figure 15.77 Select "1-Kontakt 5" in the MIDI To input chooser on track 2.

Since you have multi-selected your MIDI tracks, tracks 2 to 5 will be automatically set to 1-Kontakt 5 input across all the selected tracks at once. A great time saver!

5. Set the "MIDI To" output chooser for tracks 2 to 5 to each discrete Kontakt 5 MIDI channels 1 to 4. You should find the specific channels for your instrument

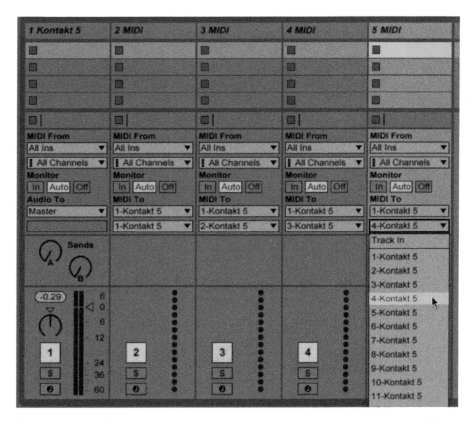

Figure 15.78 Set the MIDI To output chooser for tracks 2-5 to each discrete Kontakt MIDI input channel.

plug-in listed here. If not, check with the manufacturer and make sure your instrument has the actual multi-instrument capabilities.

6. Now open your instrument plug-in device on track 1 using the Plug-in Edit button and add four instrument presets from the Kontakt browser.

7. Set each instrument's MIDI channel input to match each of your Live MIDI track assignments on track's 2 to 5, starting with 1-Kontakt 5 and ending on 4-Kontakt 5 as in this example. (Figure 15.79.)

8. Arm track 2 and record some MIDI notes to create a MIDI clip. Repeat this across the remaining MIDI tracks until you have a Scene of MIDI clips. You'll soon see and hear your performance as a multi-instrument, all four instruments triggering on discrete MIDI channels as a multi-track performance. (Figure 15.80.)

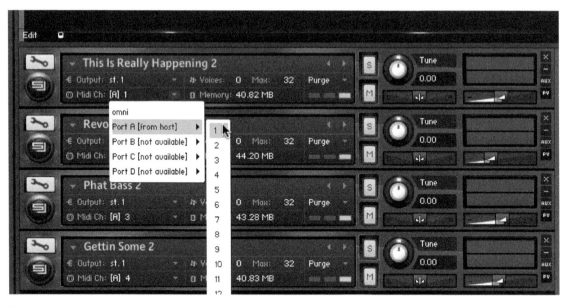

Figure 15.79 Make the appropriate MIDI routing assignments to connect your plug-in and MIDI tracks.

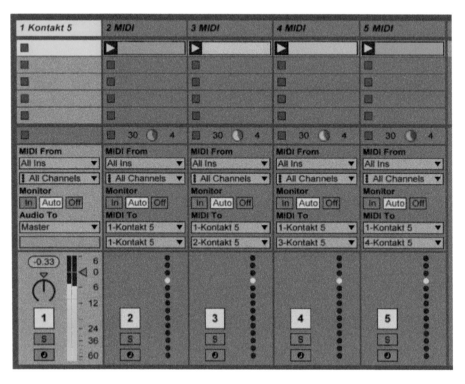

Figure 15.80
Multi-instrument
in action.

9. Remember that track 1 is serving two purposes. It acts as the host track for your dedicated plug-in instrument device and also as a dedicated monitor input. There is no need to record on this track for any reason. With the proper assignments as shown here, you have multiple tracks to record to and a very flexible way to make your productions efficient.

Figure 15.81 Track 1 is the plug-in host and the monitor source.

Device Racks

16.1 What Are Racks?

Up until now we have only looked at basic instrument devices (e.g., Impulse, Simpler, etc.) and various MIDI/Audio effects. These have all been very exciting and no doubt sparked your creative juices, but they are all examples of fairly traditional Live devices that can be created without the assistance of a Device Rack. If you thought the early examples were fun, then you are in for a treat. Instrument Racks, Drum Racks, and Effects Racks take Live devices to the next level—a good reason to graduate to Suite if you haven't already!

A Device Rack provides you with the ability to build complex devices by combining effects and instruments with flexible hands-on parametric controls into one contained device chain. This pools the power of all of Live's devices to create unique custom effects processors, multilayered synths, and intricate, effectual, and interactive performance-ready instruments. This all sounds great, but you're probably wondering what this all really means, and how you can take advantage of its theoretical power.

What is a Device Rack? Think of a Device Rack as a group of individual devices or device chains that function in tandem (parallel individual rows). Each independent device chain is self-contained and processes its own signal as it flows from left to right, not interacting with the signal of any other device chains. At the end of the Rack, the chain(s) output is summed with the other chains in the Rack, culminating at the track's output. Simply put, a Rack is a device made up of a combination of multiple chains of linked devices, although a Rack can consist of only one device chain as commonly found in Core Library presets. Now: it should pointed out that *Drum Racks* function slightly differently than the rest of the Device Racks, particularly Instrument Racks. In Drum Racks, each chain has an input/output section that allows it to be assigned to a designated MIDI note that will trigger it, whereas with all other Racks, all chains receive the same MIDI input signal except those that are affected by a specific set of rules established within the Rack. Launch to ▶ 16.4.3 for more on Drum Racks.

16.2 Rack Interface and Layout

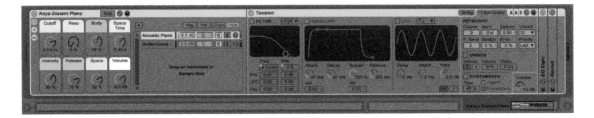

On the surface, all Racks are made up of at least three main sections: *Macro Controls*, *Chain List*, and *Devices* interface. Drum Racks include more areas, most notably *Pad View*, which we'll get to a little later. For now, let's tour the general Rack interface and layout. At the top of a Device Rack display are some standard buttons: Device Activator, Hot-Swap, and Save Preset. Next to Hot-Swap you will see a button called *Map (Macro Control Map Mode)*. This is used to map Macro knobs to parameters in the Device Rack in order to control various assignable parameters, as described in the coming section. On the left-hand side is the *View Column* containing the three Rack specific selectors: Show/Hide Macro Controls, Show/Hide Chain List, and Show/Hide Devices. Use these buttons to unfold and manage your Rack's contents and view in Device View. They will be black or yellow depending upon their view status.

16.2.1 Macros

To familiarize you with Racks, use an Instrument Rack preset that comes with Live. The first thing that sets a Device Rack apart from a non-Rack device is the set of *Macro Controls*. Macros are designed for controlling the internal parameters within a Device Rack. To show or hide Macros, *click* the *Show/Hide Macro Controls* button just below the Device Activator.

These eight controller knobs provide great power and control over the contents of a Rack. They are totally customizable. Not only can they be mapped, re-mapped, and hidden, you can set your own colors for each knob and label

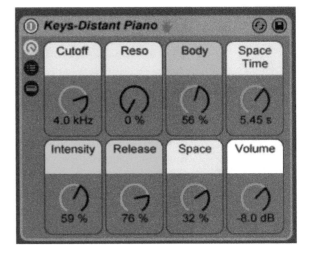

Figure 16.2 Macros for controlling internal parameters within a Device Rack.

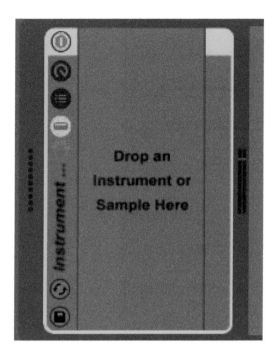

Figure 16.3 Empty Instrument Rack.

them however you wish. In the same way that Macros can control its Rack's internal parameters, they too can be assigned to MIDI/Key maps. Think of the possibilities on and off stage with unlimited controls over your instruments and effects! Refer to ▶**Scene17** for in-depth mapping. We'll come back to Macros in a moment.

16.2.2 Chain List

Racks can hold multiple chains that operate in tandem with each other, something that standard devices do not do. Signal flow is still the same as a standard instrument or effect device, except Racks employ a more complex internal routing system in order to handle multiple device chains. As mentioned earlier, each chain in a Rack (except for Drum Racks) generally receives the same incoming MIDI or audio input signal, although this is determined by the setup of the *Zone* and *Chain Select Editors*. More on that in a moment. This becomes obvious when you look at the interface and see that the Chain List houses each device chain in a horizontal layout. Not only does this represent each device chain in the Rack, but it also serves as an internal mixer and Rack view organizer for navigating

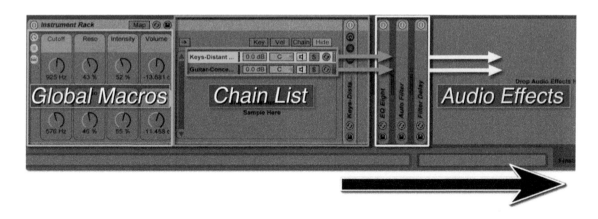

Figure 16.4 Instrument Rack: Signal flows left to right. Chains operate in parallel to one another, receiving and pass signal simultaneously through the Rack to the outputs.

each device's parameters and settings. Device Rack Chains are managed internally in a very similar way that sample zones are managed in Sampler. This is one of the great benefits of racking devices: you can assign specific conditions that determine how and when a chain within a Rack can be triggered. Each device chain in the list has a Title, Volume slider, Pan slider, *Chain Activator* button, Solo button, and Hot-Swap button, all of which can be used in conjunction with Macro controls—very powerful tools with a whole world of possibilities and real-time control.

Below the Chain List is the *Drop Area* where you can add effects, instruments, samples, and more Racks. Audio units (AUs) and virtual studio technology (VST) can be *dropped* here as well. At the top of the Chain List there are several buttons. *Auto* sets the Chain List to automatically select (highlight) a chain when it is receiving input. This way you can see when chains are triggered, allowing you to watch what is happening with your current device chain while your Set is playing.

The *Key button* brings the *Key Zone Editor* into view. This is where you can specify the Key Zones for each chain. Key zones are the areas or ranges of the MIDI keyboard from where a chain is assigned to be triggered. If a note is played outside of the assigned Key Zone, the chain will not be triggered and therefore will not play back. This is often used to split the keyboard note range so that two different instruments can be triggered via their own note range. This is similar to Sampler, if you recall: ▶ 15.3.3 .

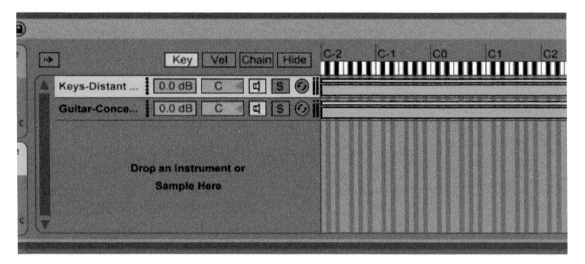

Figure 16.5 Key Zone Editor.

The *Vel* button shows the *Velocity Zone Editor*, which represents velocity values spanning from left to right. From this editor you can assign a chain to particular velocity values which, when played, will trigger that chain if the incoming MIDI note velocity values fall within the assigned range. For example, a chain assigned to velocity values 1 to 40 will be triggered when notes that fall in that value range are played. Notes that fall outside that range will not trigger the chain, resulting in the chain not playing back for that note. Velocity zones are often used for triggering various timbral layers of an instrument so that hard velocities trigger an aggressive or excited sound and softer velocities a milder, somber tone. You'll find this is not the only use for these zones. Note: Audio Effects Racks do not have "Key" or "Vel" buttons!

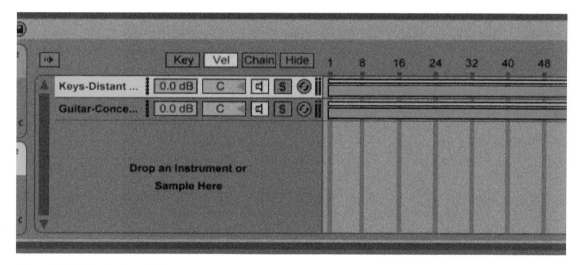

Figure 16.6 Velocity Zone Editor.

The *Chain button* displays the *Chain Select Editor*. The Chain Select Zone establishes what chains are triggered or active and when. This is similar to the way in which other zone-based assignments work. There is a possibility of 127 unique select positions or assignments that can be made. A chain can be set to one position or multiple consecutive positions similarly to the way in which a sample zone is stretched across the key map. The chain is identified as a blue horizontal line and the current select position is orange. So long as a chain falls within the orange Chain Selector, it will be triggered. This functionality is ideal for establishing the ability to choose between different chains or layers of chains to be played back when they fall within the assigned zone ranges. Theoretically, you could create 20 different chains, each on their own select position, with each representing a different instrument—like a patch list. Since each chain can be assigned to a specific chain select zone or to overlapping select zones (layered with another chain), you can select them on-the-fly by using Live's MIDI Mapping functionality to vary your device in real-time as an all-in-one multi-preset instrument or effects device. To change the position of the selected chain(s), *drag* the *Chain Selector* located in the *Chain Select Ruler* at the top of the Chain Select Editor.

Note that when any zones overlap (here or within the other Zone Editors), consider setting a *Fade Range* so transitions between zones can be smooth and less obvious to the ear. The Fade Range for each zone is located directly above the zone and can be *dragged* from either end of the zone.

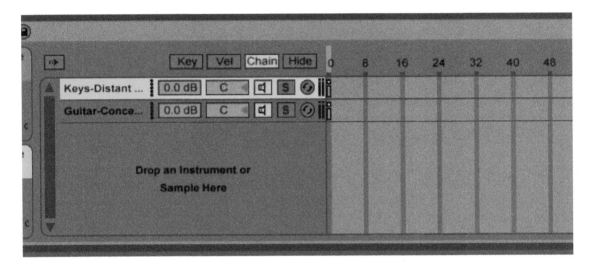

Click the *Hide* button to close out Key, Velocity, or Chain Select Zone Editors from view.

Figure 16.7
Chain Select Editor

16.2.3 Macro Mapping

Have fun with Macros, and assign them a color. To set the colors of each knob, simply *right click* or *ctrl + click* on a knob and select a color from the palette. There are two ways to assign a Macro to a control parameter and vice versa. You can use the Map button and make assignments using *Macro Map Mode* or *right click* or *ctrl + click* and make assignments using the contextual menu. To use Macro Map Mode, *click* the Map button at the top of the Rack. This is only visible when the devices are in view. *Click* the *Show/Hide Devices* selector, which will open up the Rack in Device View, then *click* Map. This opens up the *Macro Mappings Browser*

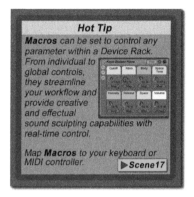

Hot Tip

Macros can be set to control any parameter within a Device Rack. From individual to global controls, they streamline your workflow and provide creative and effectual sound sculpting capabilities with real-time control.

Map *Macros* to your keyboard or MIDI controller.

in the left side of the main window of your Set. There you will see a list of all the current mappings for the selected Device Rack as well as additional parameter controls. In the Device View all of the map-able parameters and controls will now be highlighted green. To assign control, select any parameter then *click* Map on the Macro of choice. Once you're done, *click* Map again to exit Map Mode. If you prefer to use the contextual menu for mapping macros, just *right click* or *ctrl + click* on the parameter you wish to control, then choose the macro assignment from the menu. You will notice that mapped device parameters will have a green indicator when they are mapped to a macro.

16.3 Drum Racks

Although a Drum Rack is a Device Rack, it gets special attention, because it differs in its user interface, controls, and fundamental operation. The most obvious difference from an interface standpoint is the presence of *Pad View* located next to the Macros.

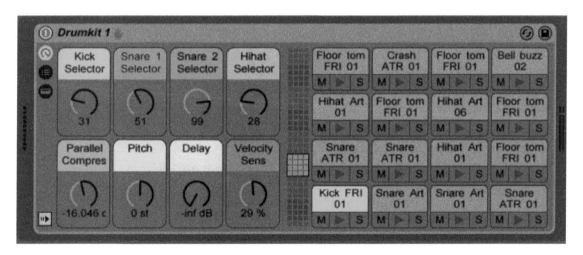

Figure 16.8
Drum Rack: Macros and Pads.

16.3.1 Pad View

Pad View is used to interface with samples and various devices. Pads are triggered like a drum pad. Each pad is assigned a unique MIDI note (1 to 128) ranging from C2 to G8. This is exactly how Live handles MIDI sliced audio clips created from the "Slice to New MIDI" command. Refer to Clip ▶ **14.3** . *Drag* samples, effects, instruments, Presets, and even audio loops onto pads. When you *drag* a sample to a pad, for example, Simpler will automatically load up as a chain assigned to that pad. If you *drag* a preset, then that preset and all of its components will load up as a chain. *Drag* effects to pads to added the effect to the current chain or create a new chain. New chains will require the addition of a device to make use of the effect. You can *drag* multiple samples at once to pads, which will then be assigned to the appropriate number of pads. Pads assignments can also be reordered: simply *drag* the pad to a different pad and the device chain will move to the new destination. Pad View is not only used to trigger chains, it also has a mute/solo feature and Hot-Swapping functionality.

Just to the left of the Pads section is *Pad Overview*. It shows the pads that are currently in view, and allows you to navigate to additional sets of pads, 16 at a time; 128 in all. Pad Overview is a vertical display beginning with the lowest Pads/MIDI notes at the bottom ascending to the highest on top. *Drag* the *View Selector* (outline box on the right over the Pad View) or use keyboard arrow keys to navigate the view in increments of an octave.

16.3.2 Internal Routing

Looking inside the Drum Rack you will once again find a Chain List along with its mixer section that has *Volume*, *Pan*, *Activator*, *Solo*, and *Hot-Swap*.

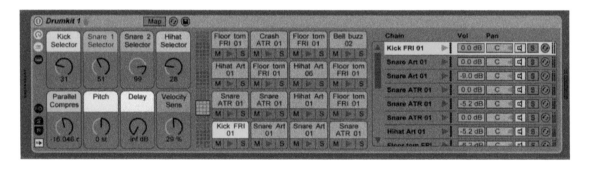

Everything is here within the Rack as expected, except that it is missing the three editors for Key Zone, Velocity Zone, and Chain Select. Additionally, the *Auto Selector* has been moved to the View Column along with three other View Selectors on the bottom left of the pads:

Figure 16.9
Drum Rack Chain List.

- *I-O*: Show/Hide Input/Output Section;
- *S*: Show/Hide Sends;

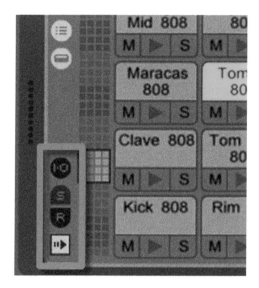

Figure 16.10 Drum Rack View Selectors.

- R: Show/Hide Returns;
- A: Activate Auto Select.

Auto Select, as mentioned, sets the Chain List to automatically select the chain currently processing MIDI input. You will see the selections change during playback. The *I-O* selector displays the MIDI Input and Output section for each chain. The *Send Selector* displays the chain's Send Level Controls and the *Return Selector* displays the Racks' return chain(s). Sends and Returns handle internal audio effects routing within the Rack, something that is unique to Drum Racks. The Returns Chain List appears directly below the Device Chain List. This is where audio effects will be added (*dropped*) to the Drum Rack. There can be up to a total of six Return Chains in all. Sends will appear in the Chain List mixer section next to volume when an audio effect has been added to the Return Chain List (creating a Return Chain).

The *Chain Send Level* slider determines how much signal is sent to another return chain in the list as indicated above the slider. Also included in the Return Chain List is a mixer section and an "Audio To" chooser for output routing.

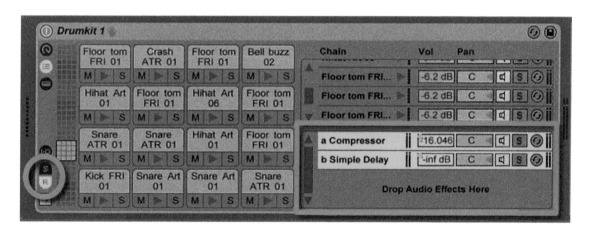

Figure 16.11 Return Chains.

As you can see, Drum Racks offer more advanced signal routing than the other device chains, whether MIDI input/output or internal effects.

Moving forward with this concept, each Drum Rack Pad responds to a unique MIDI note. In turn, each internal chain is triggered by individual MIDI input signals, although they can also receive input from multiple notes if desired. Therefore, in order to keep things fairly simple, we'll assume each chain within a Drum Rack's Chain List has been assigned to respond to a unique MIDI note. Such routing is determined by *Receive* and *Play* assignment established in the Chain List's Input/Output section. As you can see, Drum Racks are fundamentally different in how their chains process MIDI input.

Time for the rundown!

To fully wrap your head around the Drum Rack routing concept, remember that signal flows left to right in a Rack. That being said, each drum Pad is assigned to its own MIDI note (e.g., C1, C#1, D … etc.). Each Chain is triggered by a Pad. This means that when you play a MIDI note, its related Pad will trigger a related chain in the Chain List—the device chain that has been assigned to respond to it. This assignment is made via the *Receive* chooser. Whichever MIDI note is assigned here is the note that will trigger the chain, unless set to receive "All". For example, if a chain's Receive is set to MIDI note C1, then C1 will trigger the chain when it is played. If a chain's Receive is set to "All", then all incoming (played) MIDI notes trigger it. The *Play* chooser located next to Receive determines what note is sent into the device(s) within a chain when the receive note is played. To recap, Receive assigns the MIDI note that will trigger a chain and Play determines the MIDI note that is output to the device(s) in a chain.

Next to the Play is *Choke Group chooser*. Choke Groups are designed for the purpose of having one chain silence another chain when both are played in succession. Chains assigned to the same choke group will function in this way.

Figure 16.12
Chain MIDI
assignments.

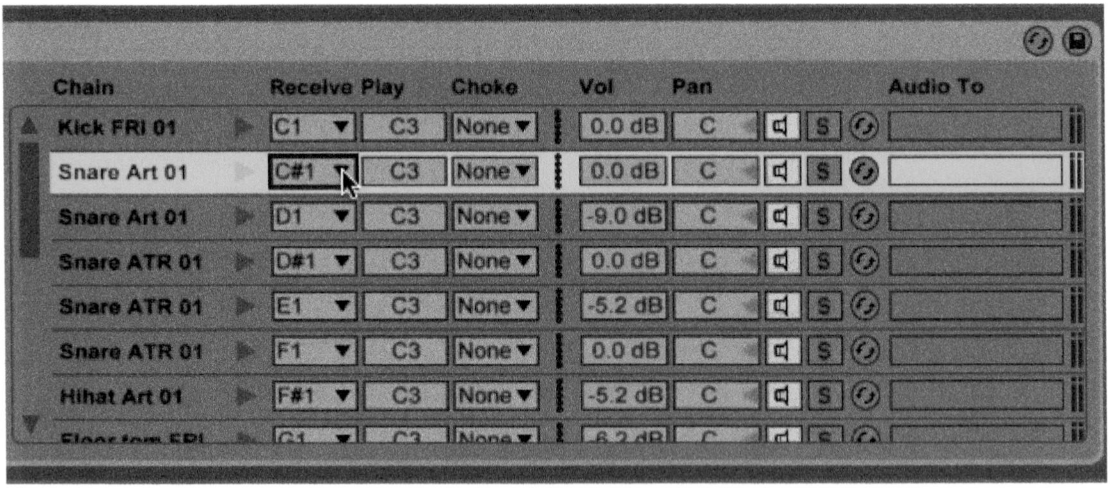

Chain	Receive	Play	Choke	Vol	Pan				Audio To
Kick FRI 01	C1 ▼	C3	None ▼	0.0 dB	C		S		
Snare Art 01	C#1 ▼	C3	None ▼	0.0 dB	C		S		
Snare Art 01	D1 ▼	C3	None ▼	-9.0 dB	C		S		
Snare ATR 01	D#1 ▼	C3	None ▼	0.0 dB	C		S		
Snare ATR 01	E1 ▼	C3	None ▼	-5.2 dB	C		S		
Snare ATR 01	F1 ▼	C3	None ▼	0.0 dB	C		S		
Hihat Art 01	F#1 ▼	C3	None ▼	-5.2 dB	C		S		
Floor tom FRI	G1 ▼	C3	None ▼	-6.2 dB	C		S		

There are 16 choke groups. This feature is designed to emulate real-world acoustic drum kit hi hats, where closed hats silence open hats.

16.4 Creating Device Racks

Fortunately, you don't have to start creating your own Drum Racks right away. There are a number of presets already designed for you to use. Just browse through the Instruments, Drums, or Audio/MIDI Effect Rack folders in the Browser Categories. To identify a Device Rack preset, look for a double-pane icon ▦ in the next to preset name. Most of the Core Library presets will be Racks. All other device presets will have a single pane icon ▦. Once you have exhausted these presets, it's time to create your own. This will bring great flexibility to your music. Let's take a look at each of the Rack types, one by one.

There are a few ways to create device Racks. You can start off with a new empty Device Rack, or you can build a device chain and then group them into a Rack after the fact. We'll start with the latter. For this demonstration, open a new Live Set.

16.4.1 Instrument Racks

OK: here's the game plan … we're going create an instrument from scratch, sculpt the timbre, add effects, group it into a Rack, add a second parallel device chain to the Rack, and then finish by mapping the Macros to the chains. Sounds fun! You should download the "Solid Sounds" Pack from Ableton's website to supplement the following exercises (*www.ableton.com/en/packs*). This will really beef up your Simpler presets! If you wish to use different instrument presets, then try your best to use ones that are similar to those described in the exercise:

1. Go to Browser *Categories>Instruments>Simpler>Bass* and *double click* or *drag* "Absence" into your Set. You should now see a new track and Simpler/Absence Spike in Device View. This is already an Instrument Rack, so we will have to *Ungroup* it.
2. *Right click* on the Device Title above the Macros and select "Ungroup". Now we can build our own Instrument Rack from scratch. (Figure 16.13.)
3. Go to *Categories >Audio Effects>Compressor* and *double click* or *drag* "Generic Compressor" to your Simpler Device View.
4. Go to *Categories>Audio Effects>EQ Eight >***Instrument** and *double click* or *drag* "E Bass Vintage" to your Simpler in Device View.

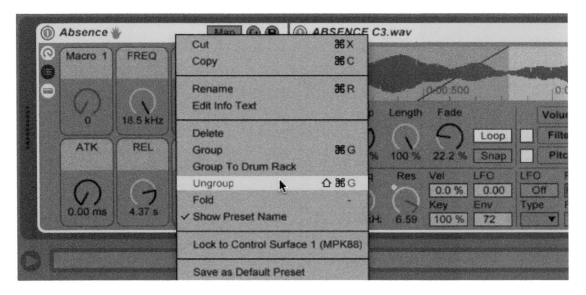

Figure 16.13 Ungroup and build an Instrument Rack from scratch.

5. Go to *Categories >Audio Effects>Saturator* and *double click* or *drag* "Warm Up Highs" to your Simpler Device View after the EQ Eight in the chain.

6. Turn the Absence track volume down to –6.0 dB.

7. Go to *Categories >Audio Effects>Ping Pong Delay* and *double click* or *drag* the default **Ping Pong Delay** to your Simpler in Device View after the compressor. Be sure not to *drag* a preset!

8. Go to *Categories >MIDI Effects>Arpeggiator* and *drag* "Arpeggiator" to your Simpler Device View and insert it before Simpler "Absence" in the chain.

9. Set the Arpeggiator "Style" to Up, "Rate" to 1/16, and "Steps" to 1.

Figure 16.14 Arpeggiator "Style" to Up, "Rate" to 1/16, and "Steps" to 1.

Now test-drive your instrument. Use the computer MIDI keyboard if you don't have a controller and *press* a few keys. Now that you have an instrument you can play with, let's group it into a Rack:

1. Select all of the devices in the chain we just created. *Click* on any device title bar in the Device View to select it and then *select all* with ([⌘+A] Mac/[ctrl+A] PC).

2. *Right click* or ([⌘+click] Mac/[ctrl+click] PC) to bring up the device contextual menu. Select "Group" to create a Rack. You can skip the contextual menu if you wish and *press* ([⌘+G] Mac/[ctrl+G] PC) instead.

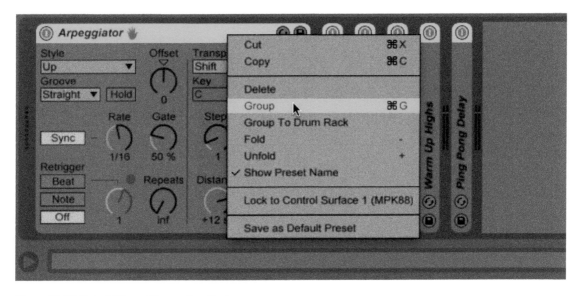

Figure16.15 Select "Group" to create a Rack.

There you have it: your first Instrument Rack. Now let's add a parallel device to the chain just for fun:

3. *Click* the Show/Hide Chain List selector to bring the chain list into view. (Figure 16.16.)

4. Go to Live *Categories>Instruments>Simpler>Bass* and *drag* "Abandon" into the Rack's Drop Area. You should now see a new chain called "Abandon" below "Absence". You have just loaded another Rack inside your Rack. This is a powerful way of building complex instruments. (Figure 16.17.)

5. In the chain list mixer section, mix the two chains so they are sonically balanced. You probably need to lower the volume of the "Abandon" bass.

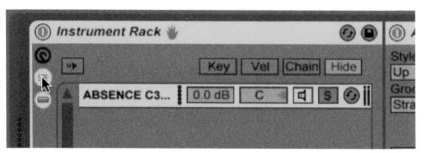

Figure 16.16
Show/Hide Chain List.

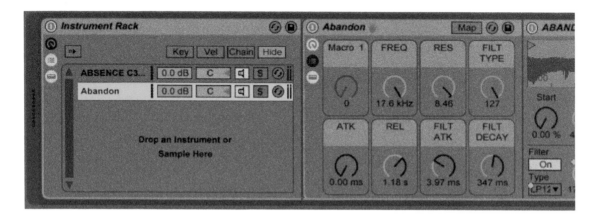

Now let's map the Macros to a few parameters within the chains:

1. *Click* on the Show/Hide Macros selector to bring the Macros into view.

2. *Click* on Map Mode to make assignments.

3. *Click* on "Absence" Chain Volume, then *click* Map on Macro 1, or *right click* or *ctrl + click* and select "Map to ... ". Do the same for "Abandon" and *click* on Macro 2. (Figure 16.18.)

4. Now is a good time to name your Macros!

5. Map "Absence's" Envelope Release to Macro 3. (Figure 16.19.)

6. Map "Absence's" PingPong Delay Dry/Wet knob to Macro 4, "Absence's" Filter Freq to Macro 5, and Res to Macro 6. (Figures 16.20, 16.21, 16.22.)

7. Map "Abandon's" Filter Freq Macro to Macro 7, and Res Macro to Macro 8. Since they are already mapped to "Abandon's" own Macro Controls, you will need to assign them directly from the "Instrument Rack" Macro to "Abandon's" Macro as opposed to assigning them from the actual Filter and Res knobs on the device interface.

Figure 16.17
Nesting Racks
inside Racks as
Chains.

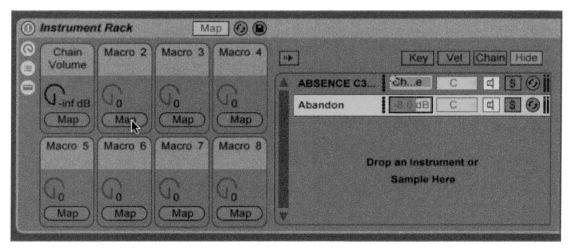

Figure 16.18 Group Rack Chains.

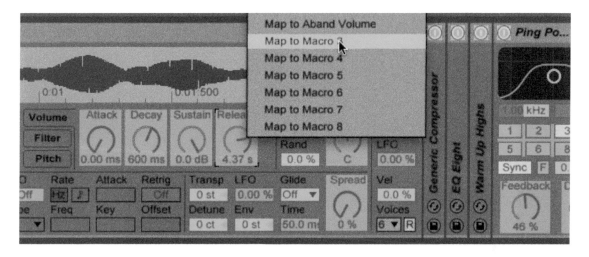

Figure 16.19
Nested Rack
Devices become a
single chain within
the main Rack
Device.

8. You will usually have to reset each parameter setting using the Macro controls after you have mapped them because Live automatically sets new Macro assignments to the value of "0" when assigned to a Macro. Feel free to mix up the parameter settings until you get something you like, or just match the example. Give your instrument a whirl and see how it sounds. Don't forget to balance your track output if you start overloading the Master track. Once you are done, take a look at some preassembled Instrument Racks on the companion website. Download "**CPP_InstRack-Pad-Lead-Arp Project**" and check out some more examples.

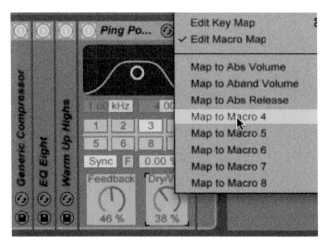

Figure 16.20
Map Macros 1 and 2.

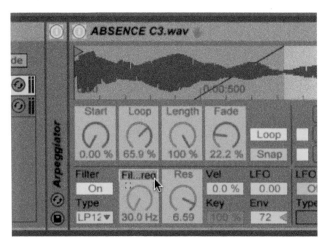

Figure 16.21
Envelope Release to Macro.

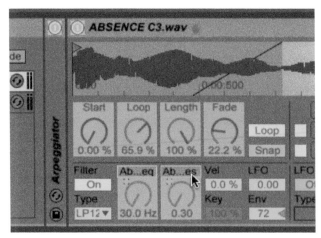

Figure 16.22
PingPong Dry/Wet to Macro 4.

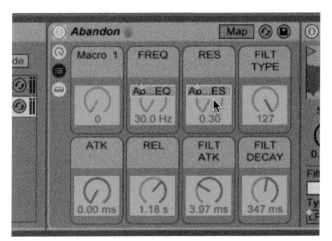

Figure 16.23
Filter Freq to Macro 5.

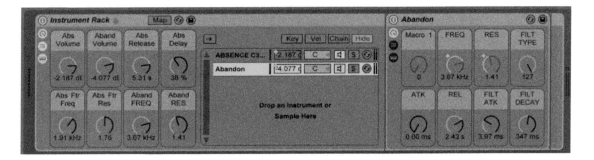

Figure 16.24
Res to Macro 6.

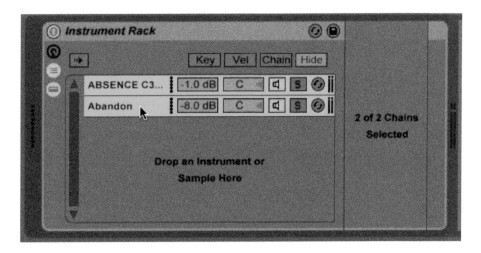

Figure 16.25 Filter Freq to Macro 7 and Res to Macro 8 via Macro to Macro mapping.

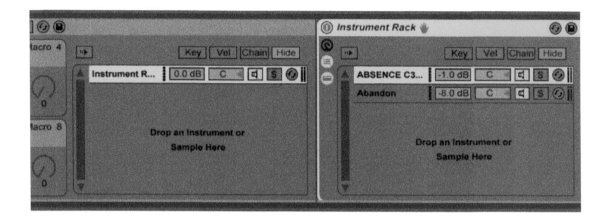

Figure 16.26
Custom Macro
assignments.

Hot Tip You'll quickly see that you have a nested Rack device within your main Rack Device. This has several advantages depending upon your needs, one being that you can create very complex multilayer instruments that are contained in one Rack.

16.4.2 Effect Racks

There are many ways to use Audio and MIDI effects Racks, so let's continue with the Instrument Rack we just created and see how we can use them. Keep in mind that MIDI effects handle MIDI information and audio effects handle audio signals. The following examples represent only a few possibilities. If you are unable locate the device presets demonstrated below, find your own suitable supplements.

Adding an Audio Effect Rack to a track device chain:

1. In Device View, hide the Chain List so you are just looking at the Macros.
2. Go to *Categories>Audio Effects>Audio Effect Rack>Space* and *drag* or *double click "Delayed Room Reverb"* to the audio effect at the end of the chain—Drop Area at the back of the Instrument Rack separated by the Level Meters. This applies the Audio Effect Rack to the output of the entire Instrument Rack and forms a new part of the track's device chain.
3. Try it out, and then delete what you just created.

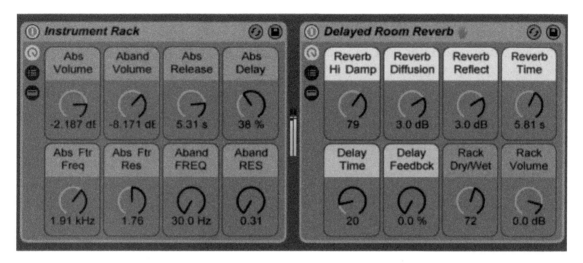

Figure 16.27 Instrument Rack, two chains, with custom Macros.

Let's try another ... adding an Audio Effect Rack to a chain within a Device Rack:

1. Select the "Abandon" from the chain list inside your Rack.

2. Go to *Categories>Audio Effects>Audio Effect Rack>Space* and *drag* "Delayed Room Reverb" to the back of the "Abandon" chain just inside of the Level Meters—not outside! This applies the Effect Rack only to this chain within the Rack. You will see a blue highlighted vertical line when you are *dragging* over a drop area. If you were to select the other chain, you would see that the Audio Effect Rack is not a part of any other chain.

3. Try it out, and then delete what you just created.

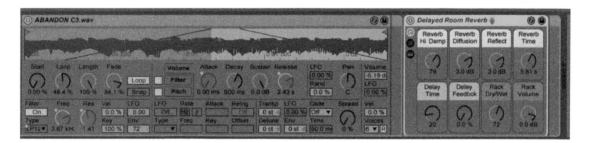

Figure 16.28 Instrument Rack and Audio Effect Rack making up a device chain.

Let's try another ... adding a MIDI Effect Rack as a track device chain:

1. In Device View, deactivate the Arpeggiater on "Absence", then hide the Chain List so you are only looking at the "Instrument Rack" Macros.

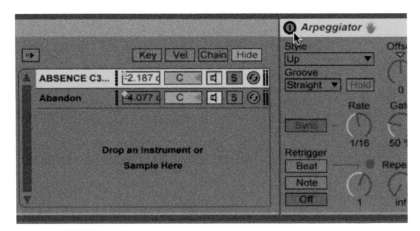

Figure 16.29 Audio Effect Rack applied to only the "Abandon" Rack chain.

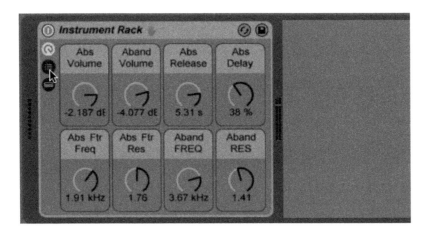

Figure 16.30 MIDI Effect Rack as a part of the track's device chain.

2. Go to *Categories>MIDI Effects>MIDI Effect Rack* and *drag* or *double click* "Chance Arp" in front of the Instrument Rack. This applies the Effect Rack to the entire Instrument Rack forming a new device in the chain.
3. Try it out. Now delete what you just created.

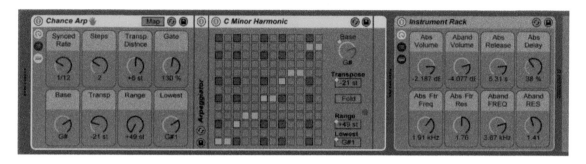

Figure 16.31 MIDI Effect Rack as a part of Device Rack chain – "Abandon" chain.

Let's try another … adding a MIDI Effect Rack to a chain within a device Rack:

1. In Device View, activate the existing MIDI Effect Arp that you deactivated in the previous example, then hide the Chain List so you are just looking at the Macros.

2. Select the "Abandon" from the Chain List inside your Rack.

3. Go to *Categories>MIDI Effects>MIDI Effect Rack* and *drag* "Chance Arp" and drop it on the "Abandon" Chain List Title. This applies the Effect Rack only to this chain within the Rack.

4. Try it out, then delete what you just created.

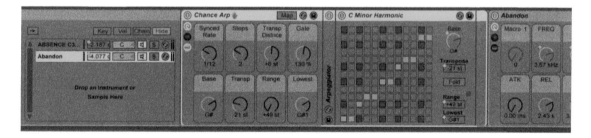

Figure 16.32 Drum Rack Pad.

16.4.3 Drum Racks

Creating Drum Racks can be as complex as you want, but let's try to keep it simple. For this example we'll use individual drum kit samples to create our own Drum Rack. On the companion website there are some additional samples and clips to accompany this exercise. Download the folder "**CPP_Drums**" and then add it as a folder into Places. Feel free to use your own kit samples or the samples from the Live 9 Core Library:

1. Create a new Set, then go to *Categories >Drums* and *double click* or drag "Drum Rack" into your Set. Be sure not to select a preset. You should now see a new track and an empty Drum Rack in Device View. (Figure 16.33.)

2. Bring the Drum Rack's Pads into view, and then go to *Places>CPP Drums* and select a kick drum and *drop* it on Pad C3. (Figures 16.34 and 16.35.)

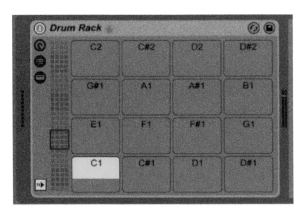

Figure 16.33
Select individual samples to load in a Drum Rack Pad.

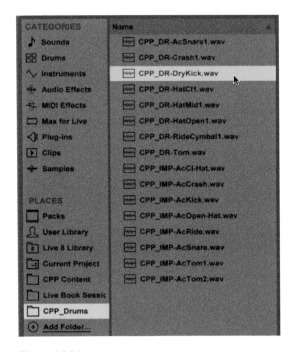

Figure 16.34
Drag + drop a sample on a Drum Rack Pad.

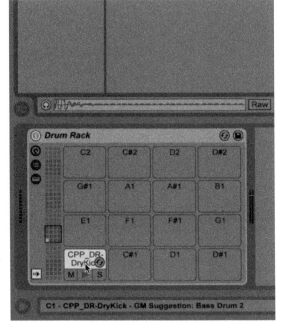

Figure 16.35
Drag in a MIDI clip to trigger your Drum Rack.

3. Choose a snare sample and *drop* it on D3. Live will automatically treat each sample as a chain and map it to the pad where you've *dropped* the sample.

4. Choose a closed hat sample and *drop* it on B3.

5. Now you need a MIDI clip. Go to *Places>CPP Loops and* select and *drag* "DR-AcDrum-Groove1.mid" to the first Clip Slot of your Drum Rack. You can also use any generic MIDI files for your Drum Rack. *Please note that for some MIDI files, you may need to drag the MIDI notes so that they trigger different Pads in order to match the MIDI note assignments in the file with the appropriate sample/pad.* (Figure 16.36.)

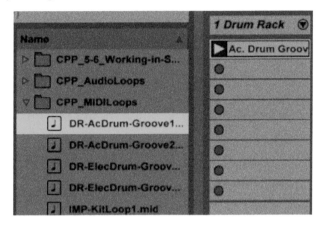

Figure 16.36 Launch your Drum Rack clip.

6. Activate the Clip Launch button and listen to the kit. Notice how the Chain List and Pad behave during playback. (Figure 16.37.)

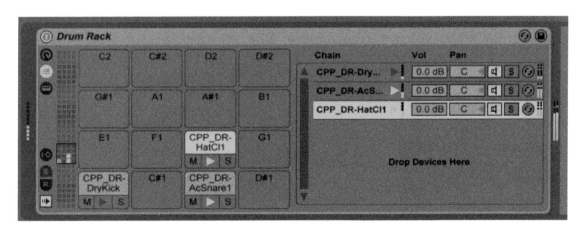

Figure 16.37 Adjust individual chain parameters to your liking.

7. While the clip is running, open up the chain list and select the snare chain. From the Simpler instrument associated with the snare sample, adjust the volume "Release" knob to extend or shorten the release so the snare sounds smooth.

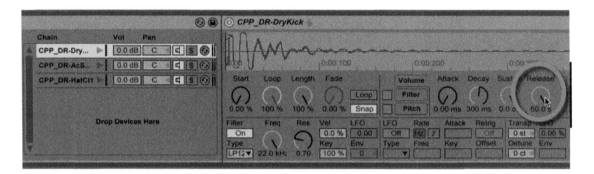

Continue to tweak, as you like.

Figure 16.38
Custom Drum
Rack.

That's all you need to get started, but if you want to take this further, you could easily add audio effects or an Effect Rack to any chain or to the entire Rack just as we did with the instrument and effects. You can even *drag* effects onto Pads in Pad view. For some preassembled Drum Racks, download "**CPP_DrumRack-Ac-Elec-Kits Project**" from the companion website.

16.4.4 Drum Racks and Audio Effects

By integrating exclusive sends and returns for audio effects, you can further the capabilities of a Drum Rack device. Let's look at how this can be achieved:

1. Load up a Drum Rack preset or insert and create a new Drum Rack Device on a MIDI track and add your own samples. The "Kit-Core 606.adg" preset from Live 9 Suite has been used for the example here. Remember you can simply *double click* on the Drum Rack in the Browser to load it into a new MIDI track without *dragging* and *dropping*.

2. *Click* on the Rack's Show/Hide Chain List Selector to open the Chain List. You'll notice the vertical row of four buttons on the extreme left-hand side of the Rack is now visible, displaying **A**—Auto, **I-O**—In/Out, **S**—Send, and **R**—Return.

3. *Click* on the "R" (Show/Hide Return Selector). The Return Chain List labeled "Drop Audio Effects Here" will open immediately beneath the standard Chain List.

4. *Drag* a Reverb audio effect (or any audio effect) directly into the Drop Area of the newly opened Return Chain List. It will now be displayed in the Return Chain list with the letter "a" in front of the effect's title.

5. Click on the "S" selector (Show/Hide Sends) on the far left of the device to open the *Chain Send Level* window for each chain. They reside directly beside the volume and pan controls in each device chain.

6. *Double click* on your audio effect in the Return Chain window to bring it into view then change the *Dry/Wet* mix knob to "100%". You may need to scroll through the Device View Selector to bring the effect into View.

7. Play some Drum Rack pads or launch a MIDI clip pattern to create some sound from the Drum Rack.

8. On one of the active device chains, increase the Send level to around "–5.0 dB". You should begin to hear the reverb effect level increasing along with your Drum Rack mix. Try this with another chain's Send value. You can also adjust the Return Chain's level from –infinity to +6 dB with the volume slider located in the middle area of the return chain.

9. Rename or Hot-Swap the Reverb or selected audio effect device at any time if needed. You can also assign the Send value to a Macro Selector in the Macro section to the left of the Drum Rack device.

10. Load another audio effect into the Return Chain Drop Area and instantly create another Return Chain just below the first one.

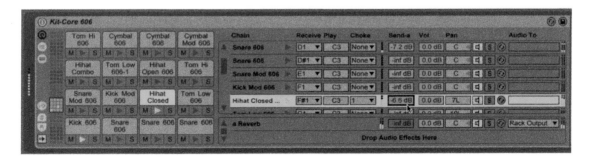

Figure 16.39 Return Chains.

You can see how your Drum Rack device comes alive using Sends and Returns with audio effects, not to mention the many creative possibilities it facilitates. Don't forget to save your Drum Rack as a new preset. You'll be able to recall it anytime with all your newly included audio effects, plus all the sends and returns you created!

If you still desire more creative power, then Group individual chains within the Drum Rack to new Racks, nesting them inside the current Drum Rack. This provides even more Macro control over each chain in the Rack. To do this, open up a contextual menu from any device Title Bar and select "Group to Drum Rack". As a matter of fact, this can also be done to any device in your set at any time if you want to convert it into a Drum Rack.

16.5 Racks in Session View

As you now know, a Rack functions as a self-contained device, made up of a combination of instruments, effects, and even more Racks. It's a nesting process, where instruments, effects, and Racks are encapsulated in a container functioning as a single device (i.e., instrument or effect). Each parallel device chain within a Rack is independent of the others, except that each chain output is summed together. The internal mix of a Rack is generally managed from its chain list, but it can also be conveniently mixed in a more traditional sense via Session View Tracks. This means that each individual chain in an Instrument Rack or Drum Rack is viewable on its own "sub-" track allowing for each chain to be viewed and submixed just like any other track type. Unfold your device chain on a track in Session View by *clicking* on the small black triangle button (*Chain Mixer Fold button*) in the track's Title Bar. When unfolded, you will see that each chain is on its own sub-track within the Device Rack track.

On each chain in Session View you'll find: *Chain Volume fader*, *Activator* (mute), *Solo*, and *Preview button* when the Mixer section is in view. This is purely a view for submixing and signal routing of Instrument/Drum Racks in the Session View. No clips can be added directly to chains, nor can you arm them. They are simply a vertical display of the Rack's chain list. The MIDI I/O choosers and Send Level sliders are exclusive to Drum Racks. (See Figure 16.44.)

One very interesting feature is that you can add Audio Effects or additional instruments to a chain just by *dragging* and *dropping* the device anywhere in the Session View over a designated chain. As soon as a new device is added in this way, you'll see the chain list and new device—for that specific chain—displayed in Device View. This helps to avoid clutter and confusion. (See Figure 16.45.)

This precise way of viewing and working with your Instrument Rack and Drum Rack Devices in Session View provides a more accessible and traditional way of mixing, especially when you group chains within a Rack. There are so many possibilities with Racks that you'd have to launch to a new book to uncover them all.

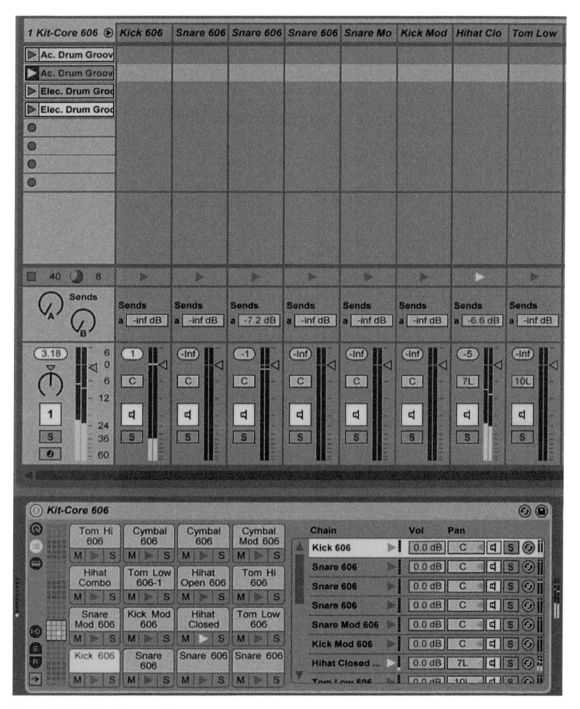

Figure 16.40 Chain in Session View.

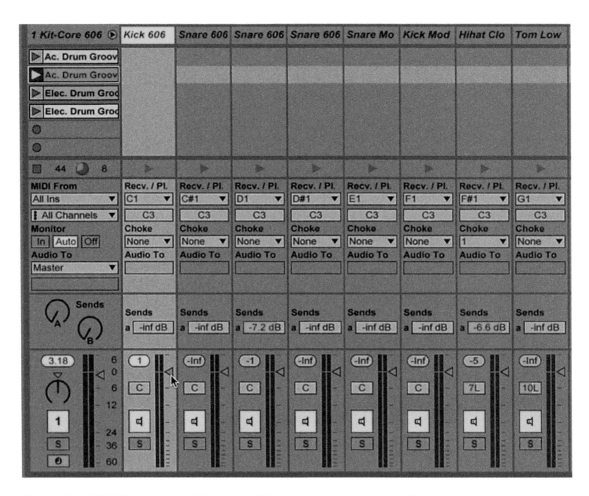

Figure 16.41 MIDI I/O choosers and Send Level sliders are exclusive to Drum Racks.

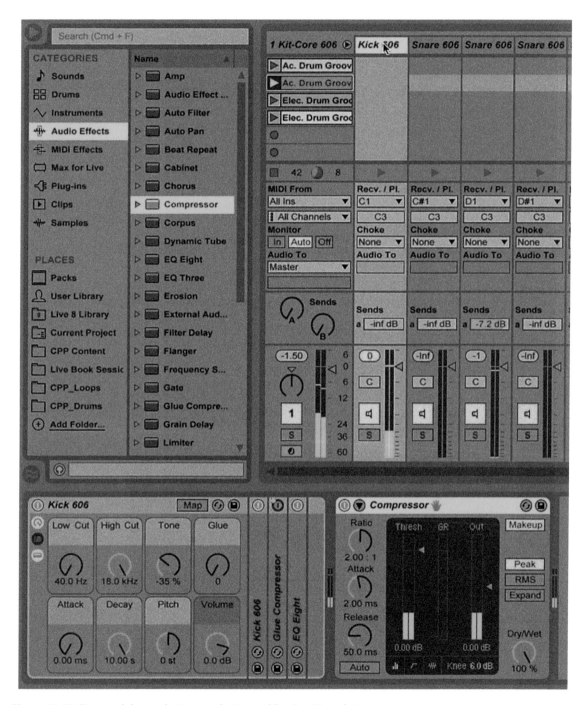

Figure 16.42 Drag and drop a device to a designated Session View chain.

16.6 Working with Racks

As with almost every feature in Live, there is a clever way to convert or isolate one aspect in order to use it independently or to build up something new. There are limitless possibilities with Device Racks. That being said, you're probably thinking to yourself … "That's great, now wouldn't it be cool if you could … ?" Odds are you can, but if not, there are numerous features and functions that are sure to strike your fancy. Take some time to experiment with Device Racks and this will become prevalent.

16.6.1 Produce: Convert Chains into MIDI Tracks

A unique way to work with chains in Session View is to turn them into discrete MIDI tracks, isolated and separated from the rest of the nested chains. Here's an example using the Rack Chain titled "Snare 606".

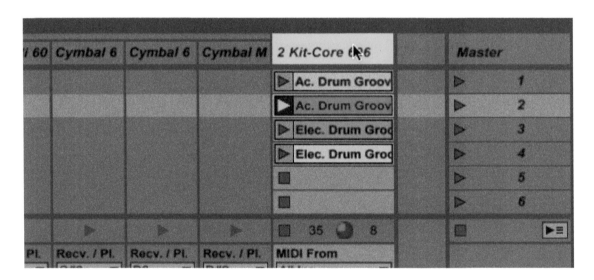

Figure 16.43 Drag a Chain to the Session Drop Area or a new empty MIDI track in Session View.

Go ahead and follow along …

Using "**Kit-Core 606.adg**" or any Drum Rack preset, unfold your Drum Rack in the Session View in order to view the chain, then *click + hold* on the "Snare 606" Chain Title Bar and *drag* it to a new empty MIDI track or into the Session View Drop Area. Live will convert your selected chain into an independent Drum Rack on a MIDI track along with its included MIDI performance—if it has MIDI assigned to trigger it. This conversion also includes any audio effects from that chain and the individual instrument sound itself from the original chain. All of the important parameters and devices themselves will be visible below in the Device View. Don't be confused by the new MIDI Track Title. Live creates the new MIDI track by copying the name of the main Rack device you pulled it from. You can easily rename it "Snare 606" or any other name you wish. By separating a single chain from the nested track chains in the Session View, you now have an isolated and editable MIDI clip that can be edited and saved for future use. Not only does this feature work with both Instrument Racks and Drum Racks, but it can also be executed in real-time while Live is playing. This is a great way to experiment with changing out the current snare, for example, with a different sound, or controlling its volume more selectively. It's important to note that when you extract a chain, it will be removed completely from its instrument. The basic design of this feature is to isolate MIDI data. One other use for this functionality is to freeze and convert the separated chain's MIDI clip to audio for warping or additional editing. Launch to Clip ▶ **8.3.1** for more on freezing and converting MIDI into Audio.

Look at this as a way to split out your MIDI notes and corresponding instruments from a complete Instrument Rack into separate MIDI tracks. Try this production tactic with a couple of your track chains and see what happens.

Controlling Your Universe

17.1 Remote Control

There are many ways to take control of your tasks in Live without the necessity of a mouse. Let's not forget that a mouse was not designed to be a musical instrument or to emulate human control. As a musician, you may feel more in control when you're not tied down or limited by the mouse. Additionally, many of you are using laptops, or are sitting/standing behind a MIDI controller device. In either case, Live caters to a humanized experience with many versatile and controllable features. In fact, almost everything in Live is can be remote controlled with either MIDI or computer keys. This includes functions such as launching, switches, buttons, input fields, and variable Parameter Values, as well as controls and parameters for the Control Bar, Clip View, Session View, Arrangement View, Mixer, tracks, clips, and Scenes—virtually everything you would want to control!

Live employs a tactile approach for mapping actions and controls to the computer keyboard or a MIDI controller. It's all about select and assign. Remote Control is governed by *Key* and *MIDI Map* Modes. Choose whichever mode/device you want to map control to—the computer keyboard or a MIDI controller. Now, before you make your mouse obsolete, take a closer look at how control surfaces and mappings are set up and assigned in Live.

Figure 17.1 Key and MIDI Map Mode switches.

17.1.1 Setting up Control

To set up Live for Remote Control, go to Live's *Preferences>MIDI/Sync tab*. You may recall looking over this tab earlier in Clip ▶ 2.3.5 when we discussed setting up your MIDI controller.

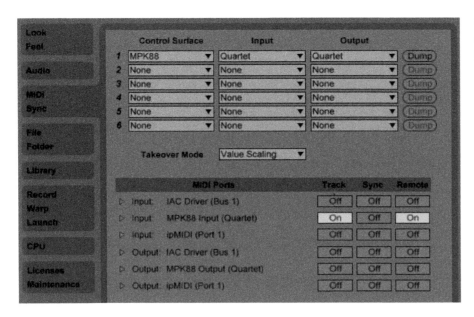

Figure 17.2
MIDI/Sync
preferences.

As far as Remote Control is concerned, you can select up to six natively supported control surfaces. If you have a controller with knobs, a slider, wheels, pads, and faders, there is a good possibility it's listed here. Choose your control surface from the Control Surface chooser menu. If your control surface is, in fact, natively supported, select it, set up the appropriate Control Surface Input/Output and MIDI Ports, and then set "Remote" to "On". If your device uses motorized faders or real-time digital displays, turn on "remote" for its output port. If your device is not supported, you can still use it; just be sure to turn Remote to "On'" for the MIDI Input Port(s) you are going to use. As an important reminder, the "Track" switch enables your device's MIDI input to communicate with Live's tracks.

Some control surfaces may require you to execute a "preset dump" (MIDI SysEx Dump) from Live if they aren't natively supported or set up to work as expected. This will allow you to install usable presets/programs that conform to Live's instruments and controls. Follow your controller's instructions, then *click* "Dump" to complete the setup of your controller (upload presets). Live will then automatically set up your device for use. These *Instant Mappings* will allow you to

Remote Control Live's various functions and parameters from your control surface based on the physical controls and configuration of your device. See your MIDI controller's documentation on how to receive MIDI dumps. Of course, you can customize and remap assignments as you see fit. For example, with an Axiom 25 controller, Dump installed three programs into the controller enabling control of Live's transport, Scenes launches, track volumes, and to trigger Impulse's sample slots. Using these presets eliminates the need to customize MIDI mappings for those specific controls and parameters.

When a control surface has been set up in Live's preferences, a *blue hand* will appear next to a device's name. This *Remote Control* icon indicates that a MIDI control surface is currently controlling the devices parameters. When you select a different device, the hand will appear with the new device, leaving the previous device. If you want to maintain control over a device even when it is not in focus (selected), you must select "Lock to Control Surface … " in its contextual menu.

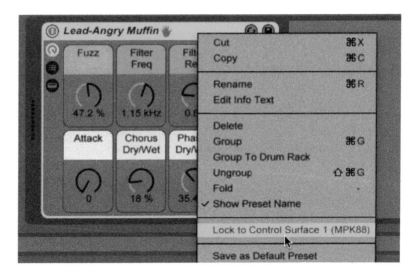

Figure 17.3 The blue hand (Remote Control) icon signifies that a device is being controlled by a control surface. Lock a control surface to the device in the contextual menu.

Takeover Modes

Before moving on, choose a *Takeover Mode*. This determines how a parameter control responds to incoming data values. There are three Modes: *None*, *Pickup*, and *Value Scaling*. Each option determines how Live behaves and reacts to sudden jumps in values sent from your controller. For example, if you assign your keyboard controller's mod wheel to control a send knob in Live, the moment you move the mod wheel it will send a new value to the send knob, thus moving it. The purpose of Takeover is to determine when and how that data is acted upon by Live. Should Live react suddenly or gradually? Take the time to determine which setting is best for your performance demands. Use Pickup for now.

17.2 MIDI Mapping

MIDI Mapping in Live is the process of assigning physical MIDI controls (continuous controller number values [CC#])—hardware faders, slider, knobs, wheel, buttons, pads, keys, etc.—to Clip Slots, Scenes, Arrangement Record, and many other various controls and parameters within Live's user interface. These assignments allow you to remote control functions and actions with your external MIDI device instead of using the mouse. Common hardware MIDI controllers are (but are not limited to) any MIDI keyboards, trigger pads, and control surfaces, as well as the *Ableton Performance Controller (APC20/40)* by Akai and Live's own *Push*. There are others out there; but these two controllers have received special attention and integration by Ableton. More than just a controller, they recreate the Live's Session View into their physical design. We'll get to an overview of these controllers later in this Scene, but if you can't wait, then Launch to ▶ 17.6 or ▶ Web .

Whether using a MIDI keyboard or a dedicated control surface, all of these devices will have some sort of physical controller(s), such as faders, knobs, keys, wheels, and pads that can be mapped to a control, instrument, or function in Live. Since a MIDI controller can also function as a traditional input device/instrument and a remote controller at the same time, MIDI mapping assignments will take priority over the default note input settings. For example, if your MIDI keyboard note C3 is mapped to launch a Scene, then it will no longer be able to input or play C3 as a note for an instrument.

To make a MIDI Remote Control assignment (MIDI Mapping), *click* on the *MIDI Map Mode switch*, located in the upper right corner of the main Live screen.

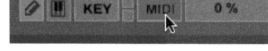

Figure 17.4 MIDI Map Mode Off.   **Figure 17.5** MIDI Map Mode On.

It will turn blue when activated. Subsequently, all map-able parameters and controls will also glow lavender/blue, seen as a highlight scheme across Live's user interface.

Figure 17.6 Session View – MIDI Map Mode.

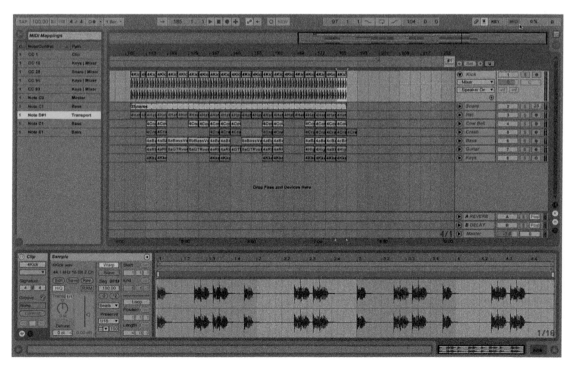

Figure 17.7 Arrangement View – MIDI Map Mode.

This signifies all of the parameters that can be mapped via your MIDI controller, control surface, or other MIDI-based hardware. Keep in mind that you can multi-map multiple parameters to the same key or MIDI control. For example, you could map a single knob on your controller to both the Pan on one track and the Send on another track, or any configuration of tracks or the same track. You'll find that such multi-map assignments allow for a MIDI control to easily control multiple parameters, which is just one way to creatively and efficiently take control of your Live universe.

Once you have entered MIDI Map Mode, *click* on any highlighted parameter and then *press/move* a MIDI knob, slider, or note on your controller to assign that control to the selected Live parameter. After making your mapping assignments, *click* the MIDI Map button again to exit Map Mode. You can assign the same physical controller to multiple parameters if needed except when it causes a direct conflict. This would be the case if you try to assign the same key, let's say C3, to clips on the same track. In this situation you must give each clip a unique assignment. A label will appear next to each assignment in the user interface indicating the specific note or control that has been assigned to remote control it—only visible while in MIDI Map Mode.

Figure 17.8
Mapping label.

To expedite mapping, use the key command ([⌘+M] Mac/[ctrl+M] PC) to enter MIDI Map Mode. After you've made your assignments, use the same key command to exit MIDI Map Mode or *click* on the MIDI Map switch again. Feel free to add more or edit the mappings as you like. If you find a prior mapping control number/message is already affixed to a parameter you want assign, *press* the Delete/Backspace key or simply assign a new one to overwrite it. This can also be done in the Mapping Browser discussed later.

Hot Tip Drag and drop audio effects in real-time without exiting Key Map or MIDI Map Mode. This proves useful when you want to quickly experiment with automating specific effect and mapping controls while remaining in Map Mode.

Once assignments have been made, the *KEY/MIDI In Indicator* will light up yellow upon remote control input activity.

Figure 7.9
KEY/MIDI In Indicator.

17.2.1 Control Behaviors

MIDI hardware controllers/control surfaces are often designed with a multitude of controllers, such as buttons, continuous controllers (knobs, wheels), keys/notes, and pads. Each type of controller sends MIDI information differently. Notes, for example, send simple on and off messages. Continuous controllers, on the other hand, send a range of MIDI data in values of 0 to 127. Some MIDI controllers also send incremental data such as +/− increments, where a button adds/subtracts a single value at a time as opposed to a free range of continuous data values. Live refers to incremental controls as Relative controllers and the continuous controllers as Absolute, but this varies amongst different controllers.

The type of controller and message data being sent determines how a control or parameter in Live will respond. With the numerous assignable controls and parameters in Live, certain MIDI controller types are more suitable for parameters and controls than others, which will have a different effect on Clip Slots, switches, and Radio buttons, buttons with multiple options, and Variable Parameters. For example, MIDI notes/keys can launch clips, activate switches, navigate Radio button options, and toggle between a Variable Parameters' min and max values (track volumes, sends, etc.). For variable parameters you will notice that a MIDI note message will toggle between the min value or the max value, but nothing in between. To better understand how your MIDI controller integrates into Live, try experimenting with various mappings with different controller types to determine which controller is best suited for the task you want to Remote Control. Don't spend too much time trying to locate each parameter type in Live. Who cares what a Radio button is?! Just dive in and slap a MIDI slider, button, key, knob, or pad onto things and see what happens.

MIDI hardware controllers are a great way to control your Set in both the Arrangement and Session View, but when combined with the MIDI keyboard, Remote Control will streamline your entire workflow, un-tethering you from your computer for hands-free manipulation of Live—well, mouse-free to be exact. One way to accomplish this is to map Scene Launches and Track Activators (track On/Off/mute buttons) in the Session View to your MIDI keyboard notes. For instance, assigning each one of the white keys to Scene launches and black keys to the track Activators (mutes) can open up a new way to work. Firing off a Scene with the press of a keyboard note is seemingly effortless and can bring forth a free-form way of working. You must try this for yourself!

17.3 Key Mapping

If you don't have a MIDI controller, never fear. Your computer keyboard can act as your controller, minus a continuous controller, of course. Key Map Mode is activated by clicking on the *Key Map Mode switch* on the Control Bar located next to the MIDI Map Mode switch or ([⌘+K] Mac/[ctrl+K] PC).

Figure 17.10 Key Map Mode "Off".

Figure 17.11 Key Map Mode "On".

When activated, the Key switch will turn orange, as will all of the map-able parameters and controls across Live's user interface.

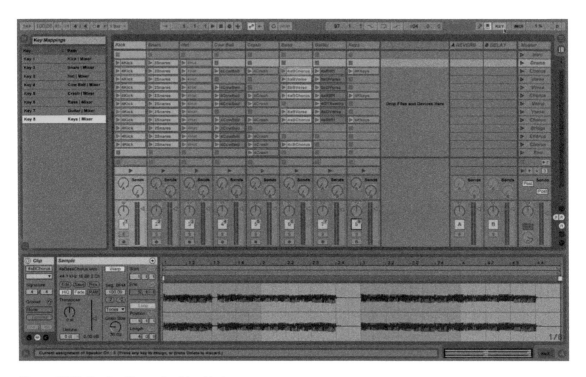

Figure 17.12 Session View – Key Map Mode.

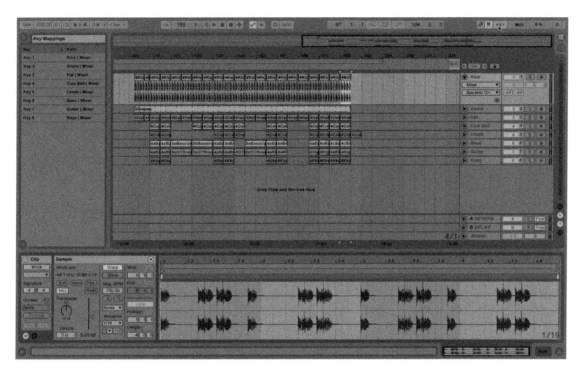

Figure 17.13 Arrangement View – Key Map Mode.

Key mappings function just like MIDI mappings. To make assignments, enter Key Map Mode then *click* on a clip, control, or parameter. After you select one, type the key on the computer keyboard that you wish to make the assignment to. In order for key mapping to work, you must make sure that you have deactivated the *Computer MIDI Keyboard*. This is the keyboard-like button located next to the Key button. When this is activated (yellow), your computer keyboard will function as a MIDI note input device rather than a remote control. In this situation you would not be able to execute your key map assignments. While activated, any keys that share both mappings and MIDI input functions will cause the Computer MIDI Keyboard button and Key to light up orange when the conflicting keys are *pressed*.

Figure 17.14 Orange here indicates conflicting KEY Map/keyboard MIDI assignments.

17.4 The Relative Session Mapping Strip

In the Session View, when you're in KEY/MIDI Map Mode, you will see on the Master track just below the Stop All Clips button an area referred to as the *Relative Session Mapping Strip*. This contains four assignable buttons that can speed up navigation and launching of Scenes. They are: *Scene Launch*, *Scene Down*, *Scene Up*, and *Scene Select*, and they will improve your workflow when working in a large Live Set:

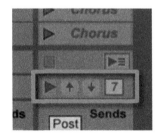

Figure 17.15
Session View – Relative Session Mapping Strip.

Scene Launch—when a controller/key is assigned to this button you can launch any selected Scene with the controller. Set the "Next Scene on Launch" preference to "On" in the Launch Preferences to further enhance and speed up navigation with this feature.

Scene Up/Down—assign a controller/key to navigate up and down through the Scenes column.

Scene Select—displays the number of the currently selected Scene. Assign a continuous controller to Scene Select and scroll through Scenes.

In the same row as the Relative Session Mapping Strip you will see that each track has an assignable *Track Launch button* located above the tracks Sends. Assign this to a Key or MIDI controller to launch a clip from that track located in the highlighted (selected) Scene. For example, whatever clip lies in the row that is highlighted will be launched when the mapped KEY/MIDI controller is used.

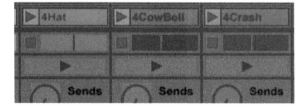

Figure 17.16 Track Launch buttons.

Let's put this feature to some good use by enabling MIDI Mapping and assigning a control to navigate and activate the Scene Launch function.

Here is a walkthrough:

1. **Enable MIDI Mapping**. You'll notice that just above the Stop All Clips on the Master track that the Relative Mapping Strip appears—*Scene Launch*, *Scene Up/Down arrows*, and *Scene Select*.

2. *Click* on and assign a separate control of your choice to *Scene Up*, *Scene Down* (Up/Down arrows), and Scene Launch so you can access and trigger them remotely.

Figure 7.17
MIDI Map Mode, Relative Mapping Strip.

3. It takes some experimenting to find the method and mapping style you are most comfortable with, but once you find it you'll enjoy mouse-free Scene navigation and launching to its fullest potential in Live.

17.5 Mapping Browser

With all of the possible mapping scenarios, Ableton has provided an organized browsing system: *Mapping Browser*. All mapping assignments are managed from this browser, which will appear on the left side of the window when in mapping mode.

Figure 17.18
Mapping Browser.

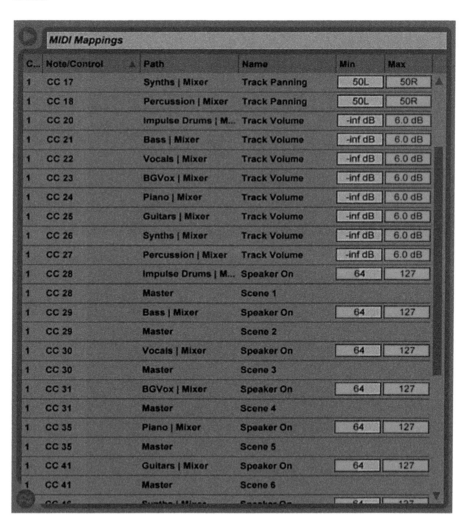

C...	Note/Control	Path	Name	Min	Max
1	CC 17	Synths | Mixer	Track Panning	50L	50R
1	CC 18	Percussion | Mixer	Track Panning	50L	50R
1	CC 20	Impulse Drums | M...	Track Volume	-inf dB	6.0 dB
1	CC 21	Bass | Mixer	Track Volume	-inf dB	6.0 dB
1	CC 22	Vocals | Mixer	Track Volume	-inf dB	6.0 dB
1	CC 23	BGVox | Mixer	Track Volume	-inf dB	6.0 dB
1	CC 24	Piano | Mixer	Track Volume	-inf dB	6.0 dB
1	CC 25	Guitars | Mixer	Track Volume	-inf dB	6.0 dB
1	CC 26	Synths | Mixer	Track Volume	-inf dB	6.0 dB
1	CC 27	Percussion | Mixer	Track Volume	-inf dB	6.0 dB
1	CC 28	Impulse Drums | M...	Speaker On	64	127
1	CC 28	Master	Scene 1		
1	CC 29	Bass | Mixer	Speaker On	64	127
1	CC 29	Master	Scene 2		
1	CC 30	Vocals | Mixer	Speaker On	64	127
1	CC 30	Master	Scene 3		
1	CC 31	BGVox | Mixer	Speaker On	64	127
1	CC 31	Master	Scene 4		
1	CC 35	Piano | Mixer	Speaker On	64	127
1	CC 35	Master	Scene 5		
1	CC 41	Guitars | Mixer	Speaker On	64	127
1	CC 41	Master	Scene 6		

The Mapping Browser displays all active note/control assignments and to what parameter they have been assigned. These are labeled by "path" (where it is located) and "name" (what it is). Min and max values are set from the browser for variable parameters. As a default, an assignment addresses the entire range of values possible for a Variable Parameter. Adjusting the range limits the selectable range of values. Use the "Invert Range" feature to reverse the direction of a specific mapped control's parameter values, inverting the Min/Max values. *Right click* or *ctrl + click* just to the left of the value slider (scalar) in the Mapping Browser. This is a great way to manipulate parameters that need to oppose one another when automating, such as EQ Bandwidth and Gain, or a low pass filter (LPF) Frequency and Resonance. Try assigning one knob control to the LPF Filter in Auto Filter and another to Resonance in Auto Filter, and then select "Invert" in the Mapping Browser for Resonance. Now you can turn both knobs inward (counterclockwise) while the value actually increases or decreases for both instead of moving in an opposite directions.

Hot Tip

Use **Invert Range** to reverse the direction of a specific mapped control's parameter value, thus inverting the 'Min/Max' value.

Right-click or **ctrl + click** just to the left of the value scalar and select 'Invert Range'.

17.6 Dedicated Live Hardware Controllers

17.6.1 The APC

The Akai/Ableton (APC 20 and 40) Performance Controller is designed to control Live's Session View in an intuitive and tactile way. The APC's custom layout of buttons is based around an 8X5 grid that emulates Live's Session View tracks, Scenes, and Clip Slots interface. Even though all the buttons, sliders, and knobs on the APC are already assigned and mapped to specific parameters in Live's Session View, they can be reassigned to initiate other control via the MIDI Map Mode function in Live. Therefore, you can assign various knobs to take over control of your favorite key functions. Launch to ▶ **Web** for more on the APC Performance Controller and also visit www.akaipro.com.

17.6.2 Push

In conjunction with the release of Live 9, Ableton launched their own controller for Live called Push, engineered by Akai Professional. It's more than a control surface; it's literally an instrument. You can create, produce, and perform beats, melodies, and much more. This inspiration is all about a total hands-on immersive environment for making music in Live 9.

Push is a true alternative controller. It may not be for everyone, but for some of you it will be the future of making music. When it comes to setting up in Live 9, it's a breeze. Push automatically integrates into Live when it is connected to your computer, Instant Mappings and all! For more on Push, launch to ▶ **Web** .

17.7 Musical Control

No matter what your end goal for using Live may be, having control is paramount. Taking control of your songs should be an effortless intuitive process that allows you to spend more time in the music than in user menus and setup. That's what Live does: it makes your experience enjoyable so you can focus on creating, producing, and performing your music does—even more so with Push.

17.7.1 Perform: Mapping Locators

When working in the Arrangement View you'll find that a *Locator* is a great way to pinpoint and start a new section of your music or a particular part of a song along the linear timeline. With KEY/MIDI mapping functionality you can Remote Control designated Locators to be triggered from your computer keyboard, MIDI controller, or control surface in order to jump to and start playback from a locator. You can also navigate to and from locators when playback is stopped in order to establish a new start location for playback. Look at this as a way of quantizing a start position while playing.

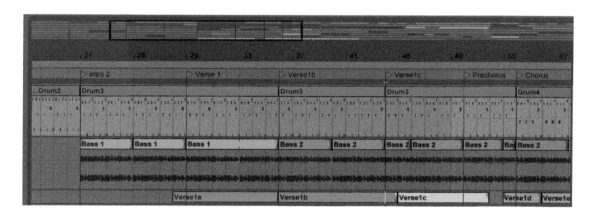

Figure 17.19 Locators along the Beat Time Ruler.

Here's an example:

1. Select and create Locator positions along your arrangement by placing your Insert Marker at the position you wish to place your Locator. *Click* the *Set* button to add a locator and continue to add as many as you need for your set. Now comes the interesting part.

2. Select the KEY Map button ([⌘+ K] Mac/[ctrl+K] PC), then *click* on any one of the new Locators to map it to a computer key.

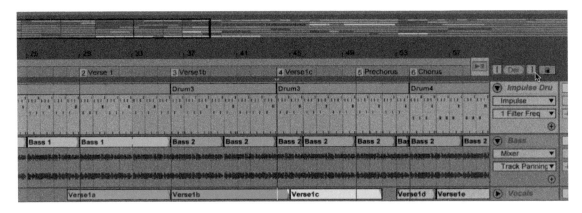

Figure 17.20 Mapping Locators.

Key	Path	Name
	Key Mappings	
Key ▲	Path	Name
Key 1	Locators	Intro 2
Key 2	Locators	Verse 1
Key 3	Locators	Verse1b
Key 4	Locators	Verse1c
Key 5	Locators	Prechorus
Key 6	Locators	Chorus
Key [	Locators	Previous Locator
Key]	Locators	Next Locator

Figure 17.21 Locators in the Mapping Browser.

Assign whichever key on your computer keyboard you wish. One tip is to select a key that will be the first in a row of keys for launching all of your locators in succession.

3. Continue to assign keys to all of your Locators and when you're done you will be able to quickly select specific locations where you want to begin playback. If you want to MIDI Map these Locators instead, just assign them to a note on your keyboard controller or drum pads on your surface controller. It's the same functionality, just MIDI Mapping instead of KEY Mapping.

4. Open the Mapping Browser to get a view of specific mapping parameters. Here you'll see all of your Locators and their names if you named them ... hint, hint.

When using mapping with Locators, try out various resolutions in the Quantize Menu for launching with Locators.

ReWiring the Digital World

18.1 Overview of ReWire

Many of you are asking: "What is ReWire?" ReWire is cross-application protocol designed by Propellerhead Software Company that allows specific software to communicate and transfer both MIDI and audio between them in real-time. In essence, ReWire streams real-time synchronized MIDI and audio signals between multiple Digital Audio Workstations (DAWs) and links their transports (play/stop), playback location pointing, and track naming information. With ReWire, Live can harness the power of other music software applications, integrating their unique synths and sequencers into your workflow. What does this mean? It means that you can use multiple music programs simultaneously in sync with Live, controlling their instruments and audio outputs while applying Live effects and everything Live has to offer to these audio signals that are all routed through your Live Set. Wow, that's a lot of stuff, not to mention all of the MIDI flowing back and forth.

Figure 18.1 Live launch window when starting up as ReWire Slave.

18.2 ReWire Modes

ReWire is a communications protocol, nothing more; therefore, it requires the use of at least two software applications. These applications must support ReWire in order to communicate. When compatible they will operate in one of two ways: as a *ReWire Host* or *ReWire Slave*. Live can operate as either, based on how the ReWire relationship is established with the other application(s) upon start up. Live refers to this relationship as *Master Mode* and *Slave Mode*. When running Live in Slave Mode, it will always be indicated as such in Live's startup screen. Whichever program you wish to be the Master, you must launch first. Launch the slave application second, after the Master has completed initializing. Once the soft wares are linked, they will begin sharing transport controls, tempo, song position, and various track information. Although this information is shared, the Master and Slave each have functions specific to their roles.

18.2.1 Master Mode: Host (Mixer) Application

The Master or "mixer" application handles the audio of both applications within its own mixer, virtually interfacing with the slave's instruments as if they were its own. Through this interfacing, you can play and control the Slave's instruments from the Master—in our case, Live. Therefore, the signals from whatever instruments are loaded in the Slave are accessible in Live. This relationship is similar to how a virtual instrument plug-in loads up in Live. Sometimes the best way to think of ReWire is that the slave is a virtual instrument; whether it's an instrument or audio tracks, it's all virtually piped back into the Master. The major difference is that the user interface exists in the slave, whereas a virtual instrument is hosted within a master. All configurations, then, must be made on the slave end, such as presets and parameters native to the slave's instrument. When Live establishes a Master connection with other software applications it can send MIDI to and receive audio from multiple ReWire slaves.

18.2.2 Slave Mode: Synth Application

The Slave or "Synth" application is controlled by the Master. It receives MIDI information from the Master, which controls its instruments. All audio is then output back to the Master, including that of audio tracks (if applicable) and the output of software instruments. The Slave can serve solely as a synth that is triggered by the master or it can also host its own MIDI sequences or audio tracks that are sent to the master's mixer via the ReWire connection. In this case the programs operate in parallel, and are held in sync and mixed by the Master. When Live is run as a ReWire Slave it can receive MIDI and send audio to a master's program, but its audio input is disabled. Other functions unavailable in Slave mode are time signature and tempo settings, sample rate settings, external MIDI sync or control messaging to externals, and ReWire hosting of other programs. All of these functions are deferred to the ReWire Master.

18.3 Live ReWired

Using ReWire is not all that difficult, but there is an order of operations that must be followed to successfully establish and route a ReWire connection. Most DAWs can only function as a ReWire Master while other programs function only as a ReWire Slave. Since Live operates as both, a ReWire Master and as a ReWire Slave, we'll look at each.

18.3.1 ReWire Master

There are many ways to route your workflow using ReWire. For the following examples we will keep it simple and use Live as the Host/Mixer of MIDI and audio. Establishing a ReWire connection is a fairly intuitive process in Live. All you will need is a program capable of being a slave. For this example, Propellerhead's Reason 6 will be used. Of course, you can use any ReWire compatible software.

Always open the Master (Live) first. Once it is open, launch the Slave application (Reason). Once you have done this, Reason will display "ReWire Slave Mode" on its I/O.

Figure 18.2 Reason ReWired as a Slave to Live.

If you see this, then you are connected. If not, close Reason, then Live—in that order—and try the process again. Once you have successfully established the ReWire connection, you are ready to start routing MIDI and audio connections. Make a note that if at any time you wish to quit the programs, you must always close the applications in the reverse order: Slave then Host.

To start out, let's take control of a Reason Device. The goal is to use the Reason instrument as an instrument within Live. To do this, we have to send MIDI out to Reason, then route the audio signal back to Live.

Controlling a Reason Instrument (Slaved) with Live:

1. In Reason, add an instrument to an empty Reason Rack. Go to the Create Menu and select "Thor Polysonic Synthesizer". By default, Thor will be routed to playback, but you may want to route Thor's Direct Out to Reason's Hardware interface outputs 3/4 if you want discrete audio from multiple Reason instruments.

Hot Tip

Order of operations: Open— ReWire Master then the Slave. Close—Slave then close the Master!

Figure 18.3 Create a Device in Reason.

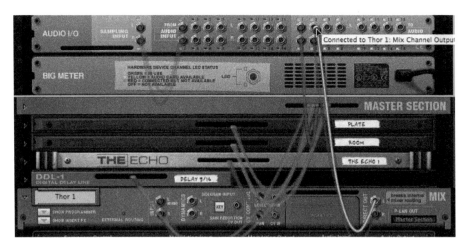

Figure 18.4 Thor routed directly to Reason's Hardware Device Interface.

2. In Live, load a new External Instrument device from *Categories>Instruments* by *double clicking* or *dragging* it into the Drop Area in the Session or Arrangement View. The external instrument device will show up as a device in Device View at the bottom of the screen as usual. *Note: if you are using Live 9 Intro, you will have to use and route a MIDI track along with an audio track in order to monitor audio from ReWire in Live. Follow the steps below, but use the track In/Out section.*

Figure 18.5
Load a new External Instrument into your Live Set.

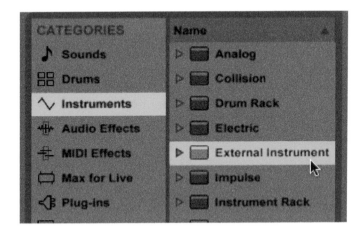

3. In Live's Device View do the following:

(a) Set the "MIDI To" chooser to "Reason".

(b) Set the "Channel Input" chooser for the Reason Device to "Thor 1".

(c) Set the "Audio From" chooser to "1/2": Mix L, Mix R.

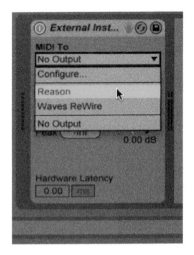

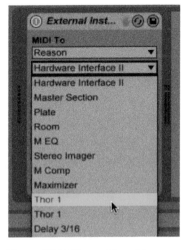

Figure 18.6 MIDI To Reason.

Figure 18.7 Channel Input to Thor 1.

Figure 18.8 Audio From to 1/2: Mix L, R, Mix.

4. Arm your Ext. Instrument track in Live and then play some keys. Now you'll be able to record a basic MIDI Clip into your new External Instrument track containing the ReWire slave's output. The great thing is that all of this routing information is contained and adjusted within the External Instrument device. Fast and simple.

When you're happy with your new clips or arrangement, you may want to record (print) the ReWire application's (Reason, in this case) audio to a designated audio track within Live for future mixing and/or reference. If you want to print Thor's output to track, then you will need to create an audio track in Live (Session or Arrangement View), then set the new track's input to Reason. Once you have all of this routed and a MIDI clip ready to play on your Ext. Instrument track, then arm your print track and begin recording Thor to a new clip in the Session View or Arrangement View. This can also be done with Freezing or Rendering, as described in ▶ **8.3.1** .

18.3.2 ReWire Slave

There are many ways to use ReWire connections, especially with Live as a Slave. In this scenario Live can be used for sequencing while another DAW is set to process the output of Live or simply host Live's mix output to two-track. You could also use Live only for its instrument devices to supplement the deficiencies of your other DAW—in this case, the Master. It's all up to you. For now, we'll focus on just making the ReWire connection. Each program is different in how you load up an instance of Live through ReWire as a Slave. Review your DAW's manual for specific instructions.

Controlling a Live Instrument Device with Another DAW:

1. Open your DAW of choice: Pro Tools 10 for this example.

2. In Pro Tools, create a new session and then create a stereo instrument track.

3. Insert a *multichannel plug-in>Instrument>Ableton Live (stereo)*. (Figure 18.9.)

    This will open up the ReWire Plug-in, set automatically to Mix L–Mix R. (Figure 18.10.)

4. In Live, load any instrument device into a MIDI track. (Figure 18.11.)

    Since you're in Slave Mode, Live's Master track output is automatically routed to output to ReWire via "Mix L/R". This means that all of your Live tracks are summed to a two-track stereo mix in Pro Tools. If you wish, you can bypass Live's Master track altogether and manually route each Live track individually to a ReWire Output. (Figure 18.12.)

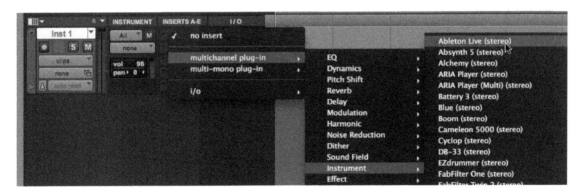

Figure 18.9 Insert an Ableton Live Stereo Instrument: multichannel plug-in>Instrument>Ableton Live (stereo).

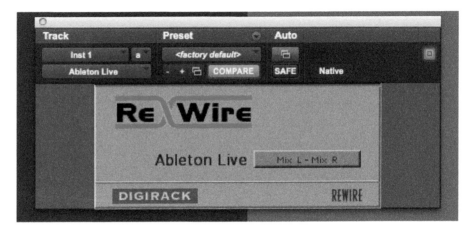

Figure 18.10 ReWire Plug-in, set automatically to Mix L–Mix R.

5. (*Optional*) If you desire to play (input MIDI data) multiple Live instrument devices via separate tracks and or channels (MIDI/instrument) in Pro Tools and have them mixed independently of each other in Pro Tools, then you must set each Live instrument track to a unique ReWire Bus output (e.g., Bus 3/4, 5/6, and so on). Set your Live MIDI track "Audio To" to "ReWire Out", and then to the appropriate Bus output to Pro Tools. Whatever output bus they are set to will also have to be assigned to a Track Input in Pro Tools. This allows you to control several individual Live instrument devices. (Figure 18.13.)

6. In Pro Tools, set the MIDI Output Selector to the instrument loaded in Live and begin playing away. You may need to arm your instrument track to monitor playback. (Figures 18.14 and 18.15.)

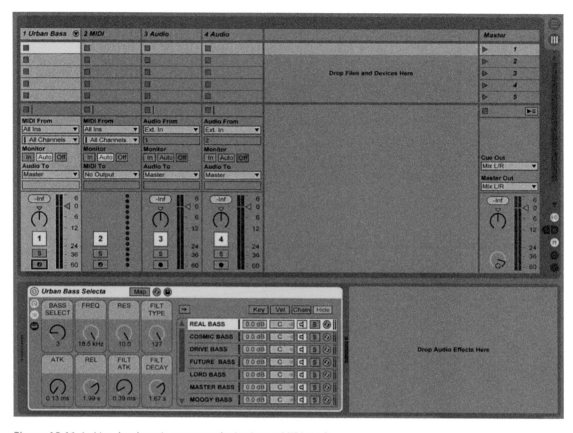

Figure 18.11 In Live, load any instrument device into a MIDI track.

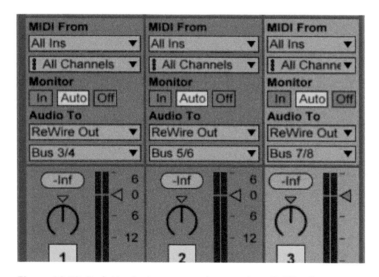

Figure 18.12 Master track output is automatically routed to output to ReWire via 1"Mix L/R".

Figure 18.13 Each Live instrument track to a unique ReWire Bus output.

Figure 18.14 Set the MIDI Output Selector to the instrument loaded in Live.

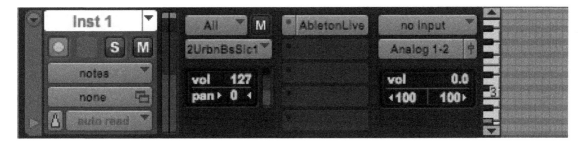

Figure 18.15 Your track should look like this.

Don't forget that you can open up multiple Live instruments and route them all into the ReWire Master (Pro Tools). Just create the necessary tracks in the Master and in the Slave. To control Live's instruments, route each track in the Master to the corresponding tracks in Live, and use a ReWire insert plug-in to monitor each track in the Master. Set your tracks to monitor your slave instruments through Mix L/R or individually via the ReWire bus options. From the Slave, send audio output to Mix L/R or on individual busses back to the ReWire Master. If you want to print Live's output to a track in the Master, you will need to send or bus Live's input signal in the Master from the ReWire plug-in track to a free audio track.

Mixing Live's Audio Output with Another DAW:

1. Open your DAW of choice: Pro Tools 10 again for this example.
2. In Pro Tools, create a new session and then create two stereo audio tracks.
3. On both tracks insert a *multichannel plug-in>Instrument>Ableton Live (stereo)*. This opens up the ReWire Plug-in set automatically to Mix L—Mix R.
4. Set audio track 1 to ReWire Bus 03 -04 and audio track 2 to Bus 05 -06. (Figures 18.16 and 18.17.)

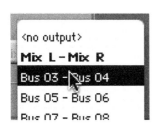

Figure 18.16
Set audio track 1 to ReWire
Bus 03–04.

Figure 18.17 ReWire Bus 03–04.

5. In Live, open up a new Set and *drop* in clip "**CPP_ Acoustic Guitar**" and "**CPP_RhythmEl-GTR1**" **audio loops** each into their own audio tracks (two tracks in all). Now you want to route these audio tracks to Pro Tools. In Slave Mode your master output will automatically be routed to ReWire output Mix L/R. For this demonstration we will bypass Live's Master track and route each track individually to the ReWire Output.

6. Set audio track 1 to ReWire Out, Bus 03–04 and audio track 2 to ReWire Out, Bus 05–06. (Figure 18.18.)

7. Launch your Live audio and you will see and hear its output in Pro Tools. (Figure 18.19.)

Of course, you can also open a whole Live Set and route, mix, and print your audio in Pro Tools or with any DAW that is capable of being a ReWire Host/Master/Mixer application.

Figure 18.18 Audio track 1 to ReWire Out, Bus 03–04 and audio track 2 to ReWire Out, Bus 05–06.

Let's review ReWire signal flow with Live. In a ReWire relationship, Live will be either the Master or Slave and will handle audio and MIDI signals. Whether dealing with audio or MIDI, the signal flow between a Slave and Master remains consistent. It is just a matter of which mode Live is set to operate in. Live is well suited to be an instrument and audio playback application, or a sequencer and mixing environment. This means that as a Slave, Live will send audio from an audio clip or an instrument device to a ReWire Master, or as a Master it will receive audio from MIDI instruments with the slave application. At this point, if you understand how to route and use Live as a ReWire MIDI instrument, then you will have no

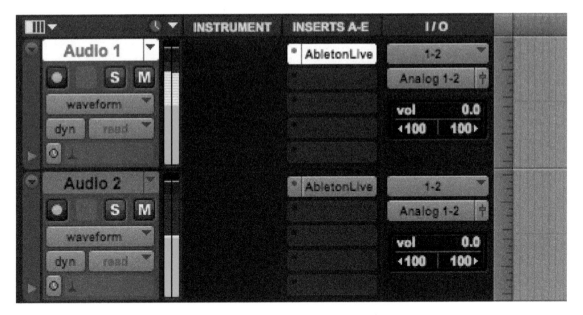

Figure 18.19 Live audio will be seen and heard in Pro Tools.

problem routing and using Live strictly for hosting ReWired audio input or slaving out audio from its audio tracks/clips to another application. (Figure 18.20.)

As a ReWire slave, Live can be used as a synth application, accepting MIDI input from the Host application via ReWire MIDI input and sending audio back to the Host via ReWire audio output. As a Master, Live sends MIDI to a synth application via ReWire MIDI output and receives audio from the synth application via ReWire audio input. The same concept applies when slaving Live and working exclusively with audio tracks/clips. The only difference is that the audio is generated and sent from audio tracks/clips as opposed to MIDI tracks/clips and MIDI instruments. (Figure 18.21.)

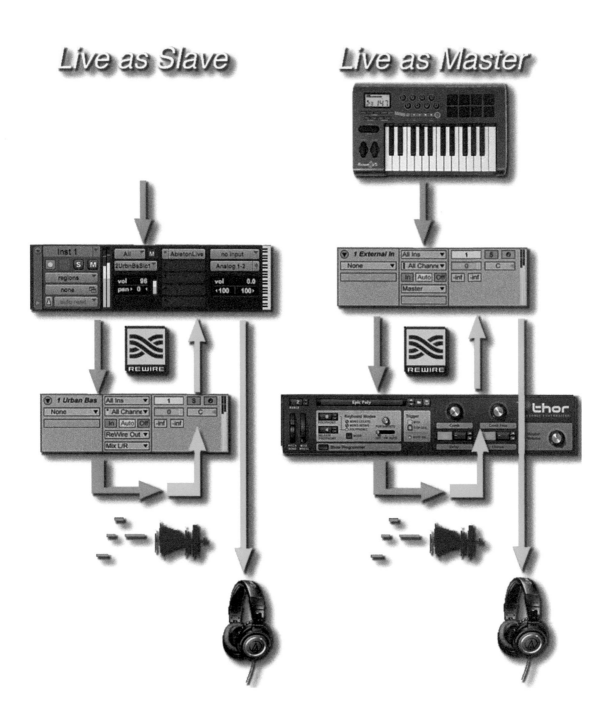

Figure 18.20 ReWire routing for MIDI.

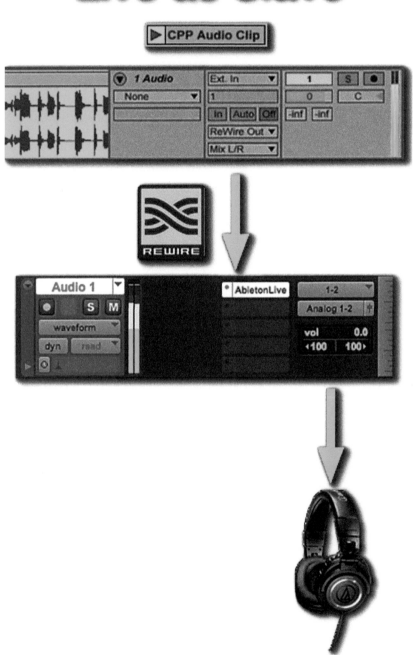

Figure 18.21 ReWire routing for audio.

Index